Estimating and Tendering for Construction Work

ESTIMATING AND TENDERING FOR CONSTRUCTION WORK

Fourth edition

Martin Brook
BEng(Tech) FCIOB

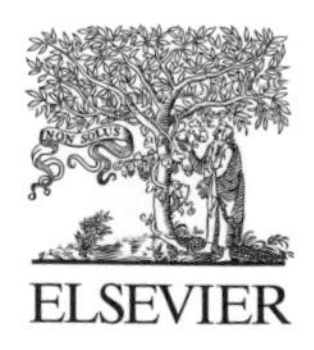

AMSTERDAM • BOSTON • HEIDELBERG • LONDON • NEW YORK
OXFORD • PARIS • SAN DIEGO • SAN FRANCISCO
SINGAPORE • SYDNEY • TOKYO
Butterworth Heinemann is an imprint of Elsevier

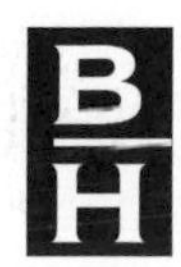

Butterworth-Heinemann is an imprint of Elsevier
Linacre House, Jordan Hill, Oxford OX2 8DP, UK
30 Corporate Drive, Suite 400, Burlington, MA 01803, USA

First published 1993
Second Edition 1998
Reprinted 1999, 2001, 2002, 2003
Third edition 2004
Reprinted 2005(twice), 2006
Fourth edition 2008

British Library Cataloguing in Publication Data
A catalogue record for this book is available from the British Library

Library of Congress Cataloging-in-Publication Data
A catalog record for this book is available from the Library of Congress

ISBN: 978-0-7506-8616-7

For information on all Butterworth-Heinemann publications visit our web site at books.elsevier.com

Typeset by Charon Tec Ltd (A Macmillan Company)
www.charontec.com

Printed and bound in the UK by MPG Books Ltd.
08 09 10 10 9 8 7 6 5 4 3 2

Contents

Preface

My aims in this book are to introduce a practical approach to estimating and tendering from a contractor's point of view, and explain the estimator's role within the construction team. The book therefore differs from other textbooks in three main ways:

1. In general it is assumed that it is the contractor who prepares estimates because in the majority of cases an estimate is produced to form the basis of a tender.
2. I have introduced many typical forms used by estimators to collate data and report to management. Most of the forms relate to two fictitious projects: a new lifeboat station and the construction of offices for Fast Transport Limited.
3. The pricing examples given in Chapter 11 have been produced using a typical build-up sheet. The items of work to which the prices relate are given at the top of each page. Estimating data are given for each trade so that students will have a source of information for building up rates. I suggest that before pricing exercises are undertaken, the first part of Chapter 11 should be read and an understanding of estimating methods should be gained from Chapter 5. The first pricing example is for a 'model rate' that gives a checklist of items to be included in a unit rate.

The estimating function has changed more in the last 15 years than at any time before. Many estimating duties can now be carried out by assistants using word processors, spreadsheets and computer-aided estimating systems. The estimator manages the process and produces clear reports for review by management.

Estimators need to understand the consequences of entering into a contract, which is often defined by a complex combination of conditions and supporting documents. They also need to appreciate the technical requirements of a project from tolerances in floor levels to the design of concrete mixes, and from temporary electrical installations to piling techniques.

The Chartered Institute of Building publishes a series of guides to good practice – the *Code of Estimating Practice* and its supplements. I have not duplicated their fine work in this book but hope that my explanation and examples show how the guidelines can be used in practice.

Contractors now assume an active role in providing financial advice to their clients. The estimator produces financial budgets for this purpose and assembles cost allowances for use during construction. Computers have been introduced by most organizations, with a combination of general-purpose and specialist software. Computers have brought many benefits during the tender period, and are seen as essential for the handover of successful tenders; adjustments can be made quickly, information can be presented clearly, and data can be transferred in a more compact form.

The changes brought about by the introduction of SMM7 and the other principles of Coordinated Project Information have reduced the number of items to be measured in a typical building contract. The item descriptions no longer provide information for pricing; the estimator must always refer to the specification and drawings. In practice this is time-consuming both for contractors and sub-contractors, and the amount of paperwork has increased immensely. Nevertheless, contractors always need a bill of quantities, whether produced by the client's quantity surveyor, by an in-house commission or by sharing the services of an independent quantity surveyor. Traditionally bills of quantities were used as a fair basis for preparing and comparing tenders, but increasingly the responsibility for quantities is being passed to contractors. It is of some concern that estimators continue to have difficulty entering bills of quantities in their estimating systems and I look forward to the time when a common approach to electronic data transfer is widely adopted.

This fourth edition has been written to reflect changes in estimating since 2004. These include:

- The Private Finance Initiative (PFI) has changed with the use of a procedure called 'competitive dialogue'. This is explained in Chapter 2.
- A substantial new section (Chapter 8) has been added for tendering on the basis of cost plans. There is an increasing trend for contractors to prepare early cost models for their clients, and develop proposals according to 'design-to-cost' principles. Other chapters have been extended to include cost-planned tenders.
- A major review of JCT contracts took place in 2005 with most of the main contract forms consolidated into new editions. Further revisions were made in 2007.
- Increases in labour and plant rates which affect rate build ups (Chapter 10), daywork calculations (Chapter 14), and pricing notes given in Chapter 11. In the South-East region of the UK, there is a shortage of skilled operatives and staff which continues to add to building costs. Across the UK, inflation continues to run ahead of the UK government measures of inflation.
- Many tables and figures have been enhanced so that they are easier to read.
- The book no longer includes procedures for incorporating nominated sub-contractors in a tender. Not only has their use declined, the practice of nominating sub-contractors is no longer recommended.
- Chapter 20, 'Computer-aided estimating' now includes electronic information systems and collaborative tools.
- Some new terminology introduced by the 6th Edition of The CIOB *Code of Estimating Practice* 1997, particularly the recommendation that structured discussions with management, are referred to as 'review' meetings, and what was the

'adjudication' meeting is now called the 'final review' meeting. This avoids conflict with the action by quantity surveyors in checking tenders which is also referred to as the 'adjudication of a tender'.

I recognize and support the role of women in construction and ask readers to accept that the use of the masculine pronoun is intended to refer equally to both sexes.

Martin Brook
2008

Acknowledgements

I wish to acknowledge the help given by Michael Hawkridge for checking the text to the first edition, and Dr Jane Brook for the cartoons.

List of figures

Abbreviations used in the text

BEC	Building Employers Confederation (changed to Construction Confederation in 1997)
BPF	British Property Federation
BRE	Building Research Establishment
BREEAM	BRE Environmental Assessment Method
BS	British Standard
BWIC	Builder's Work In Connection
CABE	Commission for Architecture and the Built Environment
CAD	Computer-Aided Design
CAWS	Common Arrangement of Work Sections
CC	Construction Confederation
CD	Compact Disc
CECA	Civil Engineering Contractors Association
CESMM3	Civil Engineering Standard Method of Measurement Third Edition
CIB	Construction Industry Board (Disbanded 2001)
CI/SfB	Construction Index – Samarbetskommitten for Byggnadsfragor
CIOB	Chartered Institute of Building
COEP	Code of Estimating Practice (Published by the CIOB)
Conc	Concrete
CPSSST	Code of Procedure for Single Stage Selective Tendering
CPI	Coordinated Project Information
DOT	Department of Transport
Exc	Excavation
FCEC	Federation of Civil Engineering Contractors (Disbanded 1996 see CECA)
FIDIC	International Federation of Consulting Engineers
HSE	Health & Safety Executive
ICE	Institution of Civil Engineers
Inc	Included
IOS	International Organization for Standardization
JCT	Joint Contracts Tribunal Limited
LAN	Local Area Network
LCD	Liquid Crystal Display
LOSC	Labour-Only Subcontractor

MB	Megabyte
ne	Not exceeding
NJCC	National Joint Consultative Committee for Building (disbanded in 1996 but documents still in use)
PC	Prime Cost
PC	Personal Computer
PQS	Private Quantity Surveyor (Consultant) also Project Quantity Surveyor
Prov	Provisional
Quant	Quantity
RAM	Random-Access Memory
RIBA	Royal Institute of British Architects
RICS	Royal Institution of Chartered Surveyors
ROM	Read-Only Memory
SMM	Standard Method of Measurement
SMM6	Standard Method of Measurement of Building Works: Sixth Edition 1978
SMM7	Standard Method of Measurement of Building Works: Seventh Edition 1988
WAN	Wide Area Network

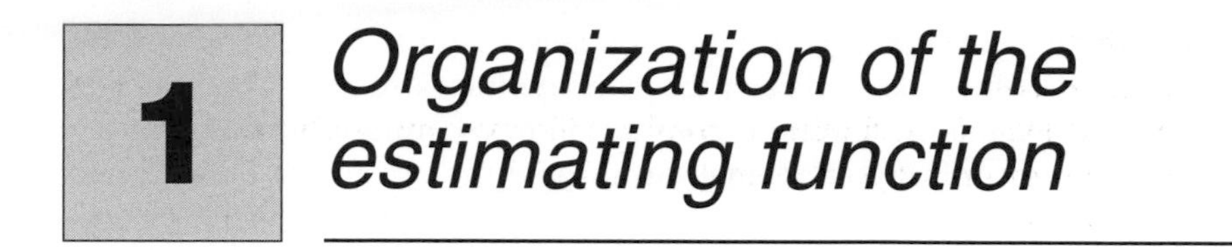

1 Organization of the estimating function

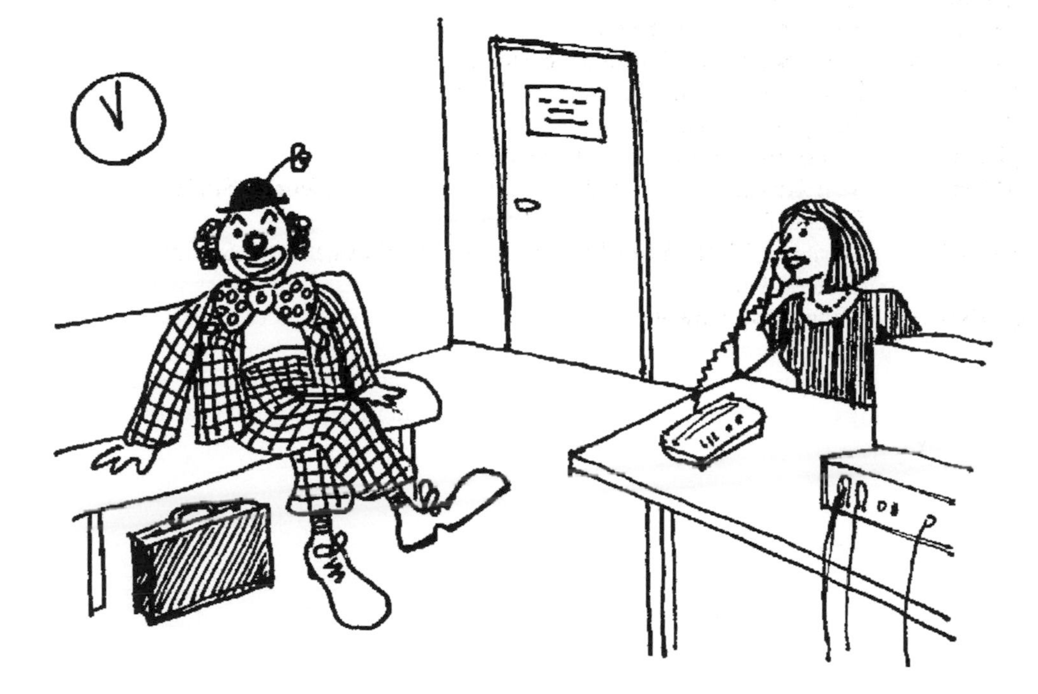

'The corporate image consultant is in reception, Sir.'

Team roles and organization

The role of the contractor's estimator is vital to the success of the organization. The estimator is responsible for predicting the most economic costs for construction in a way that is both clear and consistent. Although an estimator will have a feel for the prices in the marketplace, it is the responsibility of management to add an amount for general overheads, assess the risks and turn the estimate into a tender. The management structure for the estimating function tends to follow a common form with variations for the size of the company. In a small firm, the estimator might be expected to carry out some quantity surveying duties and will be involved in procuring materials and services. For large projects, the estimator may be part of a multi-disciplinary team led by a project manager. The estimating section in a medium-sized construction organization (Fig. 1.1) will often comprise a chief estimator, senior estimators and estimators at various stages of training.

Larger estimating departments may have administrative and estimating assistants who can check calculations, photocopy extracts from the tender documents, prepare letters and enter data in a computer-assisted estimating system.

The estimating team for a proposed project has the estimator as its coordinator and is usually made up of a contracts manager, buyer, planning engineer and quantity surveyor. The involvement of other people will vary from company to company. A project quantity surveyor is often consulted to examine amendments to conditions of contract, prepare a bill of quantities, assess commercial risks, set up design agreements and identify possible difficulties which have been experienced on previous contracts. Clients sometimes like to negotiate agreements with quantity surveyors where a good working relationship has been established and follow-on work is to be based on pricing levels agreed for previous work. A planning engineer might be asked to prepare a preliminary programme so that the proposed contract duration can be checked for possible savings. He can also prepare method statements, temporary works designs,

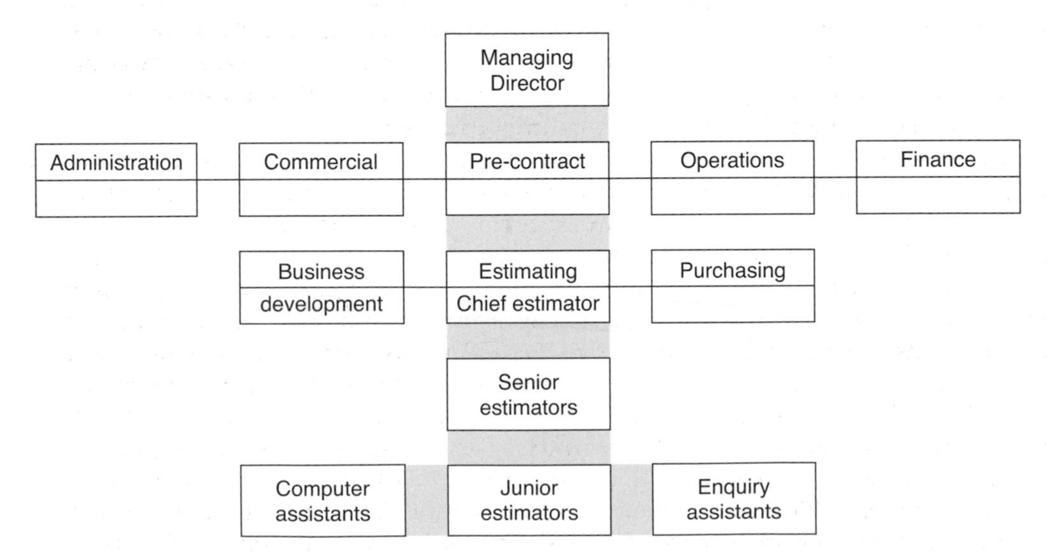

Fig. 1.1 *Estimating staff structure for a medium-sized organization*

organizational charts and site layout drawings. Some or all of this material can be used to demonstrate to a client that satisfactory systems have been developed for the project. The purchasing office will provide valuable information leading to the most economic sources for the supply of materials and plant. In many organizations today, the buyer is responsible for getting quotations from suppliers and sub-contractors. At the very least, the buyer (sometimes called 'procurement manager' or 'supply chain manager') helps prepare lists of suitable suppliers, keeps a library of product literature and advises on likely price trends and changes. A buyer can provide an invaluable service in managing enquiries and chasing quotations. His knowledge of local suppliers and current discounts is essential at the final review meeting when decisions need to be taken about the availability and future costs of materials and services.

The role of the site manager is to report on the technical and financial progress of their projects so that the estimator can learn from the company's experience on site. On completion of contracts, site staff will usually contribute to tenders for larger and more complex schemes – particularly for civil engineering and large-scale building work – where alternative construction methods have a significant affect on tender price. A site project manager is often used to lead the bid team and manage all aspects of the tender. The department dealing with business development and presentations can contribute in two ways: by maintaining close contacts with clients to ensure their needs are met, and by producing submission documents often using desktop publishing software.

The aim of the team is to gain an understanding of the technical, financial and contractual requirements of the scheme in order to produce a professional technical document with a realistic prediction of the cost of construction. The construction manager or director will then use the net cost estimate to produce the lowest commercial bid at which the company is prepared to tender. Figures 1.2 and 1.3 show the various stages in preparing a tender and the action needed with successful tenders. Figure 1.3 has additional tasks for a design and build contract.

The work flow in an estimating department is never constant; the ideal situation is to have people available who are multi-disciplinary and can deal with administrative tasks. The cost of tendering for work in the construction industry is high and is included in the general overhead which is added to each successful tender. For one-off large projects, such as PFI contracts, bidding costs can be several millions of pounds. These costs are recovered when schemes are successful but written off against annual profits when contractors fail to win. The chief estimator needs to be sure there is a reasonable chance of winning the contract if the organization is in competition with others. The decision to proceed with a tender is based on many factors including: the estimating resources available; extent of competition; tender period; quality of tender documents; type of work; location; current construction workload and conditions of contract. With all these points to consider, a chief estimator could be forgiven for declining a high number of invitations to tender to maintain a high success rate and avoid uncompetitive bids which can lead to exclusion from approved lists.

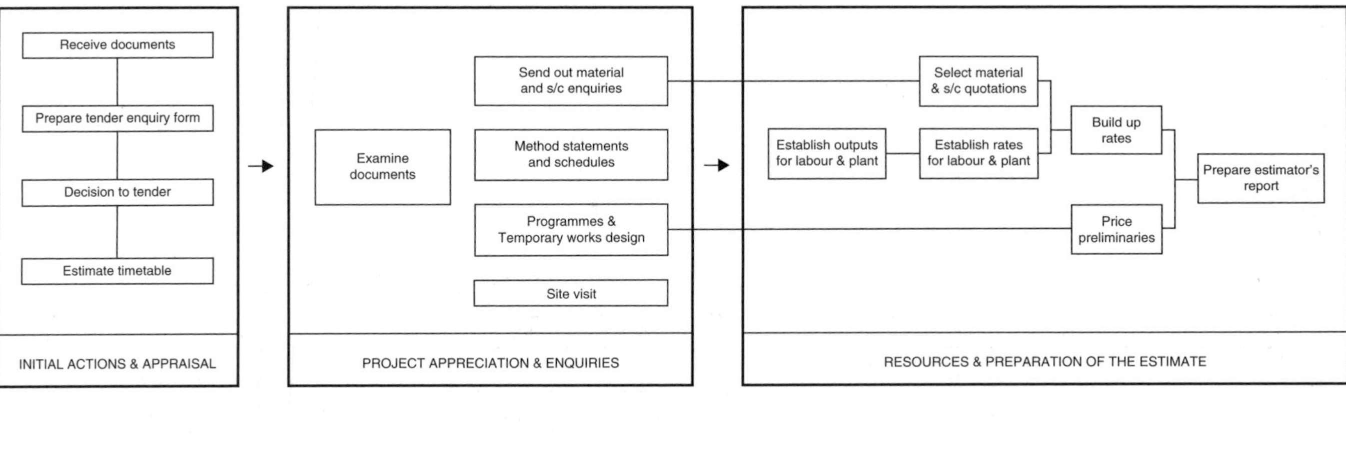

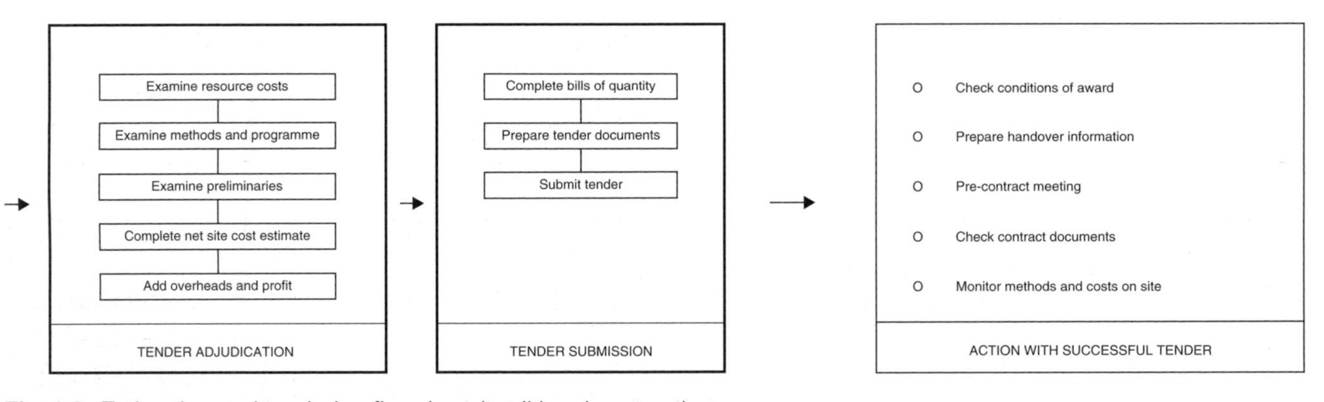

Fig. 1.2 *Estimating and tendering flowchart (traditional contract)*

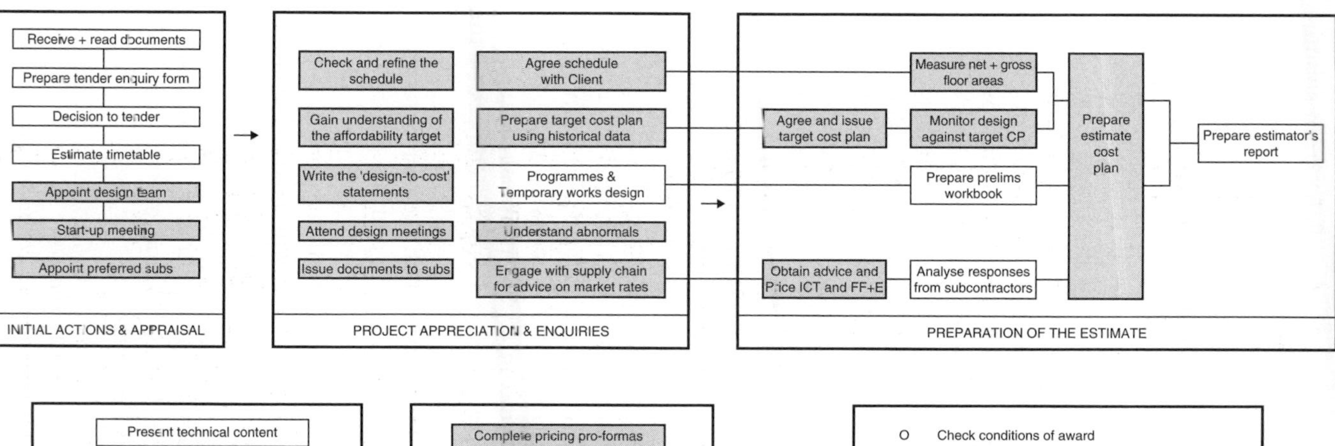

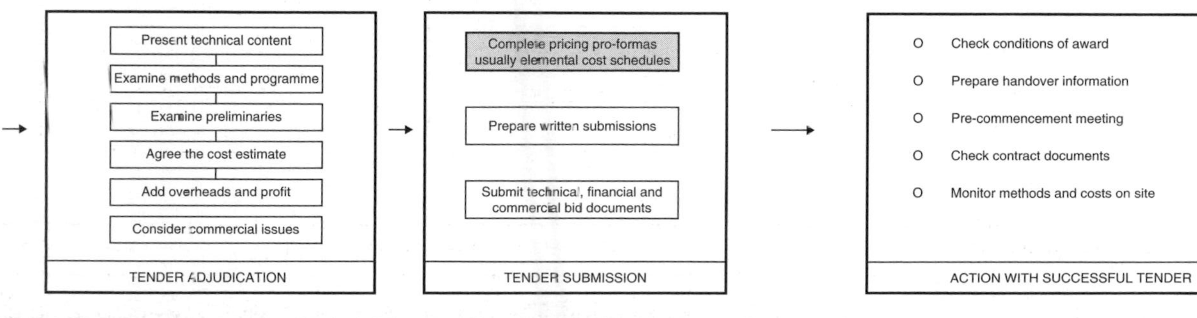

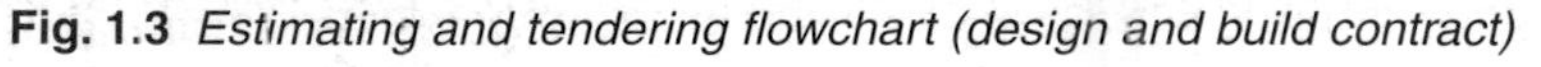

Fig. 1.3 *Estimating and tendering flowchart (design and build contract)*

On the other hand, he must recognize the goodwill which often flows from submitting competitive prices and the need to carry out work which might lead to suitable and profitable contracts.

There are several forms that can be used to plan, control and monitor estimating workload. The first is a chart to show the opportunities to tender when they have been confirmed. The information for this programme usually comes from marketing personnel, who are responsible for bringing in invitations to tender for projects that are in line with company strategy. The chief estimator will prepare a bar chart (Fig. 1.4) to show how the estimators will be assigned to present and future tenders, showing the expected dates for receipt of documents and submission of tenders. Copies are sent to heads of other departments so they can plan their input; they may also wish to attend the final review meetings. A tender register is also needed (Fig. 1.5) to record the main details of each tender such as reference number, client, price, tender date and an analysis of performance in relation to the competition.

The success ratio for a construction firm is often quoted at about 1 in 4 although it can be as bad at 1 in 6 and as good as 1 in 2 where competition is limited. Since the directors of a company are more concerned with turnover and profit, then success is better measured in terms of value, and the estimating department may be given annual targets to meet. Clearly, negotiated work and two-stage tendering can save a great deal of abortive work.

Estimators are drawn from two sources: direct from school with some good grades in GCSE subjects which suggest a potential to study to a higher technician or professional level, or from experienced staff where management has identified an aptitude and willingness for the job. In both cases a reasonable time must be spent on site to gain experience in construction methods, materials identification, use and practice. The skills that are needed are the ability to read and interpret technical documentation, the ability to communicate with clients, specialists and other members of the team, and the faculty to make accurate calculations.

Technically an estimator must have a working knowledge of all the major trades, to identify packages of work to be carried out by sub-contractors, and the direct workforce, to foresee the time and resources that will be needed. It is also necessary to have the skills needed to take off quantities from drawings, where there are no bills of quantities. When bills of quantities are provided, the estimator will need to check the principal quantities to understand how corrections to the quantities during the contract will affect the profitability of the scheme.

An estimator needs to refer to many information sources either in book form or through more modern means such as microfiche, CD-ROM and on-line databases. The following list shows some of the basic material required:

- Code of Estimating Practice (COEP) ... *pro-formas for estimators.*
- Code of Procedure for the Selection of Main Contractors ... *tendering procedures.*
- Standard Method of Measurement (SMM) ... *explanation of item coverage.*

CB CONSTRUCTION LIMITED		ESTIMATING PROGRAMME		Date	30.5.08

no.	current tenders	tender date	review date	estimator	PM	planner	QS	buyer	value £m	period wks	### June 2	9	16	23	July 30	7	14	21	28	August 4	11	18	25	September 1	8	15	22	
335	St James School extension	19.6.08	16.6.08	JE		FD		GB	2500 000	3	##	##	##															
336	Lifeboat station	3.7.08	1.7.08	GT		FD		GB	2100 000	5	##	##	##	##	##													
338	Access road	20.6.08	17.6.07	JS		BT			900 000	3	##	##	##															
341	Treatment works	10.7.08	7.7.08	JE		BT		GB	1250 000	5		##	##	##	##	##												
342	Superstore	30.7.08	30.7.08	JS	CB	FD	SS	GB	5500 000	7			##	##	##	##	##	##	##									
346	Fast transport office	22.8.08	19.8.08	GT		FD		GB	1350 000	6							##	##	##	##	##	##						
	future tenders	date due																										
	Stansford College	4.8.08							8100 000	7										##	##	##	##	##	##	##		
	Totals								**21,700,000**		###	###	###	###	###	###	###	###	###	###	###	###	###	###	###	###	-	-

Fig. 1.4 *Estimating programme*

CB CONSTRUCTION LIMITED	TENDER REGISTER	page

Tender no.	Tender title	Date received	Location	Client	QS Architect Engineer	Tender details		Tender evaluation			
						Date	Price	No. of tenderers	Rank	% over lowest	% over mean
001											
002											
003											
004											
005											
006											
007											
008											
009											
010											
011											
012											
013											
014											
015											
016											
017											
018											
019											
020											
021											

Fig. 1.5 *Tender register*

- Standard forms of contract ... *contractual obligations.*
- Standard specifications for highways and water industries ... *specifications for pricing.*
- National Working Rule Agreement ... *labour rates.*
- Definition of prime cost of daywork ... *pricing daywork percentages.*
- Daywork plant schedules ... *pricing daywork percentages.*
- Trade literature:
  (a) standard price lists
  (b) technical product information.
- Trade directory of suppliers and sub-contractors ... *lists of suppliers to receive enquiries*
- Reference data for weights of materials ... *unit rate pricing.*

Quality management

A company's quality management system must include procedures for estimating and tendering. The decisions made at tender stage will often determine the way in which the project is carried out. It is therefore important when preparing a tender to ensure that the client's requirements are understood, information is robust, and directors have approved the contractor's proposals.

Many organizations have adopted a standard approach to the process of estimating. Documented procedures are used that detail the preparation, review and submission of a tender. This is particularly useful for newly appointed staff as it provides a standard framework for the preparation of an estimate and ensures consistent records and reports for others. The preparation of documented procedures has come with the introduction of a British Standard, which provides a model for quality assurance. Now known as the BS EN ISO 9000 series, this standard was first introduced to the construction industry as BS 5750 in 1979.

The objective of a quality assurance system is to provide confidence that a product, in this instance the tender submission, is correct, is provided on time and produces the right price. This price might be defined as that which the client can afford and deems reasonable, and is sufficient for the contractor to meet his business objectives. However, it is acknowledged that tenders are always submitted on time, but owing to time and information constraints, the price may not always be the 'right price'.

The benefits of implementing quality assurance in the estimating function are:

1. *Profitability* – an improvement to the profitability of the organization.
2. *Accuracy* – a reduction of errors.
3. *Competence* – better trained staff.
4. *Efficiency* – work properly planned and systematically carried out.
5. *Job satisfaction* – for the whole estimating team.
6. *Client satisfaction* – leading to likelihood of repeat business.

Health, safety and welfare

Safety is high on the agenda of construction organizations. Estimators must understand the implications of current legislation for the design and procurement stages and include sufficient costs to carry out the work safely.

A client's professional team contributes to the writing of a health and safety file, by assessing hazards which might be inherent in the design. These hazards include possible dangers to construction operatives, staff and the public – during construction, for occupants and in carrying out repairs over the life of the building. The health and safety file tells the estimator about the project, setting out hazards associated with the design, and dangers known about the existing site.

The Construction (Design and Management) regulations demand greater responsibilities on design-and-build contractors. Their tasks will often be extended by clients to include:

1. The role of CDM Co-ordinator.
2. Vetting of designers for competence in designing safely.
3. Producing the pre-tender stage health and safety file.

Good health and safety systems ensure significant long-term business benefit, as follows:

- less staff absence
- less staff turnover
- improved productivity and efficiency
- less down time
- improved quality of work
- lower insurance premiums
- best in class.

There are some clients who remain sceptical about why they are paying for health and safety and see little benefit to their business. They are forgetting that they have a moral and legal obligation to manage the safety of the overall project, and a safety culture will affect the attractiveness of the finished product. In addition, legal action following a failure in health and safety can damage a company's reputation.

Many incidents are not covered by insurance. Also the policy excess may be greater than the individual amounts concerned. All other costs will have to be met by the contractor.

Many costs are not covered by insurance. They can include:

- investigations
- lost time and production delays

Heading	Description	£
Staff	CDM co-ordinator	35 000
	Safety manager – visiting site	15 000
	Safety manager – on site	85 000
	Temporary works design checks	30 000
	Logistics planning	25 000
Equipment	Signage external	4 000
	Signage internal	9 500
	Safety clothing – client team	8 000
	Safety clothing – visitors	2 500
	Safety clothing – staff	11 000
	Personal protective equipment	Above
	Fire equipment extinguishers etc	15 000
	First aid equipment	1 800
	First aid accommodation	10 000
	Safety lighting	5 000
	Barriers for segregated walkways	6 000
	Safety netting (priced with scaffolding)	
Training	Safety induction	3 500
	CSCS cards	25 500
Processes	Considerate Contractors Scheme	750

Fig. 1.6 *Example of the additional costs for Health and Safety for a £50m project*

- sick pay
- damage or loss of product and raw materials
- repairs to plant and equipment
- extra wages, overtime working and temporary labour
- fines
- loss of contracts
- legal costs
- loss of business reputation.

The Construction (Design and Management) Regulations impose certain requirements at tender stage. The following should be checked by the tender team:

1. A CDM co-ordinator has been appointed and attends team meetings. The CDM co-ordinator's role is 'to advise the client on health and safety issues during the design and planning phases of construction work' (HSE). In most design and build projects, the CDM co-ordinator will attend design meetings to encourage others to fulfil their responsibilities.

2. Sub-contractors have been vetted for their H&S procedures and performance.
3. Designers have been checked for the H&S procedures and performance.
4. Site accommodation, in the estimate, is of an appropriate standard to provide a safe environment for site staff and ensure a high standard of welfare.
5. Site phasing and logistics are designed to reduce the risk of accidents.
6. Designs produced are, from the start, safe to build and maintain.
7. Construction work is notified to the HSE by the CDM co-ordinator.
8. Data for the information file will be provided.
9. Planning and management of risks is improved from the start.

Health and safety is priced in various parts of the estimate but is most obvious in the preliminaries. Figure 1.6 shows part of a preliminaries spreadsheet which deals with identifiable items. Other issues are included in temporary works and plant sheets together with trade packages.

2 Procurement paths

Introduction

The Banwell Report, published in 1964, expressed the view that existing contractual and professional conventions do not allow the flexibility that is essential to an industry in the process of modernization. The report of the committee asked the industry to experiment to secure efficiency and economy in construction.

The traditional method of organizing construction work starts with appointing a consultant designer, usually an architect or engineer, or both. Other specialists may be needed, in particular a quantity surveyor is appointed to provide cost information, prepare bills of quantity, compare bids and maintain financial management during construction.

Since the early 1960s, the construction industry has experienced significant changes in the way in which contracts are managed. In some cases, contractors have been brought in at an early stage as full members of the design team; in others, such as Public–Private Partnerships (PPP), contractors have occupied the lead role. During the 1980s clients became increasingly concerned about problems such as poor design, inadequate supervision, delays and increased costs. They were also critical of the separation of design from construction, particularly between the building professions.

In an attempt to overcome some of these long-standing criticisms, the British Property Federation (BPF) published its manual for building design and construction in 1983. It wanted to introduce a new system to change attitudes and alter the way in which the members of the construction team dealt with one another. The BPF also tried to remove some of the overlap of effort between quantity surveyors and contractors without the need for the traditional bill of quantities. This system for building procurement was little used and to some extent superseded by new forms of contract such as the Engineering and Construction contract. This had the support of Sir Michael Latham in his report, *Constructing the Team* (HMSO 1994), although its implementation has been slower than Sir Michael Latham had recommended.

The design and build method has gradually grown in popularity during the last three decades by offering single-point responsibilities, certainty of price and shorter overall durations. Management contracting was used in the 1970s and 1980s for large complex projects but construction management is now seen as a more attractive choice. An alternative, which is sometimes forgotten, is the client's own in-house design team, usually led by a project manager who supervises designers, cost specialists and contractors. This method accounts for a large part of construction work because it is the one commonly used in the public sector; but even this is being replaced with new systems, in particular the Private Finance Initiative, framework agreements and Prime Contracting in the defence sector.

Clients' needs

Client organizations are divided between those in private and public sectors although this distinction is becoming more difficult to define since the privatization of many national bodies. The private sector includes industrial, commercial, social, charitable and professional organizations, and individuals. The public sector is taken to mean government departments, nationalized industries, statutory authorities, local authorities and development agencies. The experience which a client has of building procurement ranges from extensive, in the case of a client with a project management team, to none, where a private individual may want a development only once in a lifetime.

Clients will usually identify their needs in terms of commercial or social pressure to change; by an examination of primary objectives such as:

1. Space requirements: the need to improve production levels, add to production capacity, accommodate new processes or provide domestic or social accommodation;
2. Investment: to exploit opportunities to invest in buildings;
3. Identity: to enhance the individual's or organization's standing in its market or society;
4. Location: could lead to a better use of resources, capture a new market or improve amenity;
5. Politics: mainly in the public sector.

The client's experience of building will influence his expectation of the industry. Property developers on the one hand can influence their professional advisers and the contractual arrangements, and select a contractor with the right commitment to meeting project targets. The main aim is to achieve a degree of certainty in the building process. On the other hand, individuals and inexperienced clients are guided by their advisers and contractors, and will be offered what the construction team think they need.

In general a client aims to appoint a team which he or she can trust and rely on to reduce uncertainties during a building's design, construction and use. This is achieved by control of the following:

1. The design: by designing to a budget, taking advantage of the contractor's experience, avoiding excessive use of new systems, designing for buildability, safety, security,

producing a good life expectancy and low maintenance, allowing flexibility for future change and employing environmental and energy efficient designs;
2. The time: by contractors accepting more responsibility for meeting completion dates, and designers being more aware of the importance of complete information well in advance of work on site;
3. The cost: by achieving realistic cost estimates and tenders which reflect the final cost, reducing risk of contractual claims stemming from poor documentation and late receipt of information, and avoiding delays which can cause loss of revenue and costly funding arrangements.

Many clients are prepared to pay for a good service and see these objectives being met through alternative methods of contracting.

The client has traditionally occupied a passive role in the construction process. Standard forms of contract require the employer to pay for work properly executed, give possession to the site on the agreed date and appoint his professional team to design, supervise and inspect the work and account the finances. A more realistic view is that the client is the most important member of the team because, as patron for the scheme, he identifies the need for the building and he must pay everyone who is directly or indirectly involved in the construction process. This is why we now see clients taking a more active part in the control of construction work and in part explains the emergence of construction management in the UK.

Contractor involvement

During the late 1980s, clients were looking for procurement methods which could quickly produce (or refurbish) large buildings with complex designs. Clearly the contractor needed to contribute to the design phase and continue to advise on design during construction. At the same time, where projects were less complex, design and build systems were being adopted for both building and civil engineering projects.

In order to respond to these different needs, contractors have developed a wider range of construction services, sometimes setting up separate divisions within a company. These specialist services often include an interpretation of a client's brief, the development of a design to meet cost targets, creating asset renewal and maintenance regimes and the provision of hard and soft services for the life of the building.

For construction management, as seen in the USA, to flourish, contractors must accept the responsibility for producing detailed drawings and cost-effective production techniques. Whichever method is used, there will usually be a number of tendering stages that encourage the parties to harmonize their aims and develop co-operation and trust, which did not always happen in the past. If this is the way ahead, then architects and quantity surveyors will concentrate on creating an outline of the client's requirements, providing financial advice and setting up independent monitoring systems on site. Partnerships between clients and contractors provide the benefit of more open relationships based on trust and co-operation. By relaxing many of the traditional contract conditions and formalities, the parties can achieve their goals of repeat business (avoiding the cost of tendering) and a less adversarial approach.

In civil engineering, there are generally fewer professional interests, and an engineer, whether working for a client or contractor, works in a similar way. Civil engineers understand standard documentation which is used for most engineering schemes. Contractors can, however, influence the design for civil engineering work significantly, and often submit tenders with alternative bids, which can offer substantial savings to a client. Again partnership arrangements have developed in contracting, principally in process engineering, water industries and where modularization and standardization have been used.

Partnering

During the mid-1990s, partnering emerged in a number of forms, partly to reverse the suicidal fall into institutionalized conflict with appalling relationships between contracting parties in the construction industry, and more recently contractors see this as a means of securing more work by creating a competitive advantage.

Attempts to foster co-operation between contractors and clients first appeared in standard contracts with the publication of the Joint Contracts Tribunal (JCT) Management Contract in 1987. The New Engineering Contract (NEC) introduced the principles of trust and cooperation to general contracting in the early 1990s and Sir Michael Latham's 1994 report *Constructing the Team* asked for core clauses to be added to the NEC contract to establish that the employer and contractor would undertake the project 'in a spirit of mutual trust and cooperation, and to trade fairly with each other and with their subcontractors and suppliers'. Sir Michael also recommended a key objective must be 'that "win-win" solutions to problems should be devised in a spirit of partnership'.

These developments are clear attempts to get the parties to construction contracts to work together with less adversarial methods of procurement. But is a positive working spirit the same as closing the gap between design and construction?

In many ways consultants and contractors still assume their specialist roles and prejudices without having precisely the same aims. It is difficult to imagine the prescriptive method of partnering, through carefully worded contracts, being successful. It is not appropriate to draw up a contract to say; 'you will agree with each other each time an unexpected problem arises'.

Perhaps this is the reason for the growth of contractor-led partnerships. It has often been observed that contractors have developed partnering schemes in order to add a powerful ingredient to what may be a highly competitive bid. For many years alternative tenders have been submitted by civil engineering contractors based on changes to the design or specification. More recently, partnering has been used as the basis for alternative bids to combine technical innovation with an offer to look for additional savings such as sharing site staff and testing equipment; continuous improvement; ensuring quality and eliminating claims.

So which approach is better for clients? A prescriptive arrangement embodied in a modified standard form of contract or the acceptance of a contractor's proposal with an ad-hoc verbal or brief partnering agreement? In order to answer this question, and explore the expertise of tenderers, many public and influential clients are asking for elaborate pre-selection submission documents (and interviews) whereby contractors must demonstrate a proven track record in partnering with other clients.

There is some evidence that clients are satisfied with partnering arrangements particularly when an element of competition has been provided at an early stage of the scheme. It is clearly encouraging to have a team working to a set of mutual objectives which can achieve a project within the budget, no claims and completion on time. Contractors have also benefited when work has been scarce by firstly securing the work, then receiving a reasonable gross margin and finally by sharing cost savings as the project develops.

Partnering is not about the allocation of risk. Risk will depend on the contract option: design and build, lump sum or prime cost, and the nature of the works. Unforeseen ground conditions, for example, can be a risk which can be minimized by open and frank problem-solving, but are by definition unpredictable. Partnering should ensure that the team (consultants, contractor, subcontractors and suppliers) work together with what Sir Michael Latham calls a 'shared financial motivation'.

It is worth noting that clients have been prepared to pay a fair price for a good job for hundreds of years. It is for the construction industry to prove that it can deliver the service that clients deserve.

Apportionment of risk

The procurement system, and associated contractual arrangement, will dictate the financial and other risks borne by the parties to the contract. Risk cannot be eliminated by choosing a particular form of contract, but will be shifted towards one party or the other. A guide to how the risks are divided for each contractual arrangement is given in Fig. 2.1.

Lump-sum contracts based on complete pre-tender design and full documentation spread a smaller risk of cost overrun evenly between the parties. Results may be further improved by using a selective list of tenderers, avoiding nominations, checking ground conditions, and reducing the guesswork needed by contractors at tender stage.

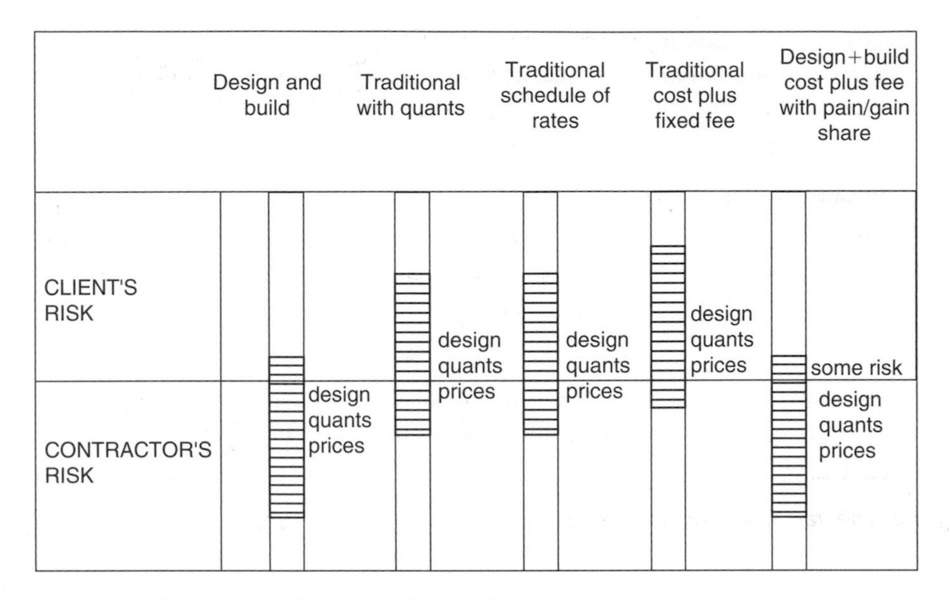

Fig. 2.1 *A guide to the apportionment of financial risk*

A contract where the price is calculated from a schedule of rates has two major problems:

1. The contractor is unable to identify the full extent of the work at tender stage, he is thus unable to plan and accurately assess his overheads, and
2. The client will not know the full price of the work until the contract is complete.

A cost reimbursement contract allows the contractor to claim all the prime cost of carrying out the work on an 'open book' basis and amounts are paid for site overheads and the management fee. Although this arrangement has the benefit of a quick start, there is little incentive to save time or costs. It would be unfair to say, however, that management contracting or construction management is more expensive than an alternative approach. All the package contracts are let competitively and the management fees are surprisingly low. It has been suggested that in the case of management contracting the management contractor makes more money by looking after the payments to package contractors. Today, for cost reimbursement contracts, costs to clients are capped by the contractor with a maximum price written into the contract.

With new framework (education sector) and prime contracting (defence sector) contracts, cost over runs between the target cost and maximum price (pain) are shared and cost under runs (gain) are also shared between the parties. The contractor is expected to tender on an early stage design, often basing the costs on historical data and agreed benchmarks.

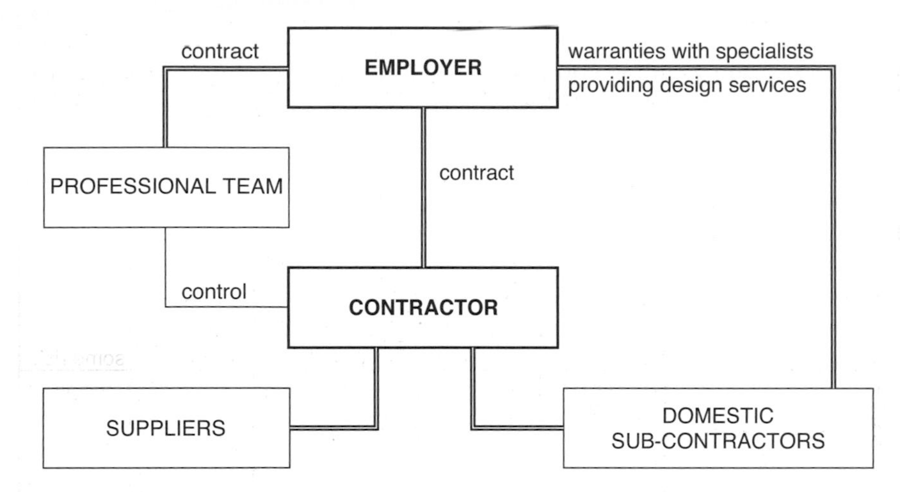

Fig. 2.2 *The traditional procurement method*

Traditional method

The traditional structure for project procurement shown in Fig. 2.2 is a sequential method because the employer takes his scheme to an advanced stage with his professional team before appointing a contractor. The consultant's role is seen as an independent one. The designer is employed to advise the client, design, and ensure the work is kept within the cost limit and complies with the standards required. A quantity surveyor can be engaged to give guidance on design costs and budgets, prepare bills of quantities, check tenders, prepare interim valuations and advise on the value of variations. Consultant structural and services engineers may be employed either by the client, or his advisers, to design the specialist parts of the project.

Separating responsibilities for design and construction is seen as the main reason for the move away from traditional contractual arrangements. The building industry suffers from the old distinctions between the professional interests and suspicion brought about by ignorance of each other's work. In civil engineering there is more freedom for individuals to move to and from consultants and contractors' organizations – there is an understanding of each other's point of view.

Instead of the direct appointment of consultants, many major building owners and developers make use of in-house project managers either to control independent consultants or to carry out all the design and financial control of the project. Project management is therefore seen as a management tool and not a procurement system. The JCT Standard Building Contract (SBC) and JCT Intermediate Building Contract (IC) with the Minor Works Building Contract (MW) are the most popular forms for building work. The ICE Form of Contract is used for most civil engineering work in both the public and private sectors, and GC/Works/1 is used for traditional civil engineering and building contracts let by central government departments. The continuing high sales of these contracts point to the commanding position of traditional methods.

Design and build

The design and build arrangement is an attractive option for clients. It simplifies the contractual links between the parties to the main contract (see Fig. 2.3) because the contractor accepts the responsibility for designing and constructing.

The benefits include: single-point responsibility, prices which reflect more closely the final cost to the client, inherently more buildable designs and an overlap of design and construction phases leading to early completion. A distinction is sometimes drawn between design and build and package deal, the latter being an agreement for the contractor to provide a semi-standardized or off-the-peg building which can be adapted to meet the client's needs. The contractual arrangement known as 'turnkey' allows a client to procure from a single contractor all the requirements of a scheme in the shortest possible time. Apart from the usual design and construction responsibilities, the agreement will often include land acquisition, short- and long-term finance, commissioning, fitting out and recruitment and training of personnel.

A design and build contractor may commission design and cost services from outside consultants or can employ a design team from within his own organization. Occasionally the client will ask the contractor to adopt a design started by his preferred

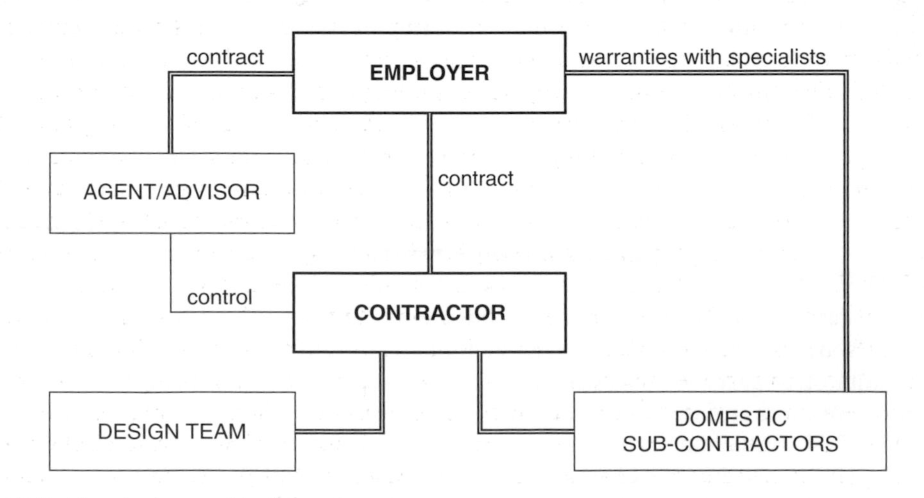

Fig. 2.3 *The design and build system*

consultant. Assigning (or 'novating') a design team in this way can arise when a client decides to switch to design and build from a traditional method, or more commonly, clients need to use a design team in order to carry out feasibility studies and gain planning consents prior to the appointment of a contractor. A designer needs a flexible outlook when he is thrust from the freedom of private practice, talking to a valued client, to being responsible mainly to the contractor, working to tight financial and time constraints, seeking solutions which satisfy the client's brief and enlarging the contractor's profits. It is generally believed that novation of consultants in design and build contracts is a recipe for friction between the parties and the contractor seldom maintains full control over his consultant team. The main drawback of novation of designers is that once the designers are fully acting in the employment of the contractor, the client finds himself without his team of advisors needed to check the tender and monitor post-tender changes and quality of work.

At tender stage the employer will introduce some competition, either open or selective tendering, which is followed by clarification of the agreement and negotiation. The National Joint Consultative Committee for Building published an advisory booklet in 1985 for private clients and public authorities planning to engage a contractor who would be responsible for the design and construction of a building project. The Code of Procedure for Selective Tendering for Design and Build stresses the importance of full and clear documents setting out the Employer's Requirements. The number of contractors invited to submit tenders in the form of Contractor's Proposals should be limited to three or four firms to reduce the high tendering costs. Each firm invited to tender for design and build work is carefully selected not only for its financial standing and construction record but its design capability and management structure for the work.

The Code of Procedure recognizes the need for longer tendering periods (often three to four months and longer on more complex schemes) and where extensive specialist work or negotiations with statutory bodies is required even more time may be needed. The employer must clearly state the form and content of the contractor's proposals and say whether the price alone will determine the offer accepted. The Code suggests that the design proposals and contract sum analysis are supporting documents which could be submitted separately. The contractor's proposals must be checked with great care because if there is a discrepancy in the employer's requirements the contractor's proposals will prevail, without any adjustment to the contract sum. The Code was replaced in 1997 by the Code of Practice for the Selection of Main Contractors, published by the Construction Industry Board. With the demise of the CIB, a Practice Note was published in 2002: 'Practice Note 6 – Main Contract tendering', which provides model forms of tender for a traditional bidding process.

Before entering into a design and build arrangement a client should consider the drawbacks. A contractor may offer a functional design which is not aesthetically appealing; he is inclined to develop a low-cost design with opportunities to increase his margins. A contractor might make a client's brief fit his own preferred solution;

the long-term life of a building might be overlooked and if the brief is vague, the client could pay an inflated price or take possession of an inferior building. A client may not realize the importance of independent professional advice. The cost of abortive designs and tendering is a heavy burden on contractors' overheads and eventually the costs will be passed on to clients.

It would be difficult to support these criticisms now that design and build is so well established. Professional contractors have taken a pride in their approach to this system that reduces conflict between the parties and gives the client single-point responsibility for design, time and cost.

In 1981 the Joint Contracts Tribunal published a new form of building contract with contractor's design, now the JCT Design and Build Contract (DB) 2007. Where the contractor is restricted to designing small discrete parts of the works and not made responsible for completing the design for the whole works, consideration should be given to using one of the JCT contracts that provide for such limited design input by the contractor and the employment of an architect/contract administrator. The new form was based on the 1980 standard form of building contract, with quantities. The contract is for a lump sum price payable in stages or monthly. In place of a bill of quantities the form provides for a contract sum analysis to assist those preparing interim valuations and valuing variations. It must be said, however, that the contract sum analysis only helps with significant variations and is of no use with day-to-day changes. The JCT published Practice Note CD/1B in 1984, which includes a useful explanation about the purpose and recommended structure for the contract sum analysis. Another reference document is 'Elements for Design and Build', published in 1996 by the Building Cost Information Service (BCIS).

With a greater number of public contracts let under the Private Finance Initiative (PFI), contractors have formed consortia to provide services which include the whole design and construction process together with responsibility for financing costs, fitting out, staffing, revenue collection, operation costs, maintenance and replacement. In this way, design and build practices are embedded in new forms of procurement such as PFI and framework agreements.

Management contracting

During the 1980s, clients were attracted to management contracting because it offered early starts to large-scale and often complex construction projects. The management contractor is appointed to work with the professional team, to contribute his construction expertise to the design and later to manage the specialist 'package' or 'works' contractors. He is responsible for the smooth running of the work on site so that the contract can be finished within time and cost. Although most major contractors have undertaken work using management contracts, there has been a feeling that it is not a final solution and a better method will evolve in the future. One

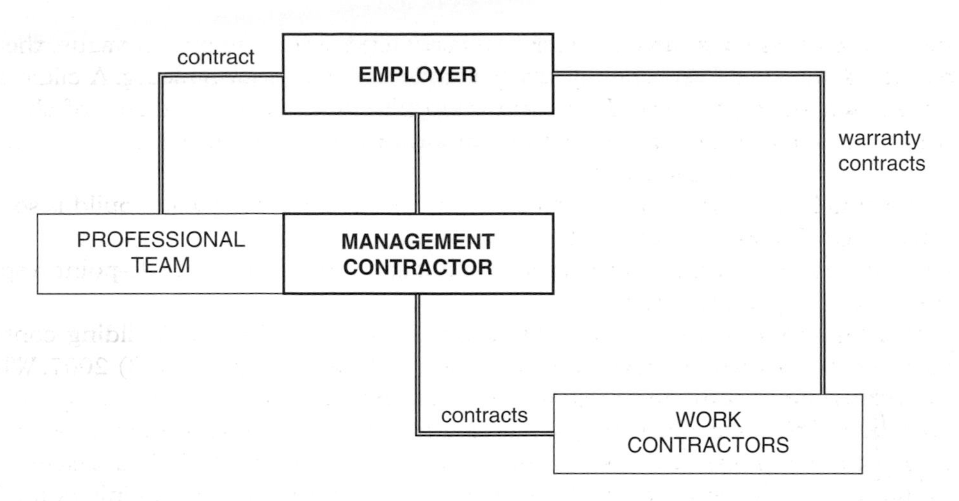

Fig. 2.4 *Management contracting*

development has been a combination of design and build and management contracting whereby the contractor produces a design and guaranteed maximum price, the work is later assigned to a number of major package contractors.

Figure 2.4 shows the system adopted by the Joint Contracts Tribunal in 1987 for its Standard Form of Management Contract, now the Management Building Contract (MC). The contract was needed to meet the growth in this procurement method and the need for standard documentation to replace the many improvised forms which had been used.

A management contractor is selected using the following criteria:

1. Experience of management contracting.
2. Quality and experience of project staff.
3. Fee.
4. Programme and method statement.

The consultants, grouped under the title of 'the professional team', prepare the drawings, specifications and bills of quantities for the various works contracts. The architect (or contract administrator) leads the professional team and issues instructions to the management contractor on behalf of the employer. The management contractor's role is in coordinating the design and preparing cost studies at the pre-construction stage. During construction his duties include placing and letting contracts with specialists, cost studies, setting out, provision of shared facilities, plant, and scaffolding, planning and monitoring the work, and coordinating all the activities on site, but not carrying out the permanent works. The management contractor's main duty is to cooperate with the professional team in the above functions.

The JCT Management Building Contract is not a lump-sum contract. The Employer pays the prime cost of carrying out the work and the fees for providing the management services. These fees will be either a lump sum or calculated as a percentage of the contract value. The recommended retention is 3% applied to both the management and works contracts, but not the fee because fees are calculated after retention is deducted. Trade discounts including the 2.5% contractor's discount are deducted from the management contractor to the benefit of the employer.

Clients are attracted to management contracting for the following reasons:

1. Construction can start before design is complete, and design can be changed during the construction phase;
2. Construction expertise is available to improve on the design;
3. Better coordination of specialist contractors through detailed planning of work packages and common facilities;
4. A contractor's knowledge of construction costs is used to maintain tight budgetary control.

Some problems have emerged, and management contracting has declined, except for a few very large projects. Contractors are less enthusiastic now that margins have fallen and sub-contractors have demanded prompt payments. For works contractors the conditions of contract are becoming more demanding with regard to the management contractor's right of set-off, liquidated damages, performance bonds and guarantees. The specialists often carry the burden of late changes to drawings and specifications which are more common when design development takes place during construction. The client cannot be sure of the final cost and will carry more risk. This is because the management contractor can pass on all the costs incurred for each trade, site staff and site facilities.

Construction management

In the USA, where the roles of the professionals are different, the client or his project manager will take a more active part in the construction phase. A construction manager is appointed as a professional consultant with powers to inspect work on site and issue instructions. There have been some spectacular building failures in the USA; a congressional inquiry in 1984 found that design quality can be impaired by excessive speed and cost-cutting exercises. Problems have been found when designers, who are often selected on a least-fee basis, pass preliminary designs to works contractors who produce the detailed drawings. This is a division of responsibilities which can lead to errors and legal action. Now that roles are far more developed, there is evidence that package contractors in the USA and Canada embrace their work in a professional manner without relying so much on being serviced by the main contractor, as so often is the case in the UK.

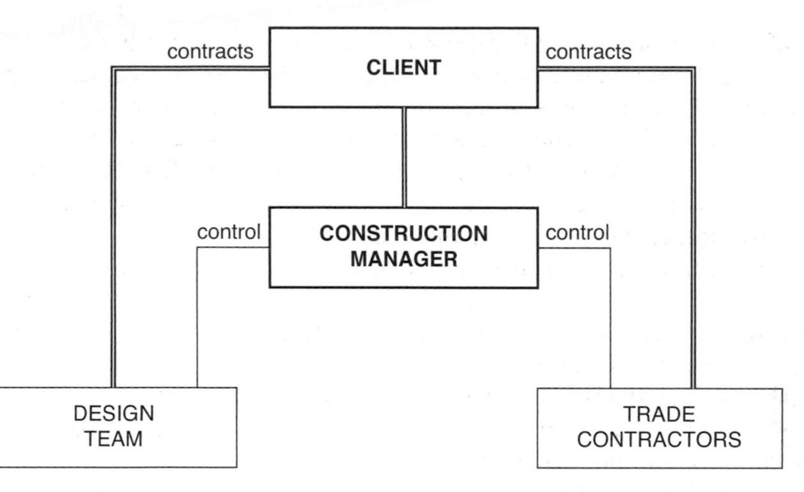

Fig. 2.5 *Construction management*

Construction management gives the client greater control over funds during construction because he has a contract with all the trade and specialist contractors (see Fig. 2.5). These contractors welcome the direct links with the client partly for the higher status this brings but more importantly because the lines of communications are clearer and payments are made sooner.

In the UK, some clients would not want to deal directly with sub-contractors or be involved in every problem of time and cost that could arise. Construction management has, however, grown steadily in the 1990s and there is ample evidence of experimentation by large property developers who want firm control of and involvement at all stages in building projects. Construction management also allows for change, and the delay of decisions until the latest possible time. In large businesses which rely on new technologies this can be important. Standard documentation has been prepared by the JCT and published as the 'Construction Management Agreement' (CM/A). Previously variants of main forms of contract have been drafted by large client bodies, and it is likely that these will continue to be used.

At tender stage, each specialist contractor receives specifications, drawings, method statements, and details of the scope of works from which each estimator can prepare his own bills of quantity. A number of onerous responsibilities are placed with package contractors, such as:

1. All risks associated with the preparation of bills of quantity which must include all work needed to complete the package whether shown clearly on the drawing or not. In some cases contractors must assess reinforcement quantities, for example, before the reinforcement is designed;

2. The need to complete elements of the design to the satisfaction of the architect;
3. Payment retentions may be kept for up to twelve months after the completion of the whole project;
4. Complex warranties for all contractors with design responsibilities.

This system of procurement is also finding favour with main contractors and their major sub-contractors. For example, a main contractor can engage an engineering services contractor as a construction manager to manage sub-packages.

Private Finance Initiative

The UK government is committed to the Private Finance Initiative (PFI) for major projects. This procurement option has been successful in delivering high-quality facilities for public services since the early 1990s. By June 2003, over 280 projects had been signed with a total value exceeding £35 billion. In other countries, there are similar new initiatives which use the alternative titles of Public–Private Partnerships (PPP) or Design, Build, Finance and Operate (DBFO).

The Private Finance Initiative provides a way of funding major capital investments, with the burden on the public purse being delayed beyond the construction stage. Private consortia are contracted to design, build, finance and in most cases operate new projects. The contract concession period is typically 25 to 33 years, during which time the building is leased by the client.

In July 2003, the Treasury report: 'PFI: Meeting the Investment Challenge' highlighted some key issues underpinning the PFI approach. The main points were:

- PFI investment in public services represents about 13.5% of total investment.
- Of 61 operational projects, 89% were delivered on time or early and all within public sector budgets.
- Benefits are achieved in new build large capital projects (small projects and schemes which are subject to rapid technological changes are less effective, and will be discouraged).
- PFI should only be used where it can be proved to be value for money.
- There is a need for the Government to ensure that value for money is not obtained at the expense of employees' terms and conditions.

There are concerns in certain sectors that there is a need to evaluate competitive interest and market capacity. In some cases there have been few bidders willing to engage in the tendering for large complex projects, especially those with a high proportion of refurbishment works.

PFI is seen to be appropriate where it provides: value for money, affordability, quality and best procurement practice. The private sector is likely to provide a better solution where the scope allows the following:

- Integrated provision of construction services, maintenance management and facilities management.
- A large proportion of risks are transferred to the contractor.
- The contractor can generate additional revenues.
- A contractor who is active in the market can bring economies of scale to the benefits to the project, for example two hospitals could have one helpdesk.
- Designs are produced which recognize whole life and operating costs.
- The project is delivered to quality standards which are linked to client satisfaction and long-term responsibilities.
- Bidders can bring innovation to the project.

There is often a considerable amount of work for bidders who follow the PFI route. The bid process can span periods in excess of two years in some cases. It is important for clients to view bid costs in proportion to the value of the project. So a £10 m college building would not be an appropriate scheme because the bidding costs could be £500–750 k for each bidder. The contractor needs to assess the likely profit in return for the investment in bid costs. This explains the higher profit needed for pre-contract risks and longer-term responsibilities for the project.

So what is different about the PFI procurement route for an estimator?

1. Understanding the requirements of the invitation to tender in order to develop the full scope of works. Examples of additional responsibilities are: creating a list of furniture and equipment that is suitable for the design; providing removal services to transfer furniture and equipment to the new facility; carrying out surveys of equipment in an existing building; and adding the cost of an extended bid period.
2. Advising on the constituents of the schedule of accommodation, checking the mathematics and monitoring drawn areas against the schedule.
3. Producing early cost plans for the finance team to test the likely unitary payment.
4. Developing target cost plans which meet the affordability and describe the scope and specification needed to achieve it.
5. Engaging with the life cycle surveyor and design team to select components and systems which give the optimum long-term solution.
6. Providing value-drawdown schedules for the financial model.

In 2005, the PFI process in the UK changed to a competitive dialogue process, shown in Fig. 2.6.

Stage	Bidders	Known as:	Comments
Information	open	Information is released to draw interest from the private sector	Memorandum of information (MOI): the document issued to those organizations expressing an interest in the project.
Pre-qualification	4–6	Pre-qualification questionnaire (PQQ)	Response to Official Journal of the European Union (OJEU).
Competitive dialogue	3–4	Invitation to participate in dialogue (ITPD)	Stage One of a competitive dialogue process ending in an interim submission.
	2–3	Invitation to continue the dialogue (ITCD)	Invitation to short list bidders following selection from the ITPD stage. Submission includes more detailed design proposals and a price for the works.
Final bids	2	Invitation to submit final bid (ITSFB)	Conclusion of dialogue and receipt of final bids.
Preferred bidder	1		The preferred bidder completes design, FM services and legal arrangements. Obtain full planning permission.
Final details	1	Preferred bidder (PB) stage	Agree final design, complete funding competition.
Close deal	1	Financial close (FC)	Contract award and often the start of construction.

Fig. 2.6 *PFI – UK competitive dialogue process*

Terminology

Awarding authority

The public sector body (department agency, NHS trust, local authority etc.) which is procuring a service through PFI.

Benchmarking

A procedure for testing whether the standard and price of services is consistent with the market standard, without any formal competitive tendering. This is usually

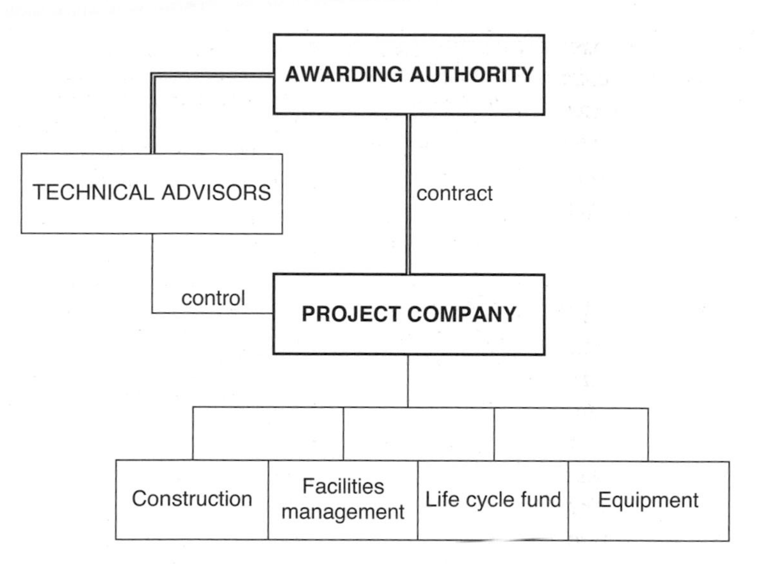

Fig. 2.7 *Private Finance Initiative*

adopted during the project concession period to ensure FM services continue to represent value for money.

Consortium

The group of private sector participants who have come together for the purpose of tendering for a PFI contract. Also becomes a Special Purpose Company (SPC) or Special Purpose Vehicle (SPV). The generic term is the Project Company, which is established by the preferred tenderer and is the contracting party for a project (see Fig. 2.7).

Market testing

A procedure for re-pricing the provision of services on periodic basis by means of competitive tender.

Output specification

The specification which sets out the requirements in non-prescriptive terms, so that the tenderers can determine how to provide the services.

Public sector comparator (PSC)

The PSC is an assessment of the scheme which includes capital costs, operating costs and third party revenues. The PSC is a benchmark against which value for money can be gauged. Clients use technical advisors to produce a reference project – sometimes called the Public Sector Scheme (PSS).

Service level specification

The specification given in the agreed project agreement setting out the standard to which the service must be provided. This is accompanied by an agreed performance monitoring regime.

TUPE

The Transfer of Undertakings (Protection of Employment) Regulations 1981.

Unavailability

The test for determining deductions from unitary payment by reference to standards for the provision of the facility.

Unitary payment

The payment by the awarding authority to the project company for the provision of the facility.

Variant bid

A bid which does not comply with the prescribed requirements of the awarding authority for a reference bid but which a tenderer is proposing as offering better value for money.

The PFI process

Step 1

The client identifies the need for a new building in the area. This is done on the basis of its own priorities. There is no private sector involvement in the choice.

Step 2

The client identifies the operational requirements that it is seeking to provide in the area concerned: for schools, the number of places required, the age group to be served, the curriculum to be taught, the methods of teaching. For a hospital the requirements might be the number of procedures to be carried out, a number of scanners, operating theatres and number of beds.

Step 3

Sites are identified for the proposed new buildings.

Step 4

Private sector firms are invited to express an interest in providing the facilities on a Public Private–Partnership (PPP) basis.

Step 5

A short list is drawn up and an invitation to tender is issued.

Step 6

The tenderers are asked to provide a site plan and building design, and to indicate the annual charge for building, equipping, maintaining and operating the new premises (including the grounds). At least 20% of the annual charge must be on a performance-related basis. Potential income from dual use of premises will need to be taken into account.

Step 7

The tender responses are evaluated on the basis of best value for money.

Step 8

The client names a preferred bidder and negotiates a contract on a PPP basis.

Step 9

The site is leased to the private contractor for thirty years. The premises are built, equipped and made available to the client from the agreed dates.

Step 10

All operational matters remain the sole responsibility of the client (governors and the council in case of a school and an NHS Trust in the case of a hospital).

Step 11

The buildings and grounds are maintained and operated to agreed standards over a 30-year life span at the expense of the private sector. The client pays an annual fee (unitary payment) or in some cases where a toll can be imposed on users the project can be financially free-standing.

Step 12

At the end of the thirty-year period the building and grounds will usually revert to the client.

Frequently asked questions

What is PFI?

'The involvement of private sector skills which offer the prospect of better value for money' – Gordon Brown, then Chancellor of the Exchequer.

Have there been any real benefits?

In February 2003 the National Audit Office published its report on building projects carried out under the private finance initiative. It found that of the 37 PFI construction projects looked at, less than a quarter came in over the original contract price. Previous experience of similar schemes indicated three-quarters of public sector schemes exceeded the price agreed at contract.

There were similar improvements for timely delivery with only a quarter of PFI projects delivered late, which compares with three-quarters of similar projects which ran over schedule previously.

What are the commercial foundations of PFI transactions?

In many cases it is a service which is being sold to the public sector over a defined period, for example the provision of computers to a government department. As an alternative, the project can be financially free-standing where the costs are recovered from private users. Examples are: the Second Severn Crossing and A69 through road tolling. There are many projects where the costs are met from public funds and partly from asset development, such as the shared use of the facility or development of other parts of the site.

People talk about better value for money. How does this happen for PFI projects?

1. Innovative and economical design calculated on whole life basis. Since the project company is responsible for maintaining the asset it is more likely to take care to secure quality of construction work.
2. Allocation of risks to the parties able to manage them at least cost. The National Audit Office Report published in February 2003 noted that some PFI contractors have actually lost money during the construction phase. This indicated that the private companies were absorbing risk that would previously have been borne by the taxpayer.
3. Greater exploitation of assets – additional income from shared-use of facilities or the sale of redundant assets.
4. Integration of design, build and service operation.

What are the real costs of all the elements of a PFI contract?

The costs are typically divided between:

- Initial construction costs (30%)
- Maintenance costs (10%)
- Hard and soft services (facilities management) (50%)
- Financing charges and project management costs (10%).

Are PFI bids tendered on a competitive basis?

Yes. The bid takes place in three stages:

1. Tenders respond to an advertisement by completing a pre-qualification questionnaire (PQQ stage).
2. Invitation to negotiate stage leads to firm bids from a short list of tenderers. A preferred bidder can then be selected.
3. Negotiation to complete the detailed contract terms with a preferred bidder.

There is also a cost check against a public sector comparator. This is a calculation showing what it would cost to provide the outputs from the private sector by a non-PFI route.

Are all risks transferred to the private sector?

Not necessarily. Sponsors need to understand which party is best placed to take responsibility for managing risks with costs being kept to a minimum, such as:

- design and construction
- planning
- routine repairs and maintenance
- demand for the facility, e.g. number of people crossing a bridge
- residual value
- technology and obsolescence
- legal requirements and regulation
- taxation
- project financing.

Risks will be different for different projects. For older buildings, for example, there might be a greater emphasis on maintenance. Furthermore, there could be defects which have been ignored for a number of years – backlog maintenance.

Who can provide the service to the public sector?

In the past, a special-purpose company (SPC) has been set up to engage in the PFI contract. This is formed by a contractor joining forces with an FM service provider and various developers and financiers.

How does a public body test the willingness of the private sector to engage in the tender process?

The simple answer is by discussing the scheme with potential operators before sending out the invitation to negotiate. For defence contracts this is achieved by holding an 'industry briefing' day. The key issues, which need to be discussed, are:

- size and length of the PFI contract
- the amount of asset provision and service delivery
- structure of the contract
- scope for transfer of risks
- management of people
- scope for shared use and alternative uses of the asset
- ownership of the asset at the end of the contract.

The estimating process for a PFI project

Enquiry documentation

The awarding authority will issue an Invitation to Tender (ITT) document to the bidders. The content will vary for different public sector schemes, but usually includes the following:

- A business case for the development.
- A fully documented public sector scheme (usually comprising drawings, cost plans, area schedules and financial projections).
- Output specifications.
- Indicative schedule of accommodation.
- Indicative room data sheets and equipment schedule.
- Lists of questions (answers to the questions become the framework for the submission documentation).
- Operational policies.

Estimating methods for PFI construction

It can be seen in Chapter 5 that the estimating method chosen will depend on the amount of information available and the design stage reached. For hospitals, there are five stages, as shown in Fig. 2.8.

Stage		Typical number of bidders	Duration	Estimating method
PQ	Pre-qualification	Over 5	1 month	No pricing required
ITPD	Invitation to participate in dialogue	3–5	4 months	Short elemental cost plan
ITCD	Invitation to continue the dialogue	2–3	6 months	Detailed elemental cost plan
ITSFB	Invitation to submit final bid	2–3	1	Detailed elemental cost plan
PB	Preferred bidder	1	8 months	Elemental cost plan with market testing

Fig. 2.8 *Stages used to procure a UK PFI project*

The first price submission, which is a formal offer, is likely to be at the ITCD stage. Since PFI projects are very large, and the design develops during the whole tender period, it is unlikely that contractor's bills of quantities can be produced in this period. There is a reliance on cost planning and approximate estimating techniques. Cost advice will be obtained from the major sub-contractors such as engineering services, civil and structural works, and cladding. It is also important to talk to specialists for demolitions, piling and asbestos removal.

The project team must read the extensive enquiry documents carefully to understand the requirements for the formal submission. In particular the financial aspects must comply with the forms to be submitted. It is sometimes difficult to assess how much the design should be developed, particularly when bidding costs must be kept under control.

At an early client meeting it would be wise to ask for a copy of the selection scoring system. Typically, marks are awarded to bidders for categories such as: technical solution (design and buildability), facilities management, asset renewal (life cycle) solution, project management, commercial and legal matters, financial proposal and capital cost.

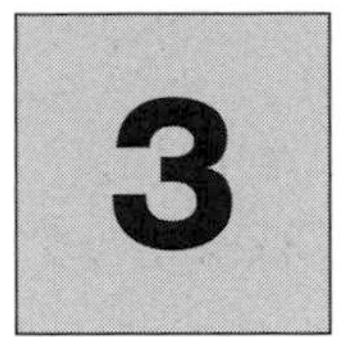

Forms of contract

Introduction

Standard forms of contract exist to identify the roles and responsibilities of the parties, and their agents; and provide rules to protect and direct the parties should things go wrong. Clients have a wide choice of standard contracts for construction work, in particular the forms used for building, which cover most of the common procurement systems. Standard conditions have been written by bodies such as the Joint Contracts Tribunal and Institution of Civil Engineers, following changing procurement methods in the industry – they seldom lead. The alternative approach would be to produce a common form of contract for all construction work whether in the public or private sector, building or civil engineering, English or Scottish Law. This idea is not new; it was one of the principal recommendations of the Banwell Report in 1964. Sir Michael Latham also addressed the problem in 1994. One of his recommendations was for public and private sector clients to begin to use the New Engineering Contract (NEC) family of contracts, in particular the Engineering and Construction contract. These ideals have not borne fruit, and there are more forms of contract published every year using different principles, terminology and apportionment of risk.

Where a standard form of contract is proposed, an estimator must carefully examine the information which will be inserted in the Appendix and note any amendments to the standard conditions so that the terms of the offer can be evaluated. An estimator should assess the cost of complying with certain terms and advise management of any onerous conditions that may influence the bid. Non-standard forms of contract are sent to the commercial department, company secretary or director so that the conditions can be evaluated before the final review meeting.

Essentials of a valid construction contract

Construction contracts are the same as any other contract, and in the end, will depend on general principles of law. A short definition of a contract is 'an agreement between two or more parties which is intended to have legal consequences'. In construction, the contract is generally for producing a building or part of the built environment, and can be entered in one of four ways:

1. Implied by conduct of the parties; a contractor may submit an offer and later have access to the site.
2. By word of mouth; typically where an offer is accepted by telephone.
3. By exchange of letters; common for small domestic works of extension, alteration or repair.
4. Using a written contract; the contract documents often include the enquiry documents, the written offer, minutes of meetings, tender-stage correspondence, a programme, a method statement and a formal contract with the agreed terms.

An estimator should keep a separate file containing all papers which will form the basis of the agreement. This is most important where negotiations take place after a formal offer has been made. If the estimator secures the work, he will need to present the contractor's undertakings to the construction staff at a handover meeting. The importance of written evidence cannot be overstressed because usually the formal documents will be the only evidence of what exactly had been agreed at the beginning of a project.

To make a contract valid and legally enforceable, certain simple rules are applied, as follows:

1. There must be an offer by one party and an acceptance by the other or others.
2. Each party must contribute something of value to the other's promise; a client is responsible for making payments and the contractor must complete the construction.
3. Each party must have the legal capacity to make a contract.
4. The parties must have exercised their own free will, without force or pressure.

A contract comes into existence when an offer has been unconditionally accepted. In construction the offer is the 'tender', 'estimate' or 'bid' and suppliers and subcontractors sometimes refer to their offers as 'quotations'. The term 'estimate' could be used in a wider context to mean a guide to how much something will cost. This ambiguity should be avoided wherever possible.

A contractor expects to receive an acceptance in clear terms from the client or his adviser. A letter of intent is often used to let a contractor know that he should prepare to start work. This statement should state clearly that all work carried out by the contractor and specialists, even if the contract does not follow, will be paid for in full.

An offer must be distinguished from an 'invitation to treat' which is an invitation for others to make an offer. In an auction sale, for example, an auctioneer invites offers which he may accept or reject. In a similar way, a client seeking tenders is not bound to accept the lowest or any bid. An offer cannot be accepted once it has terminated. Termination happens:

1. On death of either party if the contract is for personal services.
2. By the contractor withdrawing the offer.
3. After a specified time (usually stated in the tender instructions or stated by the contractor in his tender) or after a reasonable time.
4. When there has been outright rejection by the client, or where the client makes a counter offer, usually in the form of a qualified acceptance.

Although contractors and sub-contractors can withdraw their tenders at any time before acceptance, this practice can lead to many problems for the recipient. A main contractor, awarded a contract, could lose a large sum of money if a sub-contractor's offer, used in a tender, is withdrawn or changed. The main contractor should clearly state in his enquiry documents the acceptance period for sub-contractors' tenders, taking into account the requirements of the main contract and the possible delay in placing contracts. A contractor can reduce this risk by thoroughly checking quotations for sufficiency, completeness and compliance with the tender requirements. Clearly it is important to maintain up-to-date lists of reliable trade contractors.

Standard forms of contract

The standard printed forms of contract have been developed over many years to take account of the numerous events which could occur during and after a construction project. Contract law will of course deal will some of the problems, but there are many matters peculiar to construction which need clarification. Once these terms have been incorporated, they reduce the likelihood of disputes which can lead to adjudication, arbitration or litigation. Contract conditions are outlined by a reference being made to the standard conditions in the tender documents, with amendments to suit the particular project. The parties to most of the JCT contracts sign copies of the printed forms, which is not the case for the ICE and GC/work/1 forms, which could be used by reference to an 'office' copy. JCT contracts are now printed in two parts: the Agreement, which is signed by the contractor and sub-contractor, and Conditions of Sub-contract, which are incorporated by reference in the Agreement.

Some clients require a contract to be executed under seal; the standard forms have provision for this after the Articles of Agreement. A contract executed as a deed (or speciality contract) would allow an action to be brought within twelve years as opposed to six years for simple contracts. It is unwise to amend the conditions of a standard form because great effort has gone into producing a carefully drafted document with many links between clauses and other documents. Nevertheless, all contracts take effect by agreement and so standard contracts can be amended in any way the parties choose.

The standard JCT contract forms currently in use between client and contractor are:

1. Standard Building Contract – with Quantities (SBC/Q)
2. Standard Building Contract – without Quantities (SBC/XQ)
3. Standard Building Contract – with Approximate Quantities (SBC/AQ)
4. Design and Build Contract (DB)
5. Intermediate Building Contract (IC)
6. Intermediate Building Contract with Contractor's Design (ICD)
7. Minor Works Building Contract (MW)
8. Minor Works Building Contract with Contractor's design (MWD)
9. Major Contract Construction Contract (MP)
10. Construction Management Agreement (CM/A)
11. Prime Cost Building Contract (PCC)
12. Management Building Contract (MC)

Other standard contract forms currently in use between client and contractor are:

1. GC/Works/1 Government Contracts for use on major building and civil engineering projects with and without Bills of Quantities, and Design and Build.
2. DEFCON 2000 is loosely based upon GC/Works/1 but reflects Defence Estate's plans for best practice procurement in construction.
3. ICE Conditions of Contract – a family of contracts for civil engineering construction.
4. NEC Engineering and Construction Contracts are standard contracts that support the concept of partnership and encourage all parties to work together to achieve the client's objectives.
5. ACA Project Partnering Contract.

The Joint Contracts Tribunal Limited (JCT) is the company responsible for producing suites of contract documents and in operating the JCT Council. The Joint Contracts Tribunal is made up of bodies representing differing interests in building work, including the British Property Federation, Local Government Association, Construction Confederation, RIBA, ACA, RICS, and bodies representing the Scottish Building industry together with specialists and sub-contractors. The Standard Building Contract has variants that cater for contracts with contractor's design and without, with bills of quantities, without quantities and those with approximate quantities. The forms do not differ in substance, but describing and costing the work is more comprehensive with bills of quantities. Private and local authorities editions are no longer published. Each of the variants creates a lump sum contract: the lump sum is that which the contractor expects to be paid but is subject to adjustment in many carefully defined ways, mainly following the issue of an instruction. A bill of quantities is also used with the ICE Conditions of Contract Measurement Version but the conditions create a re-measurement or 'measure-and-value' contract where all the bill items will be re-measured as the work proceeds. The ICE conditions are alone in defining permanent and temporary works; the ICE form makes it clear that temporary works are solely the responsibility of the contractor except where they have been designed by the engineer. The GC/Works/1 and NEC contracts are used for building and civil engineering works. The NEC forms use non-technical language, which allows their use for a variety of construction and engineering projects. The basis for valuing work is also flexible – there are options for bills of quantity, activity schedules or a cost reimbursement basis. The GC/Works forms are also available in many variants, including major and minor works, design and build, with and without quantities, single and two-stage, construction management, mechanical and electrical, measured term and facilities management contracts.

The JCT Design and Build Contract (DB) main contract is appropriate where detailed Employer's Requirements have been prepared and provided to the contractor and the contractor is not only to carry out and complete the works, but also to complete the design. The employer will engage an agent (who may be an external consultant or employee) to administer the conditions. This contract can be used where the works are to be carried out in sections, by both private and local authority employers. Where there is a small amount of design by the contractor, the Standard Building Contract can be used.

Many new editions of most standard forms have been published in the late 1990s. This was in response to the Latham Report (1994) and many changes brought about by the Housing Grants, Construction and Regeneration Act 1996 and safety legislation. In particular many changes have been made for the following:

1. Electronic communications clause;
2. CDM Regulations;
3. Construction Industry Scheme (a contractor cannot pay a sub-contractor unless the sub-contractor has provided valid authorization in the form of a registration card or tax certificate);
4. Third party rights (contracting out);
5. Landfill tax (addition to fluctuations clause);
6. Contractor's retention bond;
7. Closer working relationships such as partnering.

The appendix section of standard forms enables the parties to insert provisions that vary from job to job, such as:

1. *Sums of money* for liquidated damages and insurances;
2. *Periods of time* for carrying out the work and making payments;

3. *Percentages* for retaining parts of the interim payments;
4. *Statements* giving the options which apply to the contract; an important example would be to show which clause has been selected for dealing with price fluctuations.

This information must be given to tenderers, otherwise they will make their own assumptions.

Sub-contract forms

Terms used

Nominated sub-contractors are persons whose final selection and approval, for supplying and fixing materials or goods, has been reserved to the architect. The use of nominated sub-contractors is very rare and the relevant clauses no longer exist in most contract forms.

Domestic sub-contractors are engaged where a contractor elects to sub-let part of the work with the written consent of the contract administrator. The ICE contract goes on to say that the contractor shall not sub-let all the works without the written consent of the employer.

'*Named sub-contractor*' is the term used in the Intermediate Form where the contractor is required to enter a (domestic) sub-contract with a firm named by the contract administrator.

Many contractors sub let large portions of their work to specialist contractors, the main exceptions being where reliable building workers are needed for difficult or small maintenance contracts. There are two arrangements for sub-letting work to domestic sub-contractors:

1. The contract administrator approves the sub-letting of the works to a firm of the contractor's choosing.
2. The contractor must choose a sub-contractor from a list of at least three names which have been included in the Specification, Schedules of Work or Contract bills.

The latter arrangement is used sometimes to replace named sub-contractors with a shortlist of specialists who may have expressed an interest in doing the work. Where large service installations are required, the quantity surveyor can send the drawings and specification to each of the sub-contractors on the list so the main contractors can avoid unnecessary duplication. The estimator just sends his enquiry letter with details of the conditions which the sub-contractor will be expected to sign.

If a single firm is named in the contract bills to carry out work that is measured, then it should in effect be a named sub-contractor and an alternative form of contract will be needed such as JCT IC or the JCT MP. If a client wishes to name a broad range of sub-contractors, then a more suitable form would be the Construction Management Agreement.

Main contract form	Ref.	Subcontract form	Ref.
Standard Building Contract	SBC	Short Form of Subcontract	ShortSub
		Standard Building Sub-contract	SBCSub
Design and Build Contract	DB	Standard Building Sub-contract with Sub-contractor's Design	SBCSub/D
		Design and Build Sub-contract	DBSub
Intermediate Building Contract	IC	Intermediate Sub-contract	ICSub
		Intermediate Named Sub-contract	ICSubNAM
Intermediate Building Contract with Contractor's Design	ICD	Intermediate Sub-contract with Sub-contractor's Design	ICSub/D
Minor Works Building Contract	MW	Short Form of Subcontract	ShortSub
		Minor Works Sub-contract with Sub-contractor's Design	MWSub/D
Major Contract Construction Contract	MP	Major Contract Sub-contract	MPSub
Construction Management Agreement	CM/A	Construction Management Trade Contract	CM/TC
Management Building Contract	MC	Management Works Contract	MCWK
Prime Cost Building Contract	PCC	Standard Building Sub-contract	SBCSub
GC/Works/1	GW	GC/Works Subcontract	GWS
		CECA Form of Sub-contract	
ICE Conditions of Contract	ICE	CECA Form of Sub-contract 'Blue Form'	
NEC Engineering Construction Contract	ECC	Engineering Sub-contract	ECS

Fig. 3.1 *Standard forms commonly used between contractors and their sub-contractors*

Most sub-contract forms are printed in two parts: the recitals, articles and subcontract particulars in one document and the subcontract conditions in the other. This could be to save money since only the former are needed each time contracts are signed.

Non-standard forms of sub-contract are sometimes used by main and management contractors to impose extra obligations and ensure that the sub-contractor is bound by the same conditions found in the main contract. The trade bodies which represent the views of specialist sub-contractors claim that their members have suffered under terms such as:

1. The 'pay when paid' arrangement, which means that a sub-contractor will be paid when the main contractor has received a payment. This practice is now negated by the Housing Grants, Construction and Regeneration Act 1996.
2. The 'discount fiddle' happens when the 2.5% discount for prompt payment is held by the main contractor, well beyond the agreed time.
3. Reduced attendances provided by main contractors, in some cases expecting sub-contractors to provide their own scaffolding, temporary services, disposal of rubbish and hoisting.
4. The sub-contractor's right to an extension of time might only be granted when the main contractor himself receives an extension
5. The main contractor can hold wide-ranging rights to take sums of money from payments, sometimes without having to prove that a loss has occurred.
6. A requirement for a sub-contractor to protect his work even when he is not present on site.
7. Badly drafted 'on demand' bonds and parent company guarantees irrespective of the size or stature of the company.

It is becoming more common for main contractors to be on the receiving end of some of these practices. In particular, some clients want set-off clauses and performance guarantees which can be taken 'on demand' and may be kept in place for a long time after the project is complete. Both main and sub-contractors when faced with such enquiries should submit their tenders with a statement asking to discuss the terms of contract with the client before entering a formal agreement.

Understanding contractual obligations

The estimator will enter the main contract obligations in the tender information form which is signed off by his directors. He has a duty to highlight any onerous conditions which will affect the obligations and risks the contractor is being asked to accept.

Contract obligations checklist

	Contract provisions	Received	To be submitted
Documents	The standard form of contract	DB	
	Contract drawings	No	1:200s
	Contract bills	No	No
	Employer's requirements	No	No
	Scope of works	Yes	No
	Contractor's proposals	No	Yes
	Contract sum analysis	No form	Yes
	Contract specification	Outline	Yes
	Priced specification or schedule of rates	No	No
	Contract cost plan	No	No
	Forms of bonds	No	No
	Health and Safety file	No – chase	No
	Activity schedule	No	No
Programme	Possession date	Yes	
	Date for completion	Yes	
	Phases or sections of work	No	
Contract	Retention	3%	
	Level of liquidated damages	£1000/wk	
	Provisions for advance payments	No	
	Stage or periodic payments	Monthly	
	Fluctuation provisions	No	
	Insurance requirements	See details	
	Requirement for PI insurance	Yes	
	'Fitness for Purpose' clause in ERs	Yes	Qualify
	Uncapped liability	No	
	Responsibility for ground conditions	Yes	
	Responsibility for reserved planning matters	Yes	Qualify
	Responsibility for possible asbestos	To survey	
	Responsibility for concept design	Yes	Qualify
	Responsibility for understanding the brief	Yes	
	Prohibited materials	None	
	Collateral warranties or third party liabilities	Yes	
	Obligations needing supply chain acceptance		
	Restrictions on use of supply chain partners	No	
	Contract with unknown project company	No	
	Access requirements for other parties	Yes	Qualify
	Provision for extension of time	Limited	
	Use of contractor's design portion in standard form		
	Collateral warranties from sub-contractors	Yes	
	Possession of site in sections	No	
	Electronic communications clause	Yes	

Fig. 3.2 *Contract obligations checklist*

MAIN CONTRACT SELECTION CHECKLIST

Procurement method	Lump sum	
	Measurement	
	Cost reimbursement	
Design	Employer	
	Part by Contractor	
	Contractor	
Cost control document	Bills of quantities	
	Schedule of rates	
	Priced specification	
	Contract sum analysis	
Payment	Stage	
	Periodic	
	Turnkey	
Roles and relationships	Client	
	Contractor	
	Design team	
	Specialists	
Time	Open	
	Fixed	
	Sections	
	Accelerated	

Fig. 3.3 *Simplified checklist for the selection of a contract*

Selection of contract forms

For many clients the choice of contract will be dictated by the type of work, size of contract and their position in society. A local authority carrying out a £2 million refurbishment contract, for example, is likely to choose the JCT Standard Building Contract, with approximate quantities. A contractor offering his services to design and build a factory unit will suggest the JCT Design and Build Contract (DB). Perhaps the most difficult decisions to be made by a client are the composition of the professional team and how financial risks will be shared. In particular he must decide whether to commission a bill of quantities or ask for tenders on a lump sum. Fig. 3.3 shows the primary elements which need to be considered. Clearly a non-construction client would need professional advice in selecting a contract that satisfies all his needs.

The Joint Contracts Tribunal publishes a guide to selecting the appropriate JCT form of contract which is available on their website: www.jctltd.co.uk.

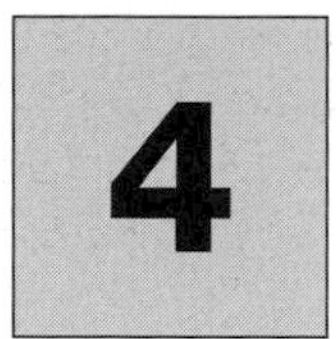

4 Tender documentation

Introduction

The key to a successful project often lies in the understanding and cooperation that is essential from all participants; each must be clearly aware of his duties and rights. The documentation is the vital link between design and construction.

Adequate and accurate drawings and specifications are indispensable if the team is going to achieve success in terms of quality, time and cost. Drawings in particular have served the construction industry well for hundreds of years as the primary means of communication. Unfortunately, poor specification writing continues to be a weak link in the information chain and leads to disputes, particularly in a competitive market where estimators will use a strict interpretation of the documents to arrive at the lowest tender. Another cause of friction is when bills of quantities differ from the drawings and specification. This often happens when the quantity surveyor is short of information from the designers.

Time spent on preparing documents, which aid the contractor's understanding of the work, will benefit the finished product. In 1964, the report of the committee chaired by Sir Harold Banwell stated:

> It is natural that a client, having taken the decision to build, should wish to see work started on site at the earliest possible moment. It is the duty of those who advise him to make it clear that time spent beforehand in settling the details of the work required and in preparing a timetable of operations … is essential if value for money is to be assured and disputes leading to claims avoided. It is also necessary for the client to be told of the need to give the contractor time to make his own detailed arrangements after the contract has been let, and of the penalties of indecision and the costs of changes of mind once the final plans have been agreed.

Tenderers will assess the quality of documentation, partly because poor information can add to the time wasted by site supervisors and partly because unreliable information can lead to claims. If the contractor has enough information he can avoid guesswork, include all the important items in his tender and will not need to add global sums for poorly defined elements of work.

For tenders based on RIBA Stage C designs, and benchmarked cost plans, design information can be limited but a different approach is used for the assessment of risk. This would be the case for PFI and framework contracts where a design and build tender is submitted prior to the preferred bidder phase. Detailed designs are developed at this post-tender stage.

Coordinated project information

The Coordinating Committee for Project Information was set up in 1979 to look for improvements in the way construction documents are produced and presented. The committee published its recommendations in December 1987 for drawings, specifications and bills of quantities for building work; and included proposals for ways in which the following problems may be overcome:

1. Missing information – not produced, or not sent to site.
2. Late information – not available in time to plan the work or order the materials.
3. Wrong information – errors of description, reference or dimension; out-of-date information.
4. Insufficient detail – both for tender and construction drawings.
5. Impracticable designs – difficult to construct.
6. Inappropriate information – not relevant or suitable for its purpose.
7. Unclear information – because of poor drafting or ambiguity.
8. Not firm – provisional information often indistinguishable from firm information.
9. Poorly arranged information – poor and inconsistent structure, unclear titling.
10. Uncoordinated information – difficult to read one document with another.
11. Conflicting information – documents which disagree with each other.

The Construction Project Information Committee (CPIC) encourages the use of CPI throughout the UK building industry. To endorse their work, Sir Michael Latham, in his 1994 report, said 'CPI is a technique which should have been normal practice years ago ... its use should be made part of the conditions of engagement of the designers'.

Drawings

Drawings are the most common means of communication for all types and sizes of project; the main exceptions being some maintenance contracts and minor works which can be scheduled or described in a written statement. The CPI initiative includes a production drawings code that gives advice on good practice for planning and producing drawings. The code stresses the need for careful co-ordination of the information, shown on drawings, with the other documents. One way to avoid mistakes is to replace specifications on drawings with reference numbers, which refer to the written specification. This could, however, lead to confusion on site if taken to an extreme case such as a drain layer asked to lay a drain R12/123 in a trench type R12/321. Would he need to be armed with the drawing and specification? Probably not; because designers understand the need for clear information for those working on site and on large-scale projects, and site engineers interpret the drawings for the operatives.

The CPI code is to be read with BS 1192:1984 'Construction Drawing Practice'. This British Standard has been rewritten during the 1980s and published in five parts. This revision was brought about by the need for international standardization of drawing practice; and many industrialized countries have taken part in the search for suitable conventions and methods. Part 5, dated 1990, is a guide for the structuring of computer graphic information. The aim of the new standard is to provide good drawing practice which will provide communication with:

1. accuracy
2. clarity
3. economy
4. consistency

between architects, contractors, civil engineers, service engineers and structural engineers.

BS 1192: Parts 1–4 have been withdrawn and replaced with CEN standards. BS ISO 128 is published in numerous parts to give general principles of technical drawing practice. BS 1192: part 5 is still published and being updated in 2007 to give catch-up with good practice in structuring and exchanging CAD information.

There are four main types of drawings commonly used in construction:

1. Survey drawings – which are based on a measured survey or an Ordinance Survey sheet; and are used to produce block and site plans.
2. Preliminary drawings – which are the designer's early interpretation of the brief.
3. Production drawings – include general arrangement drawings, layout drawings, assembly drawings, standard details such as those provided for highways drainage, schedules and additional detail drawings as necessary. They are used to go with applications for statutory approvals, to invite contractors to tender, and construction purposes.
4. Record drawings – are used to show a record of construction as it has been built and services installed. They provide essential information for maintenance staff.

Since the publication of SMM6, some drawn information can now be provided with bills of quantities. SMM6 recommended the use of bill diagrams to help describe an item of work.

In SMM7, general rule 5.3 states 'dimensioned diagrams shall show the shape and dimensions of the work covered by an item and may be used in a bill of quantities in place of a dimensioned description, but not in place of an item otherwise required to be measured'. The intention is for these diagrams to be prepared by the quantity surveyor and included in the bill of quantities. Often this has not happened with either SMM6 or SMM7. This might be because bills are produced using text-based computer systems and more drawings are now sent to contractors at tender stage.

Specifications

A specification is prepared by an architect or consulting engineer to provide written technical information mainly on the quality of materials and workmanship. The specification would be a contract document in its own right if the contractor tenders on the basis of drawings and specification only. Where bills of quantities are used for building work the specification is included with the bill of quantities as preambles. In this way the specification again becomes part of the contract documents.

There are some standard specifications published for civil engineering contracts – in particular specifications for highways and the water industry. A bill of quantities for civil engineering work will include specification clauses and a preambles section, which is used to define any departures from the standard method of measurement.

The designer notes the matters needing detailed specification clauses as he prepares the drawings. The quantity surveyor will advise on a proper format for the bill of quantities. On small contracts, where a PQS is not appointed, an architect could produce a specification which is broken down into parcels of work. The contractor would be expected to price the document to assist post-contract cost control, such as the preparation of valuations. In this context, this document is sometimes called a schedule of works or priced specification.

Another document in the CPI suite is a code for specification writing. The Project Specification Code is guide to good practice.

Many architects, engineers, quantity surveyors and contractors will subscribe to the National Building Specification (NBS), which is written in line with the Common Arrangement of Work Sections (CAWS). The National Building Specification is a library of clauses, regularly updated, using either the CI/SfB classification or the recommended CAWS method, which divides building into over 300 work sections which aim to reflect the way work is sub-contracted. In broad terms CI/SfB relates to the elements of a building and the CAWS is in trade order. Normally, only a fraction of the work sections will be used on a simple project.

Specifications are prepared by design teams using their own procedures and often vary widely in coverage and technical content. For a design and build contract, the contractor will direct the designers on the level of information needed for market testing the design and inclusion in submission documents.

It has been said that specifications have lagged furthest behind drawings and bills for quality and helpfulness. This is probably unfair where people use the NBS service. The content needs to be carefully edited and changes thoroughly researched with assistance from manufacturers and specialist sub-contractors.

There is a danger that specifications may be ignored by contractors, sub-contractors and suppliers because they:

1. contain many standard clauses which are not relevant to the job;
2. are usually too long;
3. may be a collection of protection clauses, for example 'to the best quality', because the designer is not sure what quality to specify;
4. are sometimes out of date;
5. refer to expensive standards which the contractors do not have.

Traditionally the architect has been responsible for the specification, but may delegate the printing to the PQS. The CPI initiative assumes that the designer provides more reliable specification information before tender stage. The PQS must ensure the bill descriptions do not conflict with the specification. With the introduction of SMM7, bill descriptions include cross-references to the specification, which will remove duplication.

The Project Specification Code recommends improvements, so specifications will be:

1. Complete – covering every significant aspect of the work.
2. Project specific – produced for the project, without irrelevant material.
3. Appropriate – for available materials and skills; and can be checked and standards enforced.
4. Constructive – helping all the parties to understand what is expected of them.
5. Up-to-date – using current good building practice and most recent standards.
6. Clear – economically worded.

Bills of quantities

The traditional purpose of bills of quantities is to act as a uniform basis for inviting competitive tenders, and to assist in valuing completed work. Bills of quantities are firstly designed to meet the needs of estimators, although some estimators say the bill format has changed to assist the consultants, in cost planning exercises, through the widespread use of elemental bills. On the other hand, now that contractors engage in cost planning, they benefit from having access to cost data in an elemental format. They can use elemental data to advise clients on cost plans for future projects.

A contractor can also make use of the bill of quantities in many ways, for example:

1. To plan material purchasing (note the danger in ordering from a bill: the contractor should always order materials from drawn information and the specification, making the contract administrator aware of any differences).
2. Preparing resourced programmes.
3. Cost control during the contract to ensure work is within budget.
4. Data collection during construction for bonus systems and feedback information for estimators.

Unlike drawings and specifications, there have been rules for measuring building work for many years. The first edition of the *Standard Method of Measurement for Building Works* was published in 1922 and has been a compulsory document since its incorporation in the RIBA (now JCT) contract 1933. The civil engineering methods include rules for highways and the water industry but the publication for mainstream civil engineering works is the *Civil Engineering Standard Method of Measurement* (CESMM3) now in its third edition (1991).

Bills of quantities for building are divided into the following sections:

1. preliminaries
2. preambles
3. measured work
4. prime cost and provisional sums.

There are number of formats for civil engineering bills of quantities. CESMM3 gives the following sections:

1. List of principal quantities
2. Preamble
3. Daywork schedule
4. Work items (Class A General items may be grouped in a separate part of the bill of quantities).

In both sectors of construction, the estimator prices sections 3 and 4 and the specific items described in the preliminaries, having taken full account of all the requirements in the other sections.

The preliminaries (general items) section gives general details about the project and contract conditions, as follows:

1. Description of the work, location of the site, site boundaries, names of parties, and lists of drawings.

2. The form of contract used, with any amendments clearly defined, with contract appendix details giving information such as the retention percentage, liquidated damages, possession and completion dates and fluctuation provisions.
3. Specific requirements which should be priced by the contractor as fixed or time-related items to reflect the actual costs arising from supervision, site accommodation, temporary works, site running costs, general plant, transport, client's requirements and safety.

CESMM3 and SMM7 provide for fixed and time-related items so that a contractor can show the cost of bringing plant or facilities to site, their maintenance during the job and removal on completion. The SMM7 Measurement Code suggests that prices should be split between fixed and time-related sums only if the tenderer wishes to do so. He rarely does! There should also be space in the preliminaries section of a bill for the contractor to add to the list of items to suit his particular methods of working. In CESMM these are called 'method-related charges'.

In bills of quantities for building work, the preambles contain specification clauses which provide information about the expected type and standard of materials and workmanship. They should relate to the work in the bill and so reduce the length of work descriptions.

The measured work section of the bill of quantities is divided into trade or element headings and measured according to the rules of a standard method of measurement. SMM7 defines its role by the statement 'The standard method of measurement provides a uniform basis for measuring building works, and embodies the essentials of good practice. Bills of quantities shall fully describe and accurately represent the quantity and quality of the works to be carried out.'

The JCT Standard Building Contract requires the use of the standard method of measurement where the contract includes bills of quantities. Clause 2.13.1 states 'the contract bills are to have been prepared in accordance with the Standard Method of Measurement'.

Accuracy in preparing a bill is essential because the contract conditions allow the contractor payment for any omission or error in description or quantity. Clause 2.14.1 states 'an error shall be corrected … any correction shall be treated as a variation …' and clause 2.3.1 states 'All materials and goods for the works shall so far as procurable be of the kinds and standards described in the contract bills.'

SMM7 begins with general rules for preparing bills, followed by details of preliminary particulars and about 300 work sections under 24 main headings. Rule 4.1 is an example of a rule of particular interest to an estimator:

> Dimensions shall be stated in descriptions generally in the sequence length, width, height (or depth). Where ambiguity could arise the dimension shall be identified.

Where work can be identified and described in a bill of quantities, but the quantity cannot be accurately determined, an estimate of the quantity can be given and identified as an 'approximate quantity'. This will typically occur when dealing with ground problems such as stone filling to make up levels, or maintenance work such as cutting out defective rafters.

A provisional sum in a bill of quantities is for work which cannot be described and given in items, which follows the measurement rules. SMM7 introduced two kinds of provisional sum: defined and undefined, both for work which is not completely designed. 'Defined' means the nature and quantity of the work can be identified, and the contractor must allow for programming, planning and pricing preliminaries. 'Undefined' means that the scope of the work is not known, and the contractor will be paid for all costs associated with carrying out the work, planning the work, and overheads which are reasonable.

A contingency sum is often included in a bill, as a provisional sum, for unforeseeable work, such as difficult ground conditions. The reason for its inclusion is not stated in the bill. The sum is spent at the discretion of the architect/contract administrator. SMM7 does not mention the contingency sum.

A prime cost sum is provided in a bill of quantities for work to be carried out by a nominated sub-contractor (SMM7 A51) or for materials to be obtained from a nominated supplier (SMM7 A52). Work by statutory authorities is now given as a provisional sum (SMM7 A53). SMM7 does not define PC sums to the extent found in SMM6 presumably because the form of contract deals with this. The term 'prime cost' is also used in connection with:

1. An allowance for the cost of a material such as bricks when the final selection has not been made; for example, facing brickwork PC £250.00 a thousand (the estimator must be told how to deal with waste, transport and other on costs).
2. The basic cost of labour, materials and plant in cost plus arrangements such as daywork contracts and some management contracts. SMM7 now gives dayworks as a provisional sum (A55).

SMM7 provides for certain drawings to be issued to contractors at tender stage. More detailed guidance on which drawings are needed is given at the beginning of each work section in SMM7. The following drawings are considered to be essential:

1. block plan
2. site plan
3. plans, sections and elevations.

Component drawings are required by general rule 5.2 to show the information necessary for the manufacture of components. The work sections, which require component drawings, are listed in Appendix 2 of the Measurement Code.

Bill formats

The development unit, which prepared SMM7, made some general recommendations for good practice, as follows, and included some of them in the SMM7 Measurement Code:

1. The full benefits of the CPI initiative will be gained if bills and specifications are prepared using the CAWS. By the late 1990s this recommendation had been implemented and well-established for fully documented building schemes.
2. Items for separate buildings should be kept separate, by providing separate bills.
3. Items for external works should be given in a separate bill.
4. Provisional sums, prime cost sums and dayworks should form a separate section at the end of the measured work part of the bill (avoiding confusion during the tender stage). Provisional sums inserted in the preliminaries bill cause a great deal of confusion and can be missed by an estimator expecting to find all written-in sums grouped in a dedicated section.
5. The summary should be at the end of the bills of quantities.

Estimators have a strong preference for trade bills, which separate work strictly in accordance with the measurement rules and trade headings of SMM7. This is convenient for sending enquiries to suppliers and sub-contractors, but does not help in showing the relative quantities for each building in a development.

Elemental bills relate to the functional parts (or elements) of a building, for example upper floors, roofs, and external walls. This has the benefit of helping the quantity surveyor check his cost analysis and collect data for future cost exercises, and the estimator can find the location of work. The main disadvantage is that it produces a longer bill which is not only less efficient to prepare but will add to the work of the estimator. He must bring together items for each trade from various parts of the document, which can produce a great deal of paperwork.

Sectionalized trade bills could be used to overcome the disadvantages of the elemental bill. For estimating purposes the trade order bill is subdivided into elements. If each element is printed on separate sheets, it is possible to assemble the bill in trade *or* elemental order.

Computer packages are available for producing bills of quantities and subsequent financial control. They are usually based on a library of standard items that can be called up by using codes or by accessing a hierarchal database through menus. Measurements can be entered either manually or using digitizers and the computer will sort the items before printing the complete bill of quantities.

CPI and the estimator

The CAWS was developed to align packages of work more closely with the pattern of sub-contracting in the industry. For example, SMM7 now clearly distinguishes between many cladding methods and materials in group heading H, Patent glazing, Curtain walling and many kinds of sheet cladding. Unfortunately, this fine subdivision has some awkward results. For example, an enquiry for plumbing will include measured work from Group R Disposal systems, Group S Piped water supply systems, Part T Mechanical heating, Group N Sanitary appliances and Group Y for pumps and calorifiers. Furthermore, with an elemental bill any of the 300 work categories can be repeated for each element.

Two of the objectives of SMM7 were: (1) to simplify bills of quantities and (2) to develop a method which could help with computer applications. To an extent, modern bills of quantity have been accepted by estimators because they have developed an understanding of the coding system and descriptions have not been shortened by the amount envisaged when SMM7 was published. Many quantity surveyors have avoided a total reliance on specifications; they are aware that estimators need more than an abbreviated description.

With the SMM7 there are now shorter bills of quantities. Many items have been removed where they had little cost significance. Other items have been grouped, again to lessen the number of measurable items of work. The nominal size of bar reinforcement is stated but its location is not. This means that the bill rate for 12 mm reinforcement in an eaves beam will be the same as 12 mm bars in a ground floor slab. An estimator may be able to identify the weights in each location by studying the drawings and bar schedules, but will all the estimators and sub-contractors do the same? Since SMM6 was introduced, estimators have been faced with formwork measurements grouped in height bands, which effectively mean that he does not know the actual quantity. As an example, 200 linear metres of formwork 500 mm–1.00 m high could be as little as 100 m^2 or as much as 200 m^2 of shuttering. It would take some time to measure the real area from the drawings received at tender stage.

The move towards computer-aided billing and estimating has been difficult, and many doubt whether SMM7 has helped. On first sight, the tables of measurement rules appear to be an aid to all those involved in computer-aided bill production and pricing. Unfortunately there have been some problems:

1. There are too many rogue items in an average bill, which do not match a standard coding system.
2. Libraries of standard descriptions do not use the numbering system given in SMM7.
3. Computer packages have moved away from code numbers for items, preferring to use windows of items from which relevant descriptions and resources can be selected.

It is also argued that the standard method of measurement is not a method for producing bills, nor is it an aid to pricing bills. It is purely a set of rules about how work is measured and what is to be included in an item (item coverage).

The civil engineering standard method of measurement CESMM3 states that the system of work classification adopted by the method should simplify the production of bills of quantities, making the use of computers easier. The foreword to the First Edition encourages the use of work reference numbers to identify work items. This uniform (and coded) description of work was seen as a way to standardize on the layout and contents of bills of quantities; and the engineer is recommended to use the standard method numbers in bills of quantities. This recommendation does not exist in SMM7. The use of code numbers as item references in civil engineering bills causes some confusion when inputting items in a computer system. Many estimators change the reference system to the familiar A, B, C etc. format and use the code numbers as a sort code reference.

Every estimator, whether working for a contractor or sub-contractor, must understand the coverage rules of the standard method, which applies to the contract. This is important where sundry items are now included in the main work item. For example, in SMM7 formed joints in in situ concrete are deemed to include formwork, and working space allowance must include the extra cost of work below ground water level and breaking out existing hard materials.

By making the detailed specification the central reference document under CPI, the way in which estimators work has changed. Enquiries to sub-contractors and suppliers must include all relevant specification clauses, preliminary section items and appropriate drawings, otherwise the prices will not reflect the true value of the work. Bill descriptions are shorter by adding references to the specification. The following example illustrates the problem:

Forming cavities in hollow walls 60 mm wide; wall ties spec F30:310; cavity insulation spec F30:560, 30 mm thick	115 m²

If a sub-contractor receives an incomplete enquiry, he is likely to guess what he is being asked to fix. Fortunately many quantity surveyors have recognized this problem and have enlarged item descriptions so that their meanings are clearer, for example:

Forming cavities in hollow walls 60 mm wide; stainless steel wall ties, F30:310 as System Ties Ltd, 210 mm long; Becker rigid board cavity insulation, F30:560, 30 mm thick fixed to ties with retaining clips	115 m²

Experience of pricing documents, which have been produced using the CPI guidelines, shows that some new problems have emerged, as follows:

1. Many specifications have no page numbers. The explanation is that the estimator must use the NBS codes to find the relevant clauses. The problem for the estimator is that the page numbers are needed for printing and distributing pages to sub-contractors.
2. There is confusion with the way work section numbers are used in specifications. On one page the estimator might find clause 310, which is for laying bricks, and on another page clause 310 could be for cavity wall ties. The problem is that the work section reference is missing; in the first case F10.310 and the second should be F30.310. The work category numbers must be repeated on each new page if this problem is to be solved.
3. Some sub-contractors have argued that they received the bill but not the specification; with SMM6, the bill description often had enough detail to price the work. This may be the estimator's fault but in some cases the specification references are more complex. The estimator might find the correct clause referred to in the bill but not notice that the specification clause includes references to other clauses. For example, a patent glazing specification could itself refer to a separate glazing specification, which the estimator must also send to the patent glazing sub-contractor.
4. Defined provisional sums are being used incorrectly. The tender documents should provide information about the nature of the work, a statement about how and where the work is fixed, quantities to show the scope of work, and any limitations. It is common to see defined provisional sums such as: 'drainage outfall to culvert' or 'additional dry-rot treatment'.
5. The number of drawings needed by sub-contractors at tender stage has increased dramatically. This is due to the reduction in the number of bill items; or as some would say, 'the quantity surveyor doing less work'. Many contractors have incurred an increase in printing costs since the introduction of SMM7. This may also be due to smaller margins and the need to ensure that sub-contractors will tender on exactly the same basis as the main contractor.

Now that main contract bids rely heavily on quotations from sub-contractors, the estimator must exercise great skill and care in dealing with changing procedures and new methods of measurement for bills of quantities. The PQS still has the responsibility to provide adequate information for the estimator to price. As SMM7 insists, 'More detailed information than is required by these rules shall be given where necessary in order to define the precise nature and extent of the required work.'

Documents used as the basis of a tender

The basis of the tender will dictate the way in which the contractor will be paid and the relative accuracy of the estimate. The contractor's bid will be for one of the following:

1. Fixed price contract: where the sum of money is stated in the contract as payment for work, the payment may be adjusted according to strict conditions in the contract.
2. Measurement contract: will allow the contract sum to be calculated later, usually as the aggregate of various rates submitted by the contractor. The contract sometimes includes a target price.
3. Cost-reimbursement contract: an arrangement whereby the cost, whatever it may be, will be paid by the client on the basis of the actual cost incurred by the contractor, plus overheads and profit.

Fixed price contracts

The price is fixed in advance but is subject to variation under the terms of the contract. This could include a fluctuations clause to pay for the increases caused by inflation. This definition leads to much confusion in the construction industry where 'fixed price' is the term for a price, which will not be subject to fluctuations. An arrangement which is not subject to fluctuations is better described as 'firm price'.

Lump sum contracts are the simplest type, where a lump sum offer is made by a contractor to carry out the work, which might be outlined on drawings and described in a specification but no quantities have been prepared. This is the usual form for a small job carried out by a local builder. Where a full set of working drawings and a specification are available, a drawings and specification ('plan and spec') arrangement is popular for small projects. The main advantage is the saving in time and money needed to prepare a bill of quantities. The client will also have a reasonable estimate of the total cost before the contract is signed. For small contracts where the client's requirements are clear and there are good drawings and specification, this can be a useful way to enter a contract. There are, however, some serious drawbacks. Each contractor must prepare his own bill of quantities and the employer must bear in mind the time needed during the tender stage. If construction details or specification requirements are missing, it is common to find each contractor tendering on different assumptions. The contractor must allow a contingency for the risk of making mistakes in taking off. There will be no detailed breakdown of the tender sum which would be needed for interim payments and for valuing variations. To overcome some of these disadvantages, the specification should include a description of work in a series of numbered items, each of which is to be priced.

Bills of quantities provide the most detailed basis for estimating cost. Each contractor tendering for work will be familiar with their use and can save wasteful effort in preparing quantities for the same building. They represent a clear list of items included in the contract and a schedule on which variations may be valued. Bills of quantity give a fair basis for competition, and a firm contract sum is known in advance. The main disadvantages are the time needed for the accurate preparation of bills (less of a problem with computer techniques) and the risk carried by the client for quantities. Firm bills of quantities remove the onus for correct quantities from the contractor but may inflict higher charges on the employer if discrepancies exist between the documents.

When a contractor tenders for a design and build project, he prepares his own bill of quantities (from his own drawings and specification) in order to invite sub-contract bids and arrive at a cost for direct work. Where the design team is novated to the contractor, construction drawings and specifications are usually well advanced prior to tender stage.

The contractor's proposals include a contract sum analysis. The purposes of the contract sum analysis are:

1. To value changes in the employer's requirements;
2. To value provisional sums given in the employer's requirements;
3. To allow the use of price adjustment formulae where they apply.

The contract sum analysis should be divided into sums of money for design work carried out before and during construction, and the following:

1. preliminaries;
2. provisional sums;
3. trade headings similar to those in SMM6 or SMM7.

Work in different buildings and external works is usually shown separately. Alternatively, and more commonly in practice, the client's agent produces a list of items for a contract sum analysis using elemental headings in order to compare the tender against the elemental cost plan set up as the scheme budget.

Bills of approximate quantities provide a fair basis for tendering when drawing details are not complete. The bills will represent an estimate of the quantities of work in the project. By definition, the work will be subject to remeasurement and a firm value will not be known at the start of the project. This method is commonly used with refurbishment work where the full extent of the work cannot be accurately determined. The quantities set out in bills of quantities for civil engineering are the estimated quantities and are not to be taken as the actual quantities; the actual quantities are measured during the construction phase.

Measured contracts

The total cost of a contract can be calculated by measuring the work as it advances on site and pricing the measured items using the rates given in an agreed schedule of rates or approximate bill of quantities. A schedule of rates lists all the items likely to arise, in a similar way to a bill of quantities, but no quantities are included. A schedule of rates is also used with drawing and specification contracts to value additional work. There are two principal types of schedule:

1. A standard schedule of rates, issued or published by an employer, will usually list standard items and rates, and the tenderer is asked to submit an overall percentage addition or deduction to reflect current pricing levels. Since the tender is a single figure, contractor selection is simple. Schedules of this sort enable orders to be placed before the project details are complete.
2. An ad hoc schedule of rates is a pricing document prepared for a particular job. Only those items needed for the project will be incorporated. This type of schedule is difficult to use because, in the absence of quantities, tenders are difficult to compare and the value of the project is not known at the start. An ad hoc schedule should contain approximate quantities to help overcome these problems. With all schedules of rates used at tender stage, the estimator is unable to foresee the full extent of the work. Contractors have been asked to quote for drainage trenches, for example, without knowing the ground conditions. Should the contractor assume that the ground conditions were good, and free of obstructions and other services? If he does then there is a chance that he would ask for reimbursement for additional costs for bad ground conditions.

Cost reimbursement contracts

The basis of this method is for the contractor to be repaid with the prime cost of completed work as defined in the contract, and a management fee to cover overheads and profit. The fee can be based on a percentage of cost (cost plus percentage contract) or a lump sum based on the estimated project cost (cost plus fixed fee contract). The advantages of this method are: the project can start quickly, the contractor can contribute to the design, competition can be introduced through the size of the fee, and the contractor is unlikely to cut corners. The disadvantages may be: the contractor has little incentive to save on time and resources (in some management contracts if the construction costs rise the fee to the management contractor rises), the client is unable to predict the total cost accurately, and it can be tedious to calculate costs during the construction stage. It should be remembered that most of the work is carried out by package contractors who tender for work on a traditional bill of quantities.

Formal tender documents

Formal invitation

The Code of Procedure for Single Stage Selective Tendering gives an example letter. The letter is not long because essential information is normally set out in the tender documents. The letter is needed to tell the contractor which drawings have been sent, arrangements for site visits, date for return of tender and how the tender should be submitted. If the tenderer wishes to decline an offer he should have done so at pre-selection stage. The client should issue the tender documents on an agreed date in order to enable the contractor to plan his estimating workload. A typical invitation to tender letter is given in Fig. 4.1.

Bill of quantities

If a priced bill of quantities is required with the tender then two copies should be sent to each contractor. As much information as possible should be included in the bill to reduce the need for many drawings to accompany enquiries to sub-contractors. If domestic sub-contractors are named in the bill then the consultants can send copies of the drawings and specification direct to the specialists to assist the contractors, not least in reducing the reproduction and postage costs.

Drawings

The bill of quantities will list the drawings which were used in preparing the documents. With standard methods of measurement aimed at producing shorter bills of quantities, there is a greater reliance on drawings by the tenderers. Tendering costs could be cut if copy negatives or reduction prints can be produced. Full-size drawings are clearly essential if the contractor or sub-contractor is responsible for taking-off quantities. For large projects, drawings are often issued electronically. This might be on CD-ROM, by e-mail or with tenderers downloading drawings from a secure website.

Form of tender

The form of tender is a pre-printed formal offer, usually in letter form, which ensures that all tenders are received on the same basis and should be simple to compare. The tenderer fills in his name and address and a sum of money, for a lump sum offer. It may be sent with a collusive tendering certificate and appendices that are used for

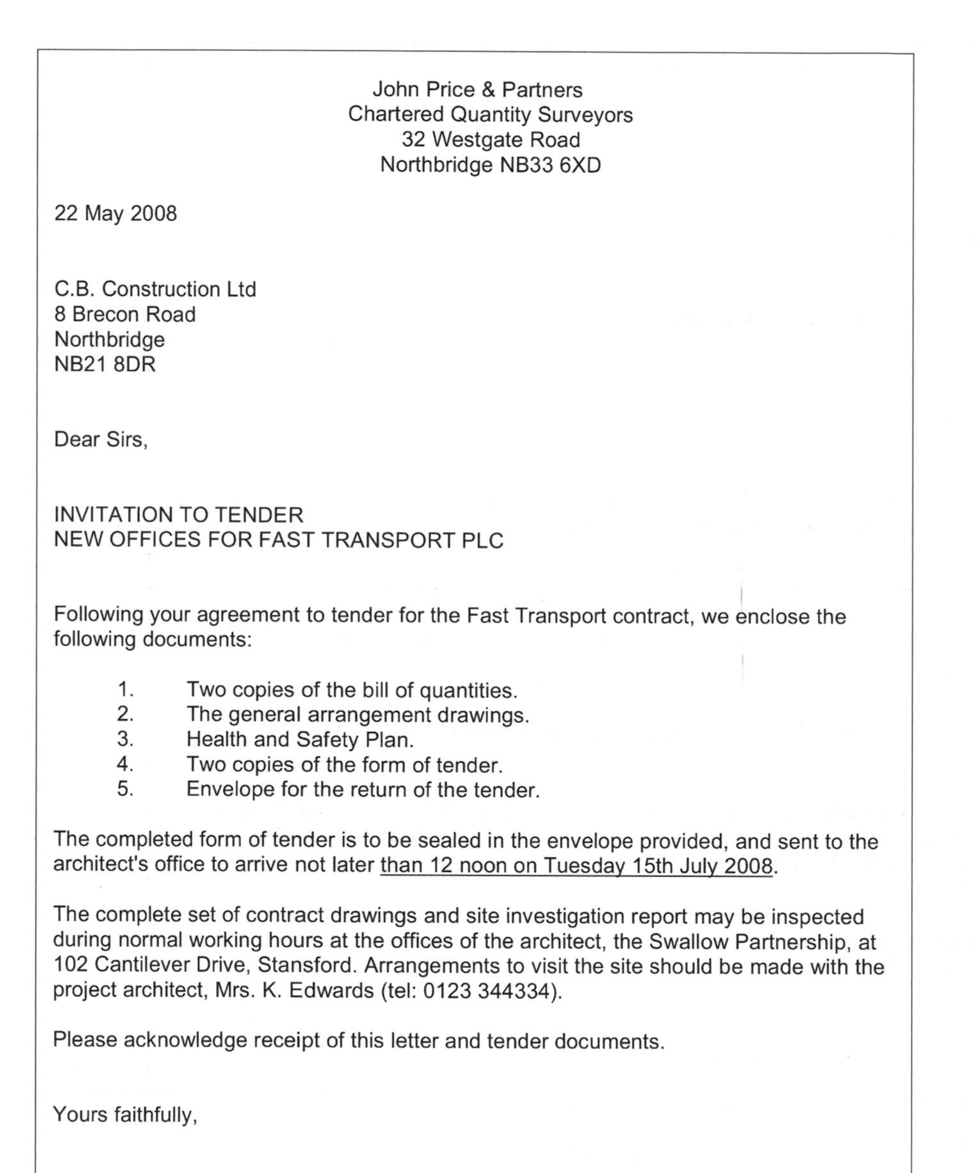

John Price & Partners
Chartered Quantity Surveyors
32 Westgate Road
Northbridge NB33 6XD

22 May 2008

C.B. Construction Ltd
8 Brecon Road
Northbridge
NB21 8DR

Dear Sirs,

INVITATION TO TENDER
NEW OFFICES FOR FAST TRANSPORT PLC

Following your agreement to tender for the Fast Transport contract, we enclose the following documents:

1. Two copies of the bill of quantities.
2. The general arrangement drawings.
3. Health and Safety Plan.
4. Two copies of the form of tender.
5. Envelope for the return of the tender.

The completed form of tender is to be sealed in the envelope provided, and sent to the architect's office to arrive not later than 12 noon on Tuesday 15th July 2008.

The complete set of contract drawings and site investigation report may be inspected during normal working hours at the offices of the architect, the Swallow Partnership, at 102 Cantilever Drive, Stansford. Arrangements to visit the site should be made with the project architect, Mrs. K. Edwards (tel: 0123 344334).

Please acknowledge receipt of this letter and tender documents.

Yours faithfully,

Fig. 4.1 *Typical formal invitation letter*

FORM OF TENDER

To: Fast Transport Ltd, Stansford

Tender for: Proposed Office Building, Stansford

Dear Sirs,

Having examined the conditions, drawings and bills of quantities, we offer to carry out and complete the works described, for the FIRM price of:

£____________________ (in words)______________________

____excluding VAT,

and complete within 34 weeks from the date of possession.

This tender will remain open for acceptance for 28 days from the date of return of tender.

We agree that should any obvious pricing or arithmetic error be discovered before acceptance of this offer in the priced bills of quantities then these errors will be corrected using Alternative 1 given in the JCT Practice Note 6 'Main Contract tendering'.

We understand that we are tendering at our own expense and that neither the lowest nor any tender need be accepted.

Signature ____________________________ Date: ______

Company: ____________________________

Address: ____________________________

Fig. 4.2 *Typical form of tender*

C.B. Construction Ltd
8 Brecon Road
Northbridge
NB21 8DR

14th July 2008

Fast Transport Ltd
Stanton Lane
Stansford

Dear Sirs,

New Offices, Stansford
Alternative Tender

Following discussions with the architect and engineer during the tender period, we have examined an alternative design which would lead to a significant saving of time and money, as follows:

1. By a small increase of plan dimensions (to the lines shown on our layout drawing F/1 attached) including some accommodation in the roof space, there would be no need for the basement construction.

2. You will see on our preliminary programme (our drawing number F/2) the contract duration can be reduced by 4 weeks to 30 weeks, with completion by 22nd December.

3. Our alternative proposals would offer a financial saving amounting to £49,552 and a tender sum of £1,013,100.

We hope that this will help you in your appraisal of the scheme and would be pleased to provide more information and discuss the work with you soon.

Yours faithfully,

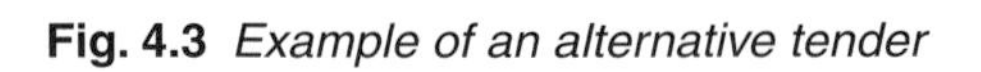

J.Lewis
Regional Manager
For CB Construction Ltd

Fig. 4.3 *Example of an alternative tender*

declarations about 'fair wages' or 'basic lists of materials'. A typical tender form is shown in Fig. 4.2; and Fig. 4.3 is an example of an alternative tender which might be produced in addition to a compliant bid. JCT Practice Note 6: 2002 recommends the use of a separate tender form for alternative bids.

Health and safety file

A pre-tender health and safety file is a requirement of the Construction (Design and Management) Regulations 2007. This document is produced by the CDM Co-ordinator appointed by the client, and included in the tender documents. At the pre-tender stage, the health and safety file will include information which the client can provide about the existing site or buildings; details of significant risks identified in the design; construction materials which could be hazardous to site personnel; and operational hazards on an occupied site.

The principal contractor is then required to develop the construction phase plan before work starts on site, and keep it up to date throughout the construction phase. The construction phase plan is prepared by the principal contractor for notifiable projects, to outline the arrangements for managing health and safety on site during construction work. The health and safety file is prepared or revised by the CDM co-ordinator. Under the legislation, the CDM co-ordinator is required to liaise with the client, designers, principal contractor and contractors. The file will contain information necessary for future construction, maintenance, refurbishment or demolition to be carried out safely, and is retained by the client or any future owner of the property. The file should be a usual and valuable document for the client.

Return envelope

Each contractor should be provided with a pre-addressed envelope clearly marked 'Tender for …'. They are to be marked so that they will be easily recognized and not opened too early or by the wrong person. Some clients insist that the contractor's name must not appear on the envelope, in order to avoid any opportunity for tampering with a particular tender.

The Joint Contracts Tribunal (JCT) published in 2002 a comprehensive list of tender enquiry documents in its Practice Note No.6 'Main Contract Tendering'.

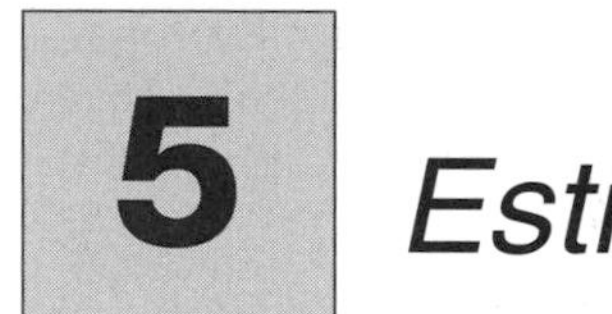

5 Estimating methods

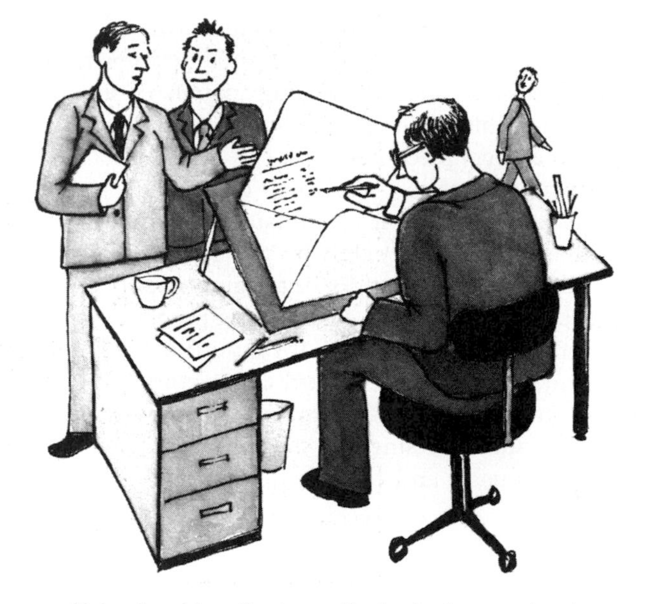

'John does his estimates on the back of an envelope'

Introduction

During the first half of the twentieth century six methods of estimating were used (Fig. 5.1). The methods are much the same today. The main difference is the current popularity of elemental cost models, which are used by quantity surveyors and contractors alike, in advising clients on their likely building costs, and helping designers to work within a budget.

Methods of estimating, used in the early stages of cost planning, depend on reli-a-ble historical cost data whereas an analytical approach to estimating is based on

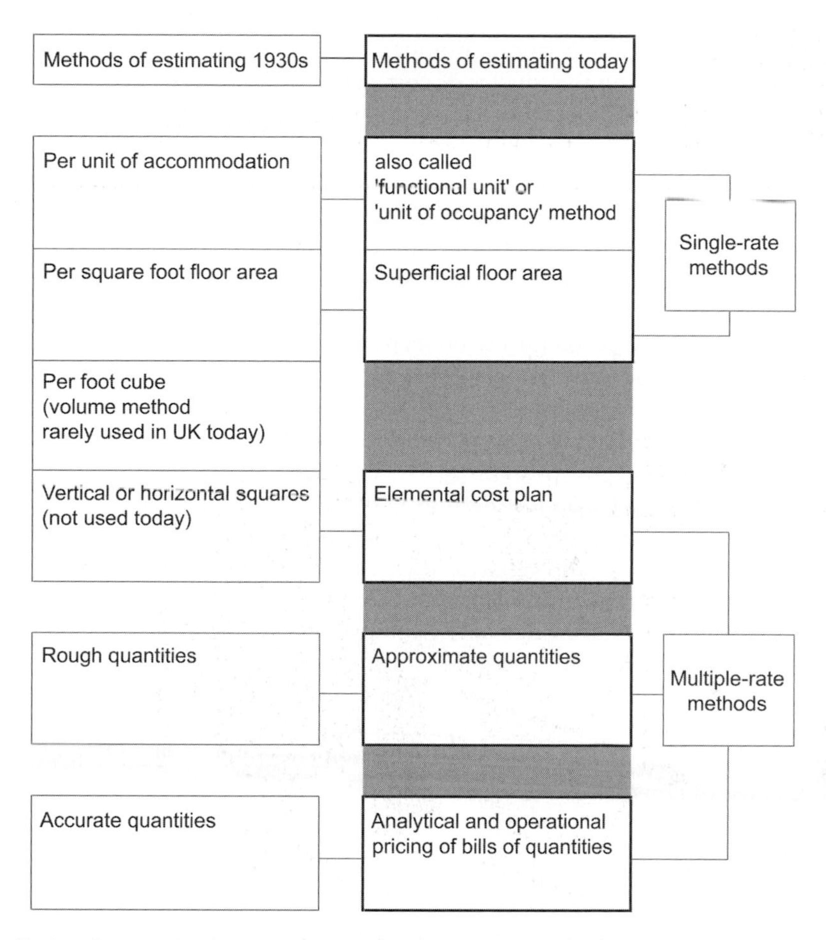

Fig. 5.1 *Estimating methods in 1930's and today*

applying current prices for resources to a well-developed design. A contractor may use a combination of estimating methods in developing a cost for a design-and-build project. For example, a client could be given a cost range for construction using the unit method and an elemental cost plan would be produced when the client's outline brief is received. Approximate (or builder's) quantities are used to produce a formal tender and when a contractor has received an order a full bill of quantities may be written for financial control during construction. The two main benefits of cost planning are:

1. To ensure tenders do not exceed the budget. This is achieved by making design decisions early with advice from the cost team. Changes made early in the design process can be accommodated without too much effect on other elements.

2. To collect cost information from a number of buildings, at various stages of development, thus improving the quality of cost data for future projects.

In some sectors of construction, *cost limits* have traditionally been applied well before a scheme is well defined. This can sometimes lead to unrealistic targets that can produce poor designs, to the detriment of the building's functionality. For example, in public schemes a great deal of effort is given to driving down floor areas. A school library could be located in a wide corridor, for example, or hospital consultants might be expected to share open-plan offices in order to hit a challenging area target.

In recognition of these problems the concept of 'value-for-money' has been adopted. In the case of new hospitals, the government has pledged more money to pay for 'consumerism', which for hospitals means more friendly spaces and more space around patients' beds. For secondary schools the aim is to transform education through the design of school buildings. Nevertheless, central government still sets challenging cost targets for public buildings which limit the scope for developing innovative designs.

The first step in cost planning is to advise a client of a budget at the inception of a project. An example of a development budget for construction costs is given in Fig. 5.2 Once preliminary drawings have been produced, a cost plan can be produced. The contractor is in the unique position of having detailed knowledge of current prices for all the resources used in construction. The PQS has the benefit of rates submitted in priced bills of quantities from a broad selection of contractors although he must be aware that rates do not necessarily reflect the actual cost of individual items of work.

The final cost of construction may be different from the forecast, for many reasons, namely:

1. The construction type: schools may be easier to predict than a bridge, the extent of repairs in a maintenance contracts can be difficult to foresee;
2. The effect of competition in the market;
3. The amount and quality of historical data available;
4. The amount of design information available;
5. The performance of the design team;
6. The nature of the site and workplace in terms of weather, ground conditions, resource prices and other uncertainties;
7. Changes introduced by the client;
8. The estimator's skill and method used.

The degree of uncertainty increases as the design stages evolve. Fig. 5.3 illustrates a diminishing cost range for a project from inception (setting a budget) to agreement of final account.

St John's Church		**Development Financial Summary**			**January 2008**
	New Church Hall (GIFA: 200 m²) and refurbishment of Church		**budget**	**actual cost**	**notes**
	Total costs		**430 551**		
1	Development costs	Concept architect	8 450		concept architect taking early retirement
		Planning consent fees	780		
		Building regulation fees	1 950		
		Additional insurances during construction	650		amount not known
		Photocopying costs	195		tender documents
			0		
2	Professional fees	Architect - pre-contract	11 700		
		Architect - post-contract	13 000		includes inspection role
		Structural engineer	5 005		includes unrecoverable VAT
		Planning supervisor	780		
		Quantity surveyor	2 925		produce valuations and value variations
		Risk assessment - fire	845		develop spec for fire alarms
			0		
3	Construction costs	Main contract	416 735		new hall and church refurbishment - excluding VAT
		Risks including inflation	20 837		allow 5%
4	Value added tax	Unrecoverable VAT	20 800		refurb portion of the works
5	Direct suppliers - fit out	Costs for sanctuary furniture	14 300		self-financing (see 6)
		Refurbish kneelers	3 900		
		Loose furniture	6 500		
			0		
6	Cost recovery	Sale of land (old hall)	−84 500		net income from sale and agent's fee
		Donations for sanctuary furniture	−14 300		£9 400 so far pledged or banked

Fig. 5.2 *Example of a development budget*

The contractor's estimator has the dual roles of forecasting the cost of construction and advising how competing organizations will bid for the same job. Although the commercial tender is the responsibility of management, the estimator must tell his managers how market trends will affect the prices, particularly where sub-contracting has a strong influence on the tenders.

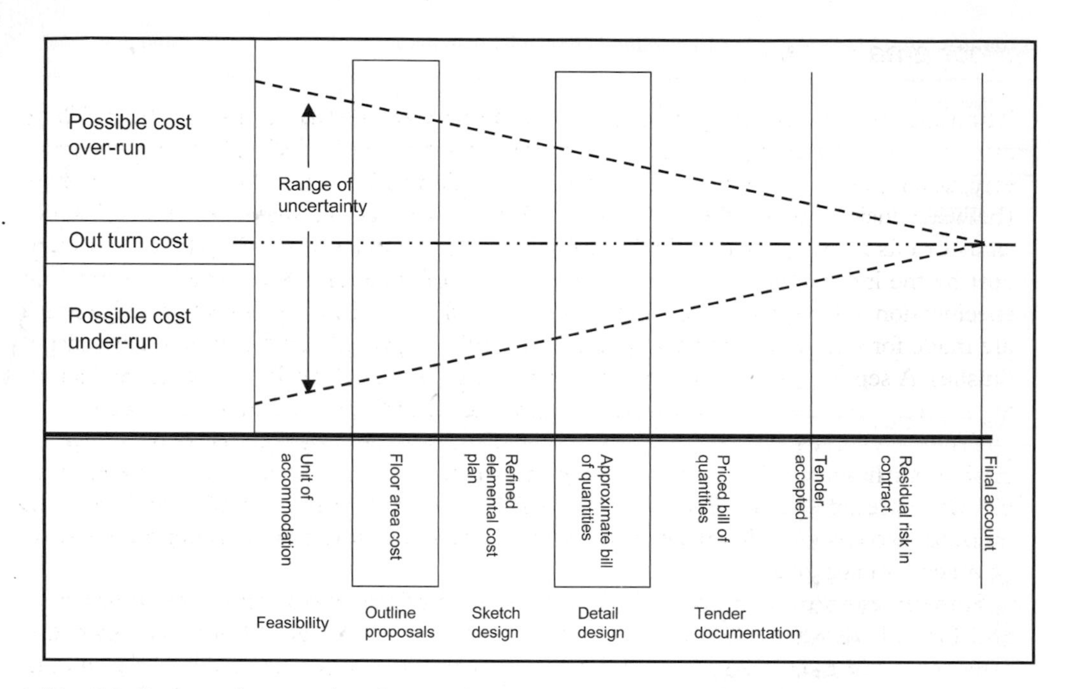

Fig. 5.3 *Degree of uncertainty for a typical construction project*

Single-rate approximate estimating

Unit of accommodation method

This method is commonly used by national bodies such as the education and health services at the inception stage of construction. If a client has an amount of money to spend (a budget) then it would be possible to consider the likely number of functional units which can be provided. From experience, it might be found that the cost of providing a study bedroom in student accommodation is £30,000. Using this figure an expenditure of £12 m would provide accommodation for approximately 400 students. On the other hand if the number of units is known, a budget cost (usually expressed as a cost range) can be calculated.

Providing there are recent comparable data available, the unit method is useful where a simple and quick cost range is needed in the early stages. It is difficult, however, to adjust the costs for specific projects, in different locations, with varying ground conditions and so on.

Floor area method

The main reason for the popularity of the floor area method is its simplicity. There are few rules to remember and the cost per square metre is well understood by property developers and contractors. A proposed building is measured at each floor level (between inside faces of external walls); no deductions are made for internal walls, stairs or lifts zones. Previous similar building costs are used by dividing the construction cost by the internal floor area. Adjustments can be made for location and inflation, but specification adjustments are much more difficult to estimate. Subjective judgements are made for size, shape, number of storeys, services, ground conditions and standard of finishes. A separate assessment should be made for external works, demolitions, incoming services and drainage, which can be significantly different for similar buildings.

There are many buildings where the unit of accommodation method is impracticable, such as warehouse projects or open-plan offices. In these cases the superficial floor area method is found to be reliable with an accuracy of 10% to 15%. This method also works well with certain external works contracts such as concrete paving or macadam surfacing.

Sometimes contractors are asked to quote for building work using sketch drawings and limited information about the site. The request might be: 'If you can build this 5000 m^2 office building for £1250/m^2, the contract is yours.' It is unlikely that a contractor would risk signing a contract on this basis. First, a clear scope of works would be needed together with a site survey and soil investigation report. Then there are planning conditions and the costs for bringing services onto the site which can vary greatly from site to site. On the other hand, it would be tempting to produce a tender using good 'design-to-cost' principles (see Chapter 8).

Building volume method

There are several methods which use the volume of a building as the cost yardstick, but they are not widely used today. In some European countries, architects and engineers are familiar with building costs expressed as cubic metre prices. In Germany, there are publications which list typical building costs in terms of their volume, and the procedure for calculating volumes is given in a DIN standard.

Multiple-rate approximate estimating

Elemental cost plans

There are two kinds of elemental cost plan. One is produced to show how a client's affordability target can be broken down into sufficient detail to control the development of the design ('top-down'). The other is a summary of a more detailed cost plan which has been produced using approximate quantities and advice from specialist suppliers ('bottom-up').

An elemental cost plan is a list of the elements of a building such as substructure, frame and upper floors, each with its share of the total budget cost (see Fig. 5.4). It follows a standard range of elements which are defined by the Building Cost

CB Construction Limited, Northbridge

Proposed Workshop for Fast Transport Limited

GIFA (m^2)	2 310

	Element	Cost £/m^2	Element cost
1	**Substructure**	85	196 573
2	**Superstructure**		
	Frame	86	199 264
	Roof coverings	63	146 025
	Roof drainage	5	12 285
	External walls	42	98 033
	Windows	17	38 415
	External doors	7	15 405
	Internal walls	9	19 761
	Internal doors	8	17 603
3	**Internal finishes**		
	Wall finishes	14	32 313
	Floor finishes	5	11 067
	Ceiling finishes	5	10 589
4	**Fittings and furniture**	2	5 187
5	**Services**		
	Sanitary appliances	3	7 865
	Internal drainage	0	inc
	Hot and cold water	0	inc
	Heating	16	37 128
	Electrical installation	12	28 145
	BWIC	1	1 976
6	**External works**		
	Site works	50	116 383
	Drainage	14	32 682
	External services	4	9 776
7	**Preliminaries**	52	120 705
8	**Contingencies**	21	48 295
9	**Budget total**	522	1205 474

Fig. 5.4 *Elemental cost plan for a portal-framed building*

Information Service (BCIS) in its publication, *Standard Form of Cost Analysis*. It is essential that an estimator collects historical data with full knowledge of what constitutes each element. For example, floor finishes include screeds, skirtings and floor coverings, but exclude finishes to stairs and structural screeds which are in other sections.

A Contract Sum Analysis used in design and build contracts is also structured in the form of building elements. The BCIS publishes guidance for structuring a cost plan for design and build contracts in its publication, *Elements for Design and Build*, BCIS 1996.

The forecast cost of each element can be calculated in two ways:

1. By measuring the approximate quantity of each element and applying a unit rate;
2. By calculating the proportion of total cost for each element on a similar building and using this ratio to divide the budget for the proposed building into its elemental breakdown.

The second method is better shown by example. If a contractor has built some portal-framed factories he will know the costs of each sub-contract package and can express this information as costs for each element as a rate (£/m^2 of floor area). Fig. 5.5 illustrates a typical analysis for a factory building. The site team has been asked to feed back cost information to the estimator by converting package values to elemental costs.

A cost plan for another similar factory can be generated by multiplying each rate by the new floor area. Fig. 5.6 shows the second factory which the contractor will further adjust for inflation, and significant specification changes. Typical examples would be the number of sanitary appliances, internal doors, roller shutter doors and ground improvements. In this example the contractor was confident about this approach because he found the floor area and wall-to-floor ratio to be similar to the earlier factory.

If a budget is wanted for another factory with a much smaller floor area, say 1200 m^2 for example, then a different approach would be needed, since the wall/floor ratio will be greater. The estimator should look at some elements such as external walls and apply a rate/m^2 (of the element itself). The preliminaries cannot be assessed using the floor area either. An allowance for preliminaries should be calculated using the cost per week of time-related costs for a similar factory and multiplying by the duration for the new scheme. In this way, a combination of historical data (the cost of elements per m^2 of floor area) and calculated costs for certain elements is used.

Contractors and PQSs are becoming more adept at using this method and have adapted the basic principles for computer systems. A spreadsheet template can store the information shown in Fig. 5.6 and the effect of changes can be seen immediately they are made. Spreadsheets are now used to produce sophisticated budgets for clients at the early stages of design.

	CB Construction Limited, Northbridge	***COST FEEDBACK***	
	Factory for Hitech Cables Limited	**GIFA (m^2)**	**3 120**
	Element	**Element cost**	**Rate £/m^2**
1	**Substructure**	242 385	78
2	**Superstructure**		
	Frame	269 633	86
	Roof coverings	156 468	50
	Roof drainage	14 976	5
	External walls	125 554	40
	Windows	31 135	10
	External doors	21 554	7
	Internal walls	11 414	4
	Internal doors	19 942	6
3	**Internal finishes**		
	Wall finishes	23 218	7
	Floor finishes	13 065	4
	Ceiling finishes	7 748	2
4	**Fittings and furniture**	9 425	3
5	**Services**		
	Sanitary appliances	9 633	3
	Internal drainage	inc	0
	Hot and cold water	inc	0
	Heating	33 215	11
	Electrical installation	47 931	15
	BWIC	4 719	2
6	**External works**		
	Site works	164 515	53
	Drainage	43 173	14
	External services	6 656	2
7	**Preliminaries**	187 915	60
8	**Contingencies**	73 164	23
9	**Budget total**	1517 438	486

Fig. 5.5 *Elemental cost plan for building under construction*

Approximate quantities

There are many ways in which approximate quantities are used, depending on who uses them and for what purpose. A PQS may want an alternative estimating technique

CB Construction Limited, Northbridge

		COST FEEDBACK		*NEW PROJECT*
		Hitech Cables		**Pluto Blinds**
		GIFA	3 120	2 860
	Element	**Element cost**	**Cost £/m²**	**New budget**
1	**Substructure**	242 385	78	222 186
2	**Superstructure**			
	Frame	269 633	86	247 164
	Roof coverings	156 468	50	143 429
	Roof drainage	14 976	5	13 728
	External walls	125 554	40	115 091
	Windows	31 135	10	28 540
	External doors	21 554	7	19 758
	Internal walls	11 414	4	10 463
	Internal doors	19 942	6	18 280
3	**Internal finishes**			
	Wall finishes	23 218	7	21 283
	Floor finishes	13 065	4	11 976
	Ceiling finishes	7 748	2	7 102
4	**Fittings and furniture**	9 425	3	8 640
5	**Services**			
	Sanitary appliances	9 633	3	8 830
	Internal drainage	inc	inc	inc
	Hot and cold water	inc	inc	inc
	Heating	33 215	11	30 447
	Electrical installation	47 931	15	43 937
	BWIC	4 719	2	4 326
6	**External works**			
	Site works	164 515	53	150 805
	Drainage	43 173	14	39 575
	External services	6 656	2	6 101
7	**Preliminaries**	187 915	60	172 255
8	**Contingencies**	73 164	23	67 067
9	**Budget total**	1517 438	486	1390 985

Fig. 5.6 *Elemental cost plan for similar factory building*

to check cost forecasts before tenders are returned. Measurements will be concentrated into as few items as possible for grouped work components. A simple example is a cavity wall measured and priced with both skins included in the unit rate. The rate will include forming the cavity, wall ties, plastering and pointing. Rates for composite items can be found in price books, calculated from rates in priced bills of quantities or calculated from first principles.

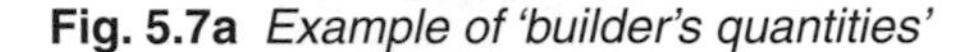

	CB Construction Limited				
	Builder's quantities for Pluto Blinds				
	Description	Quantity	Unit	Rate	**Total**
A	Excavate to reduce level	332	m^3		
B	Excavate for foundations ne 1.0 m deep	248	m^3		
C	Excavate machine pits ne 4.0 m deep	112	m^3		
D	Disposal of surplus from site	445	m^3		
E	Backfilling with selected exc material	247	m^3		
F	DOT type 1 under slab; 400 mm thick	1 330	m^3		

Fig. 5.7a *Example of 'builder's quantities'*

A contractor needs to produce bills of approximate quantities when tendering for work based on drawings and specifications. He will seldom allow the entire ancillary and subsidiary work items found in the standard method of measurement; but must be careful to tell sub-contractors the assumptions made. There is a strong case for attaching a preamble on the rules of measurement used, so any misunderstandings and disputes will be reduced (Fig. 5.7).

The accuracy of this method is related to how far the design has developed. At least the quantities are based on the planned construction and not a previous job and realistic allowances are made for plan shape, height of building, type of ground, quality of finishes etc. For these reasons it is widely used and being developed with computer systems using database and spreadsheet software to produce standard bills for repetitive building types. The danger is that the cost calculated using approximate quantities can appear to be as accurate as a full bill of quantities based on working drawings. It is more likely to be an underestimate of the cost of construction unless a generous contingency is added for small components, fittings, fixings and design development.

In common with all approximate estimating techniques, there are some difficulties which need to be recognized when advising clients. Some of the difficulties to be faced are:

1. The reliability of historical data must always be questioned.
2. Preliminaries are usually unique to a particular job and should be calculated whenever there is deviation from an identical scheme.
3. Incoming services are seldom the same on different sites and can only be assessed after detailed consultation with service providers.
4. Contract conditions can vary markedly between projects; the requirements for bonds, insurances and liquidated damages can be particularly onerous.
5. The contingency sum for design development must be estimated for each job.
6. At RIBS stage C design, there is often incomplete survey information, and so further site surveys are needed to confirm the prices. For example, an intrusive asbestos survey will be needed before a firm price can be confirmed.

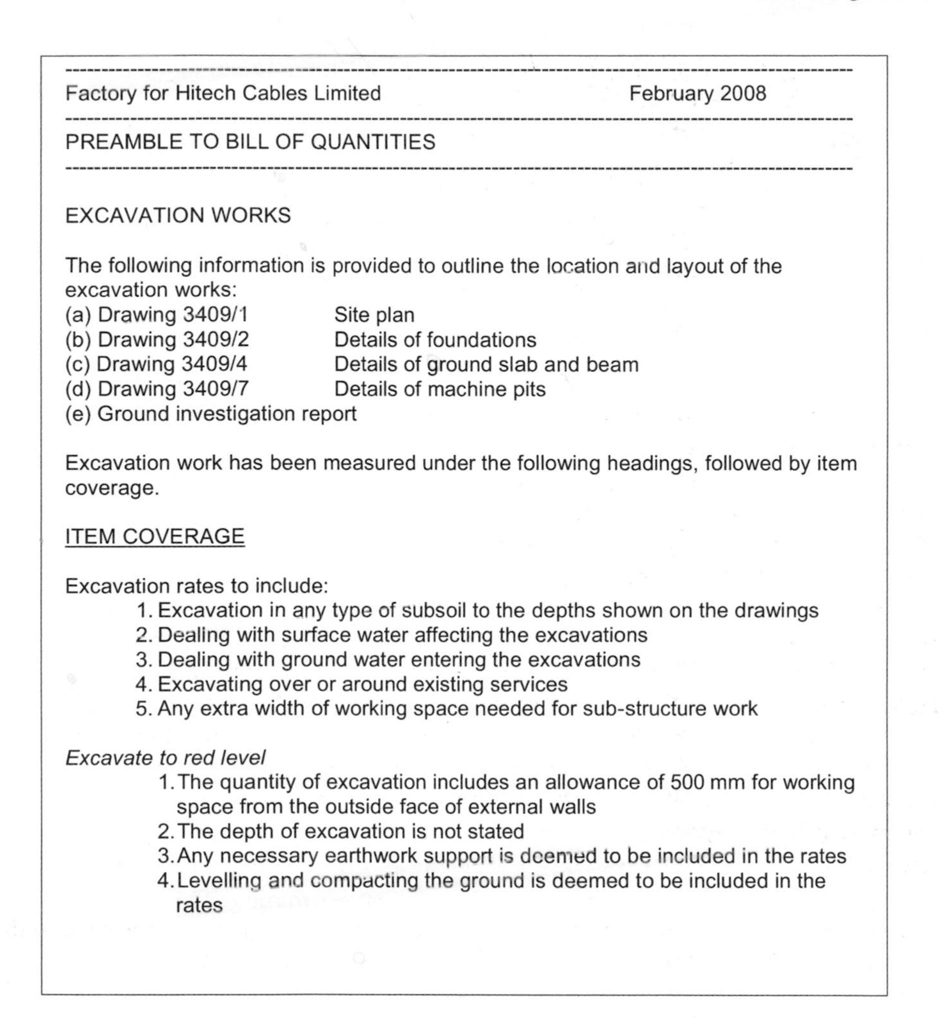

Factory for Hitech Cables Limited February 2008

PREAMBLE TO BILL OF QUANTITIES

EXCAVATION WORKS

The following information is provided to outline the location and layout of the excavation works:
(a) Drawing 3409/1 Site plan
(b) Drawing 3409/2 Details of foundations
(c) Drawing 3409/4 Details of ground slab and beam
(d) Drawing 3409/7 Details of machine pits
(e) Ground investigation report

Excavation work has been measured under the following headings, followed by item coverage.

ITEM COVERAGE

Excavation rates to include:
1. Excavation in any type of subsoil to the depths shown on the drawings
2. Dealing with surface water affecting the excavations
3. Dealing with ground water entering the excavations
4. Excavating over or around existing services
5. Any extra width of working space needed for sub-structure work

Excavate to red level
1. The quantity of excavation includes an allowance of 500 mm for working space from the outside face of external walls
2. The depth of excavation is not stated
3. Any necessary earthwork support is deemed to be included in the rates
4. Levelling and compacting the ground is deemed to be included in the rates

Fig. 5.7b *Example of preamble for 'builder's quantities'*

Analytical estimating

Analytical estimating is a method for determining unit rates by examining individual resources and the amounts needed for each unit of work. This method for pricing bills of quantities is described in the CIOB Code of Estimating Practice, in four stages:

1. Establish all-in rates for the individual resources in terms of a rate per hour for labour, a rate per hour for items of plant and the cost per unit of material delivered and unloaded at the site.

2. Select methods and outputs to calculate net unit rates to set against items in the bill of quantities.
3. Add to the net cost preliminaries, contingencies, inflation and risk.
4. Summarize resources and prepare reports for management.

The ability to analyse unit rates is an important skill for all those engaged in construction. Quantity surveyors and architects may need to value variations according to the rules of a contract. Standard contracts have wording such as: 'where the work is not of similar character to work set out in the contract bills the work shall be valued at fair rates and prices'. This presumably means a properly built-up unit rate either derived from first principles or one taken from a respected price book.

Contractors rely on pricing carried out by their sub-contractors for an increasing share of the work. A contractor's estimator should be able to recognize the correctness of such pricing and have the ability to build up rates for work which is to be carried out by his direct workforce.

Analytical pricing of bills of quantities is more than just applying resources to items of work to produce a unit rate. The constituents of a rate are inserted in the bill, and totalled for each page, each section, and carried to the summary, so that the contractor has a complete picture of the resource costs at the final review meeting. Fig. 5.8 shows a typical printout from a contractor's bill where the rates and totals are show between the item descriptions. Fig. 5.9 is an example of a contractor's bill of quantities priced analytically using a spreadsheet package.

The benefits of analytical pricing of bills of quantities are:

1. The total cost of labour is a product of analytical estimating. This is needed to calculate the cost of insurances, transport of operatives, small tools and equipment, and workforce levels.
2. The breakdown of resource costs is needed to calculate the inflation allowance for firm price tenders.
3. Labour and plant totals for elements of the work are used to calculate activity durations for the tender programme.
4. A breakdown of prices is needed in each trade to make comparisons between direct work and labour-only sub-contracts.
5. The costs of resources are needed to calculate the cost commitment cash flow forecast.
6. Adjustments can be made to any part of the estimate right up to the submission date.
7. The resource breakdowns will be used on site for post-tender cost control, bonus systems, monitoring and forward costing.

Factory for Hitech Cables Limited **February 2008**

	Description			quant	unit	rate	total
	Breakdown	*Lab rate*	*Plt rate*	*Mat Rate*	*Sub rate*		
		LAB	*PLT*	*MAT*	*SUB*		
a	Excavate to reduce level			332	m^2	5	1 726
		1	*4*				
		492	*1 234*				
b	Excavate for founds ne 1.0 m deep			248	m^3	7	1 693
		2	*5*				
		551	*1 141*				
c	Excavate machine pits ne 4.0 m deep			112	m^3	9	965
		3	6				
		333	632				
d	Disposal of surplus from site			445	m^3	23	10 361
		7	16				
		3 303	7 058				
e	Backfilling with selected excavated material			247	m^3	6	1 394
		1	4				
		366	1 028				
f	DOT type 1 under slab 400 mm thick			1 330	m^3	26	34 742
		1	4	20			
		1 971	5 533	27 238			
	Total to summary						**50 881**
	Breakdown	*7 017*	*16 626*	*27 238*			*50 881*

Fig. 5.8 *Contractor's bill of quantities priced analytically*

Most contractors know the benefits of analytical estimating but sometimes have difficulty finding time to apply the technique to all tenders. The two main problems are that *all* the rates must be priced analytically for the system to work, and many extra calculations are needed to extend the rates to totals. A computer estimating system is designed to overcome these difficulties and will produce the resource summaries automatically.

Unit rate pricing of a bill of quantities is carried out to certain conventions: those which are expected by the client's representative, and those which the contractor has

Weighbridge foundation for Hitech Cables Limited **15.2.08** **Analysis of rates**

	Surface weighbridge (15 m long)	quant	unit	rate	total	lab	plt	mat	s/c	LAB	PLT	MAT	S/C
A	Excavate to reduce lev ne 1.0 m dp	23	m^3	12	267	5	6			119	149		
B	Excavate for thickening & downstand	9	m^3	32	291	26	6			233	58		
C	Load and remove to tip on site	17	m^3	10	176		10				176		
D	Backfill with selected material	15	m^3	7	110	3	4			45	65		
E	Level and compact	79	m^2	1	76	1	0			59	17		
F	Earthwork support	6	m^2	4	25	1	3			8	17		
G	Hardcore (Free Issue)	18	m^3	7	132	3	4			54	78		
H	Blind with dust (Free Issue)	62	m^2	1	53	0	0			27	27		
J	Soil stabilization mat	79	m^2	2	137	1		1		43		93	
K	Concrete grade 40N in foundation	17	m^3	125	2 124	30		95		513		1 612	
L	Concrete grade 40N in ramps	15	m^3	129	1 939	34		95		517		1 422	
M	Concrete grade 40N in upstands	6	m^3	168	1 008	65		103		388		621	
N	Concrete grade 40N in plinths	1	m^3	157	157	54		103		54		103	
P	Rebar 12 mm (upstand & downstand)	1	t	1 390	834	463		926		278		556	
Q	Rebar 16 mm	0	t	1 293	155	431		862		52		103	
R	Fabric A393	280	m^2	9	2 470	2		7		512		1 957	
S	Dowel bars; 25 mm	30	nr	4	129	2		2		65		65	
T	Form plinths 900×900×250 mm high	4	nr	42	168	28		14		112		56	
U	Sawn formwork to sides of founds	47	m^2	45	2 126	30		15		1 418		709	
V	Sawn formwork to sides of upstands	17	m^2	50	842	32		17		549		293	
W	Cast in service duct	1	nr	22	22	13		9		13		9	
X	Grouting baseplates on return visit	1	item	334	334	269		65		269		65	
Y	Steel bumper stops	284	kg	4	1 162	1		3		247		914	
				TOTALS	14 737			check	14 737	5 575	585	8 577	-

Fig. 5.9 *Contractor's spreadsheet for weighbridge foundation*

developed. The notes at the beginning of a bill of quantities usually include instructions such as:

1. All rates shall be inclusive of labour, materials, transport, plant, tools, equipment, establishment and overhead charges, and all associated costs, margins and profit.
2. All items shall be priced; the value of any items unpriced shall be deemed to be included elsewhere in the bill of quantities.

The contractor, on the other hand, is likely to produce bill rates which *exclude*:

1. General overheads and establishment charges;
2. Profit, which can only be calculated after the net estimate is complete;
3. Restrictions, which apply to more than one item such as difficult access, difficult handling and protection;
4. Plant, which is common to several activities such as compressors, hoists, mixers, dumpers and cranes.

Contractors may include a nominal mark-up or 'spread' to the rates, which can be supplemented by sums in the preliminaries part of the bill when the true overheads and profit are known after the final review meeting. A computer-aided estimating system would allow some of the overheads and profit to be spread over various parts of the bill of quantities. For example, a contractor might want to add 20% to the earthworks rates. This could improve the cash-flow position of the project but would put the contractor at risk if the extent of earthworks reduced.

There are many PQSs and civil engineers who would want to introduce analytical bills. This would be a bill format with extra columns for labour, plant, materials, subcontractors, and overheads/profit, which would be submitted by the lowest tenderer before entering into a contract. The client's consultants argue that although contractors would not take on this extra work, and reveal confidential information, the idea has the following advantages:

1. There would be a clearer basis from which to value variations.
2. The settlement of final accounts could be based on an examination of which elements had changed, and the effect on the programme may be clearer.
3. The design team could see where they had chosen designs which were labour-intensive.
4. If the analysis was extended into valuations, the contractor could use the data for his own cost-monitoring systems without doubling his effort.

In order to comply with the instructions to tender, contractors may insert all their rates in the sub-contract column, and argue that the work will be sub-contracted.

In a recent example of this form of pricing, the contractors tendering for a framework agreement were asked to produce a full analysis for all their rates. The exercise became unworkable when estimators had no means of splitting rates for engineering services and specialist equipment such as a lift installation.

Operational estimating

Operational estimating is a form of analytical estimating where all the resources needed for part of the construction are considered together. For example, an estimator pricing manholes using the Civil Engineering Standard Method of Measurement needs to gauge the time taken to build a complete manhole, whereas a building estimator is expected to price all the individual items for excavation, concrete work, brickwork etc., measured under the rules of the appropriate work sections.

The following examples show some of the many other situations where work is priced as whole packages:

1. Excavation including trimming, consolidation and disposal;
2. Placing concrete in floor slabs including mechanical plant, labour, fabric reinforcement, membranes, isolation joints and trowelling;
3. Formwork to complex structures including a unique design, hired-in forms and falsework;
4. Drain runs including excavation, earthwork support, bedding, pipework and backfill;
5. Repairs which often involve more than one trade or a multi-skilled operative;
6. Roof trusses including the use of a crane, a suitable gang of operatives and temporary works.

It must be said that building estimators have become skilled at applying production outputs to units of work and then occasionally employing operational estimating techniques to check the results. Civil engineers, on the other hand, usually examine methods and durations before pricing the work. This is because different construction methods for civil engineering can have a significant effect on costs. There is also a greater reliance on the specification, the drawings and preambles which give the item coverage.

The term 'operational estimating' is often applied to methods that rely on a forecast of anticipated durations of activities, and a resource levelling exercise. The estimator must start with an appraisal of the details on the drawings, the extent of the work described in the specification and bill, and a study of the site conditions. Next, the sequence of work will be found by considering the restraints brought about by site layout, client's requirements, the design, time of year, and temporary works. The critical operation at each stage of the construction can then be plotted and the rest of the activities sketched in. Labour and plant schedules can be drawn up for direct work, and specialist sub-contractors will be asked for their advice about their work. It may be necessary to change the programme if there are any unwanted peaks and troughs in the resources needed on site. The estimator will then have a list of resources for each operation from which to calculate costs. This approach will often produce a cost based on a particular method for carrying out the work. If this has brought about a saving in costs the estimator will prepare a method statement so site staff can understand the assumptions made in preparing the estimate.

When a building estimator uses operational estimating with a traditional bill of quantities he has great difficulty dividing the cost of a piece of work amongst all the related bill items. Where, for example, should an estimator put the rate for casting a concrete floor which includes a DPM, fabric reinforcement, power floating and sealer? The PQS often insists on rates being inserted against items that have a value, so there is better financial control during construction. Clearly this is not a problem with a contractor's bill of quantities produced for design and build or plan and specification projects, because there is no bill of quantities submitted. Another solution to the problem for building estimators would be to rough price the bill early on in the tender period and adjust the rates when operational methods highlight greater or lower costs. This is commonly done during the final review stage, and the rough pricing technique is popular with those using computer systems.

The advantages of operational estimating are:

1. Activities are examined to select those methods that are practicable.
2. Outputs are based on a programme, which includes holiday breaks, time of year, idle time, facilities available on site etc., giving a more realistic guide to the time needed for labour and plant.
3. Alterations and repair work are usually measured as global items which can be overpriced if all the possible trades are examined separately.
4. In a competitive market, the estimator may only look at the labour and plant needed for the core item of work, such as the brickwork in a manhole assuming the bricklayer can fix the cover while finishing the brickwork and the excavator can dig the pit when it digs the pipe trench.

Figure 5.9 is a contractor's bill of quantities for a weighbridge foundation priced analytically. The estimator used an all-in rate of £15.00/hr for all his labour and applied his usual labour outputs from his tables of constants.

The site manager has kept records from previous similar jobs which show that this type of weighbridge foundation usually takes two weeks to construct with four men, and a return visit is needed for two men to grout in the equipment. A backacter and roller costing £35.00/hr is needed for three days. This gives the following net cost for labour and plant:

Labour	4 nr	×	2 weeks	×	45 hours	×	£15.00	=	5 400.00
	2 nr	×	2 days	×	9 hours	×	£15.00	=	540.00
							Total	=	£5 940.00
Plant	1 nr	×	3 days	×	8 hours	×	£35.00	=	£840.00

It can be seen from the comparison that when the project is assessed as a whole, the net cost of labour and plant is more than the total from the unit rate analysis (Fig. 5.9). The estimator may have used his normal constants for labour and plant without checking whether there is a continuous flow of work for labour and plant resources. Perhaps the site manager should next look at materials wastage that he has experienced, in particular blinding concrete and fabric reinforcement, which could be significantly higher for such a small contract than an estimator would normally allow.

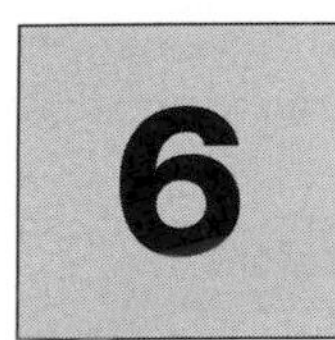

6 Contractor selection and decision to tender

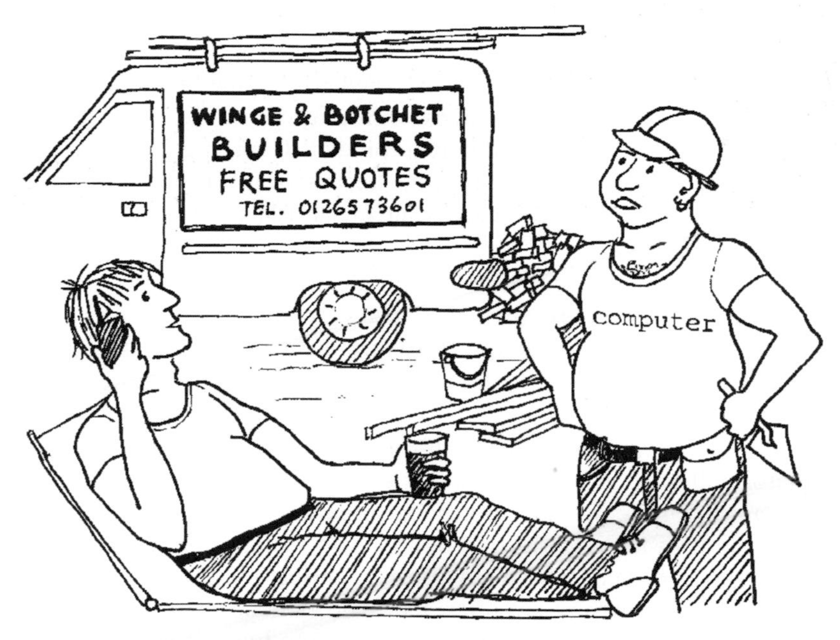

'let me run that through our computer'

Introduction

How does a construction organization maintain its turnover? Some enquiries arrive 'out of the blue' arising from hearsay, the *Yellow Pages*, or advertising. Others are sent on the strength of earlier successful contracts or following a direct salesman approach. New markets can be entered by replying to invitations for open tenders; some opportunities can be created by speculation. The greater part of work carried out in the construction industry is secured through a process of tendering which is intended to be an unbiased means of selecting a contractor to carry out work. The client, through an evaluation of his needs, determines the criteria for selection. The aims of selection are to find a contractor who can supply a product for a competitive price, and can demonstrate the following:

1. A reputation for good quality workmanship and efficient organization.
2. The ability to complete on time.
3. A strong financial standing with a good business record.
4. The expertise suited to the size and type of project.
5. An understanding of the requirements of the scheme in terms of the type of work, the quality expected and the need to achieve target completion dates.

The construction industry is rarely concerned with providing off-the-shelf products; most projects involve unique designs, with purpose-written specifications to be finished in a time which is often difficult to predict. Construction clients must balance the importance of cost, quality and time because it is rare for all three to be satisfied. A client can reduce the technical and financial risks in two ways:

1. By fully designing the project before selecting contractors.
2. Using a full construction service which includes interpreting the clients requirements, carrying all design responsibilities, and meeting the funding allowances.

It is not only clients who need to establish the financial standing of the other party. The contractor will have to be satisfied that the client has the ability to pay, and on time. In the past, contractors have not been so careful about selecting their clients. This has changed with the introduction of bonds and guarantees which are now used by both parties to contracts. The word 'trust' is unfortunately absent from conventional agreements, and lawyers are often the main beneficiaries. Recent partnership agreements have been developed using large-scale modifications to standard forms, or occasionally the NEC Engineering and Construction contract that is plainly written with mutual understanding at its core. The report of Sir Michael Latham in 1994 expresses concern that endless changes to the existing conditions will not avoid confrontation.

Clearly tendering in a competitive marketplace is the norm and will remain the basis for procuring most construction work. But contractors and clients both see the need for longer-term relationships. Since the mid-1990s, partnering between the parties to a construction project has emerged as a route to better communications and a means to improve business performance. There are many forms of partnering, ranging from improved interaction in a traditional contract to long-term relationships using common objectives throughout the supply chain in order to deliver continuous improvement over time. As a result it should be possible to secure lower costs, improved quality and a reasonable profit for everyone.

Bundling projects

There are many examples today of bundled contracts, which allow a client to reduce tender costs by awarding a number of similar projects using some sample schemes for the competitive selection stage. So, for example, a city council could invite tenders for a number of secondary schools. Outline designs and cost plans would be submitted for some 'sample' schemes which exhibit features of the broader building stock. It is common to include a new secondary school on a green-field site, a school which provides facilities for students with special needs and a mixed development which includes some retained buildings needing refurbishment.

The advantages of bundling are:

- Procurement costs are reduced by entering a competitive process only once.
- Bidding costs are reduced because follow-on projects are priced by only one contractor.
- Time for procurement stages is reduced.
- Contractors can offer continuous improvement by building up the expertise to build a well-known product. Specifications can be developed and improved from building to building and supply chain members can engage in the process.
- There are long-term commitments to work collaboratively in teams with common objectives.

If bundling is within the Private Finance Initiative (PFI) sector, the continuous improvement objectives extend to finance arrangements, risk transfer, life-cycle management, and maintenance regimes.

There are three main disadvantages:

- Bundling sometimes excludes smaller construction firms, or at best put them in a sub-contractor position.
- Client organizations may lose their in-house expertise because some roles are transferred to the contractor. This is necessary if the savings are to be realized.
- Non-sample schemes need to be benchmarked against the tendered sample projects. This and the subsequent negotiation exercises can take some time.

Contractors also bundle contracts with their suppliers. Again using schools as an example, it would be advantageous to obtain doors from one suppler for a batch of schools. There would be economies of scale, known specifications and a lot more cooperation and understanding between the parties.

Competition and negotiation

Contractors may be selected by competition or negotiation and sometimes by a combination of both. Open competition is an arrangement where an advertisement in local newspapers or trade journals invites contractors to apply for tender documents. A deposit is usually required to ensure that only serious offers are made; presumably it is needed to cover the cost of copying the documents. Local authorities have been advised against open tendering because it often leads to excessive tender lists where the cost of abortive tendering is considerable. There are instances of selection criteria being applied after the tender has been submitted, so a bid could be rejected if a contractor does not belong to an approved trade association, for example, after he has submitted his tender. They argue that this method allows new contractors to join the market and increases the chance of gaining a low price. Regional and national contractors avoid this method because they can see no reason to compete against anyone who asks to be included on the tender list and later be subjected to the further hurdle of contract compliance clauses.

Selective tendering consists of drawing up a list of chosen firms and asking them to tender. It is by far the most common arrangement because it allows price to be the deciding criterion; all other selection factors will have been dealt with at the prequalification stage. A more enlightened approach is for the price to be fixed by the client; the selection criteria would become design and deliverability. There are three ways in which selective tendering lists are drawn up:

1. An advertisement may produce several interested contractors and suitable firms are selected to tender.
2. The consultants may contact those they would wish to put on an ad hoc list.
3. Many local authorities and national bodies keep approved lists of contractors in certain categories, such as work type and cost range.

Contractors who ask to be included on select lists of tenderers are usually asked to provide information about their financial and technical performance, particularly about the type of work under consideration. Sir Michael Latham in his 1994 report recommended a single qualification document for contractors wanting to tender for public sector work. Recommendations for the use of a single qualification document were published by the Construction Industry Board in its 1997 document 'Framework for a National Register of Contractors'. However, with the demise of the National Joint Consultative Committee for Building (NJCC), and the CIB, the remaining guide for good practice is The JCT Practice Note 6 (2002).

This JCT Practice Note recommends that the following documents are used to decide which potential contractors to include on the tender list.

For many years, the building industry has used the 'Code of Procedure for Single Stage Selective Tendering' published by the NJCC. It was replaced in 1997 with the CIB Code of practice for the Selection of Main Contractors. The JCT Practice Note 6: 2002 'Main Contract Tendering' has in turn replaced the previous guidance notes, with far less detail. These procedures follow a number of well-defined stages for pre-selection and tender stage actions. Their success relies on complete designs before tenders are invited and the use of standard forms of contract but can be used with other procurement systems. The following points illustrate the coverage of the codes:

1. Preliminary enquiry – contractors are given the opportunity to decide whether they wish to tender by receiving a preliminary enquiry letter, 4 to 6 weeks before the despatch of tender documents.
2. Number of tenderers – the recommended number of tenderers is a maximum of six (three or four for design and build) and further names could be held in reserve.
3. Tender documents – the aim of the documents is that all tenders will be received on the same basis so that competition is limited to price only.
4. Time for tendering – normally at least four working weeks should be allowed, and more time may be needed depending on the size and complexity of the project.
5. Qualified tenders – tenderers should not try to vary the basis of their tenders using qualifications. Queries or unacceptable contract conditions should be raised at least 10 days before tenders are due. The consultants can then tell all the tenderers of their decisions and if necessary extend the time for tendering. A contractor should be asked to withdraw significant qualifications or else face rejection. This is necessary to ensure tenders are received on a like-for-like basis.
6. Withdrawal of tenders – a tender may be accepted as long as it remains open; a definite period is usually stated in the tender documents. The tenderer may withdraw his offer before its acceptance, under English Law.
7. Assessing tenders – the tenders should be opened as soon as possible after they are received. Priced bills may be submitted in a separate envelope by all the contractors, or more likely only the bills of the lowest tenderer will be called for and submitted within four working days. Once the contract has been let, every contractor should be issued with a list of tender prices; alternatively, tender prices should be given in ascending order and the names listed in alphabetical order.
8. Examination and adjustment of priced bills – the PQS will treat the information in the tender documents as confidential and report errors in computation to the architect and client. There are two methods for dealing with errors: Alternative 1 gives the tenderer the opportunity to confirm his offer or withdraw it; Alternative 2 allows the contractor to confirm his offer or amend it to correct genuine errors. If the contractor amends his offer with a revised tender which is no longer the lowest, the tender of the lowest will be considered.
9. Negotiated reduction of tender – the code of procedure recognizes the need to look for savings in the cost of a project where the tender exceeds the employer's

JCT Model Form of Preliminary Enquiry	In the form of an enquiry letter asking for confirmation that the contractor is going to submit a bona fide tender and that the rules of tendering will be those given in Practice Note 6.
Preliminary Information Schedule	• Description of the works • Location • Approximate cost range • Extent of contractor design • Anticipated start date and duration • Phasing and access to the works • Tender issue date and period for tendering • Employer and professional team details • Form of contract and particulars • Other contracts and collateral warranties
Questionnaire	• The contractor's recent experience with the type of work in the proposed time scale • Technical skills and capabilities for this type of work • Management structure and personnel resources • The proposed management team • Policies for health and safety and staff development • Liaison between parties • Capacity to carry out the work with current and expected workload • Financial standing of the company

Fig. 6.1 *Preliminary Invitation to Tender*

budget. This can be achieved by negotiation with the lowest tenderer, or the next lowest if negotiations fail.

Two-stage selective tendering may be adopted as an alternative to single-stage selection when a contractor's assistance is needed during the design stage. The first stage will produce a competitive tender based on approximate bills of quantities using preliminary design information. The contractor selected at the first stage helps with design, programming, and cost comparisons, and submits a final tender for the works, without competition, based on the original pricing levels.

The National Joint Consultative Committee for Building (NJCC) published codes of procedure for two-stage selective tendering and selective tendering for design and build. The principles were the same as those described for single-stage tendering. For design and build schemes the client must ascertain the design and build experience of each contractor and limit the number of tenderers to three, or four at the most, because there are large costs involved in preparing designs and cost proposals. Contractors must be told the basis for assessment where the price is not the sole basis for the award. The code suggests that the relative importance of cost, quality and time for construction should be included in the employer's requirements. An employer could, for example, state the target cost and timescale in his tender documents so the principal criterion for selection will be the quality and appearance of the building.

When a contract is negotiated, a contractor is often selected on the basis of past performance, recommendation, familiarity with the work, or from previous experience with the client or his advisers. In certain circumstances only one contractor may be able to provide the service required as in the case of system building. It is more difficult for those in the public sector to negotiate because EC directives insist that projects over a specified value must be subject to competition. Negotiation allows early contractor selection where the extent of work is not fully known and time is of the essence, and more time would be wasted in preparing full tender documents.

The process of negotiation starts with an outline design and a pricing document such as a bill of approximate quantities or cost plan. The contractor will insert rates which will be agreed by negotiation between the PQS and contractor's QS or estimator. Without competition the initial price may be higher than would be gained by other means, but this may not be a serious problem. An employer is often looking for other factors such as confidence, reliability, speed and experience of working with a known contractor.

Serial tenders allow a number of similar projects to be placed with a particular contractor and thereby provide the incentive of a continuous flow of work. The contractor is normally selected using a priced master bill or sample cost plans. Separate contracts for each individual project can then be arranged using the priced bill or cost plan as a basis for pricing levels.

Abuse of tendering procedures

The various codes, and practice notes, have encouraged all those involved in tendering to use fair and efficient methods which are the best and most professional techniques in use today. The prime aim is to select the right contractor who will give the client good value for money. Unfortunately, individual interests and lack of time can stand in the way of good practice, and the parties to a contract are often unclear about the true nature of the agreement. Some of the problems faced by the estimator are given below.

Large tender lists

With high costs of tendering in mind, many reputable contractors will not willingly take part in pricing a project where there is a large number of bidders. With an emphasis today on design and build contracts, the cost of tendering can be very high. It could also be argued that it is very difficult to build a close working relationship with a client when there are so many contractors tendering for the work.

Short tender periods

The time for tendering should be determined by four factors: the size of the project, the complexity of the project, the standard of the enquiry documents and the amount of information needed to accompany the bid. In practice the enquiry documentation is often late, thus eroding the time available for the estimate. A 'rough' estimate could be produced quickly but a contingency sum would be needed for unknown risks. In order to reduce the uncertainty, contractors would prefer to examine the project, the site, and the documents and agree methods with the contract staff and subcontractors, prepare a programme and prepare a detailed estimate. In fact the longer the tender period, the more likely it is that the contractor will find savings which would increase the possibility of winning the contract and may produce a better price for the client. The estimator will try to respond to such short tender periods by telephoning his enquiries to suppliers and sub-contractors, making use of information from previous jobs, manually or with the help of a computer. The depth of analysis will be reduced, there is a greater risk of errors, and the price is likely to be greater to reflect such problems.

Tender documentation for traditional contracts

The estimator should receive enough drawings to understand the nature and scope of the works. The minimums needed are elevations and floor plans to measure temporary works (such as scaffolding), site plans to consider materials access and distribution and component drawings where non-standard elements are to be priced.

References to brand names and specialist suppliers should include current telephone numbers and addresses. Information must be provided about any restrictions which might affect the contractor's choice of method. The site investigation report (or extracts) should be sent to each contractor. With design and build projects, problems have arisen when all contractors have been expected to carry out their own site investigations – clearly an enormous waste of effort and a further burden on the already considerable costs of tendering.

Perhaps differences between documents might be expected at this stage, and so the bills of quantities are used to specify the amount and quality of the works. Discrepancies between the bill descriptions and specification clauses do cause problems but should reduce with the use of coordinated project information. There will always be people who want to change the agreed conventions. The estimator needs to be alert to traps such as 'earthwork support shall include all means of holding up the sides of excavations including sheet piling' (normally measurable) or 'hack off external render where necessary and renew' (*where necessary* could be small isolated sections or the whole wall if the contract administrator so decides). Amendments to the tender documents should be avoided but can be allowed early in the tender period. Once quotations have been received from suppliers and sub-contractors, changes will be difficult to build into the bid.

Estimating without bills of quantities is much more time-consuming, not only because so much time is needed to take off quantities but enquiries to sub-contractors are delayed and the risk of errors is greater.

Asking for tenders when the work is unlikely to proceed

There is a tradition in the construction industry for estimates to be given without charge to the client. This can be at great cost to unsuccessful contractors. Some have reported that it costs about 0.25% of the tender price to prepare a bid for a traditional lump-sum form of contract: a design and build tender can cost as much as 2%. Contractors will continue to accept this financial risk providing they are submitting tenders to clients who use selective tendering and eventually award a contract to one of the bidders.

Qualified tenders and alternative bids

A tenderer should submit his bid without adding conditions to his offer. All contractors must consider the terms of their offers, and sometimes will not be able to comply fully with the instructions of the client. On the other hand they should recognize the need for a common basis from which the best bid can be selected. Contractors may produce an improvement to the design or see a method for completing quicker, and often can calculate an alternative price. Providing an offer is made which complies with the original brief, alternative tenders are considered by employers.

Failure to notify results

A contractor can monitor his tender effectiveness when he receives information about his performance in relation to other tenderers. Tender prices should be published if a contractor is to review his suitability for the type and value of projects. Clients are becoming increasingly reluctant to publish figures because the lowest tenderer could attempt to recover the difference in value between his tender and the second lowest, either before the contract is awarded or later during the construction period. Contractors commonly ask for a briefing on their performance, but will not be told the other tender sums.

Late receipt of tender documents

Estimators do their best to deal with requests for tenders sometimes at short notice, but when tender documents arrive later than promised their programme of work will be affected, and other opportunities to tender may be harmed. It has become common practice for tender submission dates to be held firm regardless of how late the tender documents are despatched.

Decision to tender

All employees of the firm should be made aware that they have a part to play in capturing the opportunities that arise. Senior management will feed back knowledge of projects gained from conversations with prospective clients and partners in related professions at business and social events. Equally, a job surveyor may well gain knowledge picked up while having a pint with his opposite number from the PQS office. All such snippets of information should be fed to the central source and recorded. A list of expected tenders may become a formal report in bigger organizations so resources can be used effectively. Invitations to tender arrive at a contractor's office in a variety of ways and it is important that they should be channelled to a central source for collating and monitoring. Where the organization has a marketing section then this may be the most suitable location. Alternatively, they can be held within the estimating department.

Formal invitations to tender are normally communicated by either letter or telephone. It is to be hoped that the enquiry follows the format laid down in the JCT Practice Note 6 and communicated by letter or facsimile. Compliance with the recommendations given in the Practice Note should be honoured by all parties. The client's professional adviser should provide in good time basic information about the project and ask the contractor if he wishes to be considered for inclusion on a selective tender list. The contractor then has the opportunity to decide, knowing there will be a limit on the number of tenderers.

Regrettably, telephoned enquiries persist! The person answering the phone needs to ask for all the information he would have if a preliminary enquiry had been sent.

He is sometimes asked for a decision immediately, which is usually 'Yes' because he knows that his boss can reverse the decision when the documents come in. It is suggested that a pad of forms be available by the telephone of all those likely to accept a call asking if the firm is willing to tender (see Fig. 6.2). The form contains some basic

CB CONSTRUCTION PRELIMINARY TENDER ENQUIRY			
Job title :			
Employer :		**Relationships**	
Architect/PM		**Estimated value**	£
QS		**Funding arrangements**	
Brief description :			
Competition	Competitive Negotiated	No of bidders	Single/two stage
Contract type : **PFI** **Framework** **D+B** **Traditional**		**Bills :** **Fluctuations :** **Bond :** **Damages :**	yes / no firm / fluct yes / no £ per
Programme :	Tender due in : Tender due out :	**Start date :** **Duration :**	
Action taken :			
Comments :			
Signed :		Date :	
Approved :		Date :	

Fig. 6.2 *Preliminary enquiry information form*

headings as an *aide mémoire* to those receiving the request. An abstract from the forms and formal letters of invitation could be kept on a weekly report form.

The decision to tender should be made by the chief estimator or general manager using the following points:

1. How strong are our relationships with the client? If they are not yet tested, is the client willing to build a partnering approach?
2. Is the work of a type which the contractor has experience of, both in winning tenders and completing profitably? Does it conflict with the company's objectives and future workload?
3. How many contractors will be invited to tender?
4. Has the contractor the necessary supervisory staff and labour available? (He may not wish to recruit untried and unknown personnel in key positions.)
5. Will the estimating department have staff available with suitable expertise for the type of work to be priced?
6. Does the location of the proposed site fit the organization's economic area of operation?
7. Are there too many risks in the technical and contractual aspects of the project?
8. Does the project have opportunities to build future business with the client?
9. Will suitable documents be produced for tender purposes? A busy estimating office may give priority to work that has been measured. Poor documentation might give clue to the standard of working documents during construction.
10. Has enough time been given to prepare a sensible estimate?
11. What will be the cost of preparing the tender? A contractor might limit the number of design and construct tenders, for example, in order to limit his exposure to cost. For unsuccessful tenders, these costs are usually not recoverable.

As a client needs to establish the contractor's financial standing, so in turn the contractor will need to be satisfied that the client has the ability to pay, and on time. Similarly, as the contractor is investigated for performance on similar work, whether his management structure is satisfactory and his present resources can cope with the added workload, so too the contractor will need to consider experience of working with the client and his professional team.

Contractors should not feel that they are under pressure to tender for a contract. Most invitations to tender include a statement such as: 'Your inability to accept will in no way prejudice your opportunities for tendering for further work under my/our direction.' This is a plea for the contractor to give an honest answer without fear of being penalized in the future.

When the full tender documents arrive, a contractor will again review the risks associated with the contract and should consider whether he wishes to confirm or decline to tender.

If the invitation to tender is to be declined, the client's adviser should be told immediately, preferably by phone, giving the reasons, and the documents must be returned quickly so that another bid can be invited from a firm on the reserve list. If the decision is to proceed, the estimator should acknowledge the safe receipt of all the tender documents and confirm that a tender will be submitted.

Inspection of tender documents

The arrival of the tender documents within a contractor's office invariably causes a stir; everyone is eager to have a look. The documents should be passed directly to the estimating department. The CIOB Code of Estimating Practice (COEP) states that they should be inspected by the person who will be responsible for preparing the estimate. This should be the head of department – the individual who takes the responsibility for the estimate – not the estimator who will later be appointed to deal with the task. It is most important that the early inspection is carried out by a person experienced in current procedures and documentation, and capable in decision-making and effective in communicating with others.

The documents should first be checked that they accord with those listed in the letter of invitation, normally:

1. Two copies of the bills of quantities.
2. Two copies of the general arrangement drawings.
3. Two copies of the form of tender.
4. Addressed envelope for the return of tender (and priced bills if applicable).

If the documentation is not complete the fact should be reported immediately by telephone. If reference is made in the letter that certain sections of the bills will follow shortly and the tendering time is as stated in the preliminary invitation, an appeal should be made for a revision of time for tendering. It is good practice that the time for tendering should be calculated from the date of issue of the last section of the documents.

The preliminaries sections of the bills need to be examined carefully at this stage, particularly the general and contractual particulars called for under A10–A37 SMM7. The drawings from which the bills were prepared should be listed in accordance with A11 and the drawings set out in General Rule 5 should be enclosed. Drawing number references should match those recorded, for example if the bills state drawing No. 90/3/2910C and 'D' is supplied, then it must be questioned.

If further information is needed (to find the extent of temporary works, for example) more drawings may be sought. It is important that the estimator works from actual full-scale prints rather than making a visual inspection at the consultant's office. As a general rule, clients issue all information that is relevant and available at the tender stage.

The contractor may already have a guide to the value of the project; if not, he could get a rough guide to the tender figure by applying approximate rates to the principal quantities. The initial inspection of the tender documents is completed by producing a tender information (enquiry record) form which is similar to the Preliminary Enquiry form but with more details of the estimated cost and contract details. The COEP provides a form for this purpose. The Tender Information form is a valuable source of information because it provides management with a summary of the tender which is being prepared and can be kept for all previous tenders whether successful or not.

Competition legislation

The Competition Act 1998 is designed to make sure that businesses compete on a level footing by outlawing certain types of anti-competitive behaviour. The Office of Fair Trading (OFT) has strong powers to investigate businesses suspected of breaching the Act and to impose tough penalties on those that do.

All businesses, no matter how small, need to know about the Act – to avoid becoming a victim, and to avoid breaking the law. The Act should not be viewed in isolation. The Enterprise Act 2002 among other things introduces a cartel offence under which individuals who dishonestly take part in the most serious types of anti-competitive agreements may be criminally prosecuted.

In addition, as a result of amendments to the Company Directors Disqualification Act 1986 under the Enterprise Act 2002, company directors whose companies breach competition law (including the prohibitions in the Act) may be subject to Competition Disqualification Orders, which will prevent them from being concerned in the management of a company for a maximum of 15 years.

Prohibiting anti-competitive agreements

The Competition Act 1998 came into force on 1 March 2000. It prohibits both informal and formal arrangements, whether or not they are in writing. So an informal understanding where Companies A and B agree to match the prices of Company C will be caught in the same way as a formal agreement between competitors to set prices.

Although many different types of agreement are caught by the prohibition, the Act lists specific examples to which the prohibition particularly applies. These include:

- agreeing to fix purchase or selling prices or other trading conditions;
- agreeing to limit or control production, markets, technical development or investment;

- agreeing to share markets or supply sources;
- agreeing to make contracts subject to unrelated conditions;
- agreeing to apply different trading conditions to equivalent transactions, thereby placing some parties at a competitive disadvantage.

Key aspects of the legislation are:

- anti-competitive agreements, cartels and abuses of a dominant position are now unlawful from the outset;
- businesses which infringe the prohibitions are liable to financial penalties of up to 10% of UK turnover for up to 3 years;
- competitors and customers are entitled to seek damages;
- the Director General of Fair Trading has new powers to step in at the outset to stop anti-competitive behaviour;
- investigators are able to launch 'dawn raids', and to enter premises with reasonable force;
- the new leniency policy will make it easier for cartels to be exposed.

The intention is to create a regulatory framework that is tough on those who seek to impair competition but allows those who do compete fairly the opportunity to thrive.

Professional fee scales, which used to require that professions charged particular fees for particular jobs, are now illegal because they are effectively price fixing. If they were now used they would render participants liable to both company and individual penalties. Another example in construction is the practice of sharing bid costs with a competitor. These may be fees for design work or taking off quantities. Provided that bid cost sharing is not prohibited by the tender arrangements, and provided that there is no other collaboration or collusion, this is likely to be acceptable. However, it would not be acceptable for bid costs to be shared without declaring this to the customer, or if the sharing in any way meant that the prices to be charged were agreed or discussed with the competitor. It would also be illegal for either competitor to agree not to bid.

Cartels

In its simplest terms, a cartel is an agreement between businesses not to compete with each other. The agreement is usually verbal and often informal.

Typically, cartel members may agree on:

- prices
- output levels
- discounts

- credit terms
- which customers they will supply
- which areas they will supply
- who should win a contract (bid rigging).

Cartels can occur in almost any industry and can involve goods or services at the manufacturing, distribution or retail level. However, some sectors are more susceptible to cartels than others because of the structure or the way in which they operate. For example, where:

- there are few competitors;
- the products have similar characteristics, leaving little scope for competition on quality or service;
- communication channels between competitors are already established;
- the industry is suffering from excess capacity or there is general recession.

Cartels are a particularly damaging form of anti-competitive behaviour – taking action against them is one of the OFT's priorities under the Act. A business could be a victim of a cartel or could be breaking the law. Either way, it is vital that people know how cartels can affect their business.

A member of a cartel could be fined up to 10% of its UK turnover for up to three years. As a result of the Enterprise Act 2002, participation in cartel agreements may expose individuals responsible for those agreements to criminal sanctions. However, if a business ends its involvement and confesses to the Office of Fair Trading, it can be granted immunity or a significant reduction in any fine.

If there is a compliance programme in place this may be taken into account as a mitigating factor when calculating the financial penalty. The precise circumstances of the infringement, and in particular the efforts made by management to ensure that the programme has been properly implemented, will be carefully considered.

The Enterprise Act 2002

The Enterprise Act received Royal Assent on 7 November 2002. It covers a range of measures to enhance enterprise through strengthening the UK's competition law framework, transforming the UK's approach to bankruptcy and corporate rescue, and empowering consumers.

The Act builds on the progress made by the Competition Act 1998. The substantive consumer and competition provisions of the Act came into force on 20 June 2003.

The measures in the Enterprise Act will empower consumers, modernize the insolvency regime so that it supports enterprise, and help to make UK markets more competitive. The main reforms in the Act are: criminal sanctions with a maximum penalty of five years in prison to deter those individuals who dishonestly operate hardcore cartels – agreements to fix prices, share markets, limit production and rig bids. The offence will be tightly defined, ensuring that honest businesspeople will have nothing to fear. US research shows that cartels raise the prices of the affected goods and services by 10% on average.

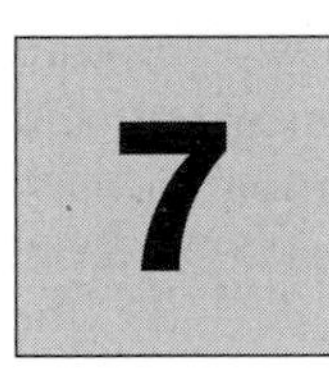

7 Project appreciation and enquiries to suppliers and sub-contractors

Introduction

Following management's decision to tender, the tender documents are given to the estimator to prepare the estimate. He should read the documents to gain an overall understanding of the project. A decision can then be made about the help needed from other departments for project management, planning, procurement and commercial appraisal.

If a bill of quantities is available, enquiry schedules can be drawn up immediately, and documents will be prepared for suppliers and sub-contractors (see Chapter 8). Enquiries need to be sent promptly so that specialists have enough time to prepare their quotations.

Once the enquiries are under way, the estimator will broaden his understanding of the project by scheduling principal quantities and PC and provisional sums; he will undertake visits to site and if necessary the offices of the consultants.

Estimate timetable

For most tenders there is an absolute requirement to meet the submission date. The estimator must programme the activities needed to produce a tender to show how the deadline can be met and explain to other members of the team their part in the plan. Each project is different, and some dates such as those for the return of quotations and a design freeze (design and build) require firm action to maintain the programme dates.

Time allowed for tendering is usually limited by the client's need to start a project quickly. If the design stage has been delayed it is often the tender stage that is shortened. Flexibility is needed to concentrate on the critical parts on the estimate preparation. The estimator can press on to complete his work with a day or so to spare for reconciling and checking the estimate. Much of the early part of the tender period is given over to the dispatch of enquiries to suppliers and sub-contractors and setting up job files when a computer system is used. Figure 7.1 shows a simple

CB CONSTRUCTION	**ESTIMATE TIMETABLE**

Project :	Lifeboat station		
Ref. No:	T384	Date :	16/6/08

	June													July							
	14	15	16	17	18	21	22	23	24	25	28	29	30	1	2	5	6	7	8	9	12
Documents received	e																				
Decision making	e																				
Study documents		e																			
Mark up enquiries		e	e																		
Dispatch enquiries			b	b	b																
Date for return of mat prices																					
Date for return of s/c prices																					
Computer entry			a	a	a	a	a														
Visit site					e																
Visit consultants					e																
Risk workshops					a												e				
Study methods, temp works						e p	p	p	p	p											
and programme																					
Main pricing								e	e	e	e	e	e	e	e	e	e				
Extend bill rates and													a	a	a	a	a	a			
chase quotations																					
Project overheads																p	p	e	e		
Summaries and report																		e	e		
Final review																					
Prepare docs for submission																					
Submission																					

Key : e = estimator, b = buyer, p = planner + project manager, a = estimating assistant, c = commercial manager

Fig. 7.1 *A typical estimate timetable*

timetable for producing an estimate and tender. This programme is simple to produce using a 'blank' standard form because many of the activities are common to all tenders.

The estimator is responsible for preparing the estimating timetable showing the key dates for him and the other members of the estimating team. As the team leader, the estimator will need to coordinate the other people in the team. For large projects the bidding team is normally led by a project manager. The COEP gives a typical checklist for a coordination meeting (chaired by the chief estimator), which is presumably for large-scale or complex projects.

Pricing strategy

Procedures for estimating are well understood by most contractors, and formal plans are not needed for every tender. On the other hand, large-scale projects, such as hospitals and office blocks, need to be planned from the start by agreeing pricing methods and strategies at a 'start-up' meeting.

A typical pricing strategy document, shown in Fig. 7.2, lists the aims, explains the actions, records progress and identifies the person responsible for completing the actions. This is an important document to agree with management – before recourses are allocated to the tender. It would be reckless to wait until a mid-tender review meeting to discover that the approach is not in line with management's expectations.

Schedules

The estimator should list all the prime cost and provisional sums to identify the work which will be carried out by other contractors. A summary of costs written into the bill will become part of the estimator's report for management at the final review stage. The example given in Fig. 7.3 shows the structural steelwork and electrical sub-contractors have been chosen by the client and no enquiries will be sent by the contractor at tender stage.

The summary of PC and provisional sums can also be used to show the attendances required by each named sub-contractor. SMM7 (A51.1.3) gives a list of the items of special attendance which must be given in a bill of quantities if required. The summary can include these items in the form of a checklist. The CIOB Code of Estimating Practice provides an alternative form, which encourages the estimator to produce a breakdown of attendances into labour, plant, materials and sub-contractors. Most of the costs of providing these attendances are evaluated when pricing the preliminaries because they can be considered in relation to the project as a whole.

CB CONSTRUCTION LIMITED **Pricing Strategy**	Project	**Ashbury College**		
	Type	**Design + Build**		
	Ref. No:	**T384**	Date:	**27.6.08**

Ref	Aims	Actions	Progress	Owner
1	To research similar schemes and pricing data	Obtain data from Morton College and office database		JM
2	To agree a target cost	Ask Brian to speak to Client about affordability. Use Morton College to produce elemental target cost plan		JM
3	To influence the design of the scheme	Take cost plan to first design meeting. Advise architect about target for each element of the building		JM
4	To quantify the work	Send drawings and specs to Joe Clarke. Ensure quants are received within two weeks. Quants to be in spreadsheet format		PC
5	To monitor the design – ensure it develops within the target cost plan	Attend design meetings. Get Joe to check quants on all drawings issued		JM
6	To obtain quotations for at least 85% of the value of works	Richard send enquiries to all main trades including groundworks		RH
7	To programme the works	Rachel produce programme with optimum solution and one which meets the client's end date. Also programme the preferred bidder phase.		RE
8	To quantify and price project overheads	Plant, temporary works, scaffolding and supervision to be quantified by planner. Allow subsistence costs for project manager		JM
9	To complete the estimate for review by management	Mid tender review to be on Tuesday 15 July. Final review Wed 6 August 2008		JM
10	To identify risks and opportunities	Phil arrange risk meeting on Mon 12 July. Check for overlaps with trade contractors		PC
11	To comply with submission requirements and follow-up by contacting client	Ask James Barker to produce/edit submission document		JM/JB

Fig. 7.2 *Typical pricing strategy for a 'design and build' project*

The estimator needs to abstract 'direct' work items that will be carried out by the main contractor, such as excavation, concrete work, brickwork and drainage. The trade abstract shown in Fig. 7.4 brings together all the pages to be priced under each trade heading, and helps the estimator to assign pricing duties when more than one estimator is working on the tender.

CB CONSTRUCTION LIMITED	PC & Provisional Sums	Project	Lifeboat Station		
		Ref. No:	T384	Date:	13.6.08

Bill Ref:	Description	Prov Sums	Prime Cost Sums Gross	Discount	Nett	Notes for pricing preliminaries
	PC SUMS					Special attendances
6/1a	Structural steelwork Named by client		23 000	575	22 425	good access roads and hardstanding
6/1e	Electrical installation Named by client		15 600	390	15 210	scaffolding covered storage
6/2a	Fire doors Named by architect		3 840	192	3 648	
	PROVISIONAL SUMS					Prelims for defined prov sums :
6/2m	Contingencies	5 000				
6/2n	Drainage to sump	1 000				
6/2p	Glazed roof over entrance	3 500				scaffolding protection and cleaning
6/3	Daywork – labour add 110%	1 000 1 100				
	Daywork – materials add 15%	500 75				
	Daywork – plant add 60%	500 300				
	Totals :	12 975	42 440	1 157	41 283	

Fig. 7.3 *List of PC and Provisional Sums at project appreciation stage*

The estimating team

The roles of the members of the estimating team (Fig. 7.5) will vary from company to company and will depend on the size of the job. Some companies prefer to hand a copy of the documents to the buyer for sending out enquiries and others elect to keep

CB CONSTRUCTION LIMITED	Project:	Fast Transport		
Trade abstract	Ref:	T354	Date:	20.6.08

Trade		Bill pages	Spec pages	Estr	Quant	Unit	Lab	Plt	Mat	Total
D20	Earthworks	3/1-5, 4/35-37	2/1-12	JM	1 075	m^3	7 800	7 980	5 700	21 480
R12	Drainage	5/1-34	2/56-65	JM	823	m	8 650	6 520	9 520	24 690
E10	Concrete work	3/5-7, 4/38-40	2/13-15	JM	956	m^3	12 520	3 750	51 840	68 110
E30	Reinforcement	3/11, 4/43	2/15-16	JM	76	t	13 680		24 700	38 380
E20	Formwork	3/7-10, 4/40-42	2/17-22	JM	2 150	m^2	30 950		16 530	47 480
E40	Concrete sundries	3/7,11-13, 4/44, 45	2/14	JM			2 150		3 180	5 330
F31	Precast concrete	4/46	2/35	PC			1 850		8 250	10 100
F10	Brickwork	3/19-22	2/24-28	PC	76	th	18 460		19 520	37 980
F11	Blockwork	3/20	2/27-32	PC	3 100	m^2	18 880		17 450	36 330
F30	Brick sundries	3/22, 23	2/24	PC			4 520		5 250	9 770
G20	Timber	3/28-31	2/39-44	PC	4 210	m	5 150		6 210	11 360
P20	Joinery	3/31-38	2/39-46	PC			2 380		5 310	7 690
G12	Metalwork	3/29, 39	2/47-48	PC			2 680		8 450	11 130
P31	BWIC	3/55		JM			2 110		2 030	4 140
	Attendances	6/1		JM			410		385	795
					totals		132 190	18 250	184 325	334 765

Fig. 7.4 *Trade abstract for sections to be priced by the contractor*

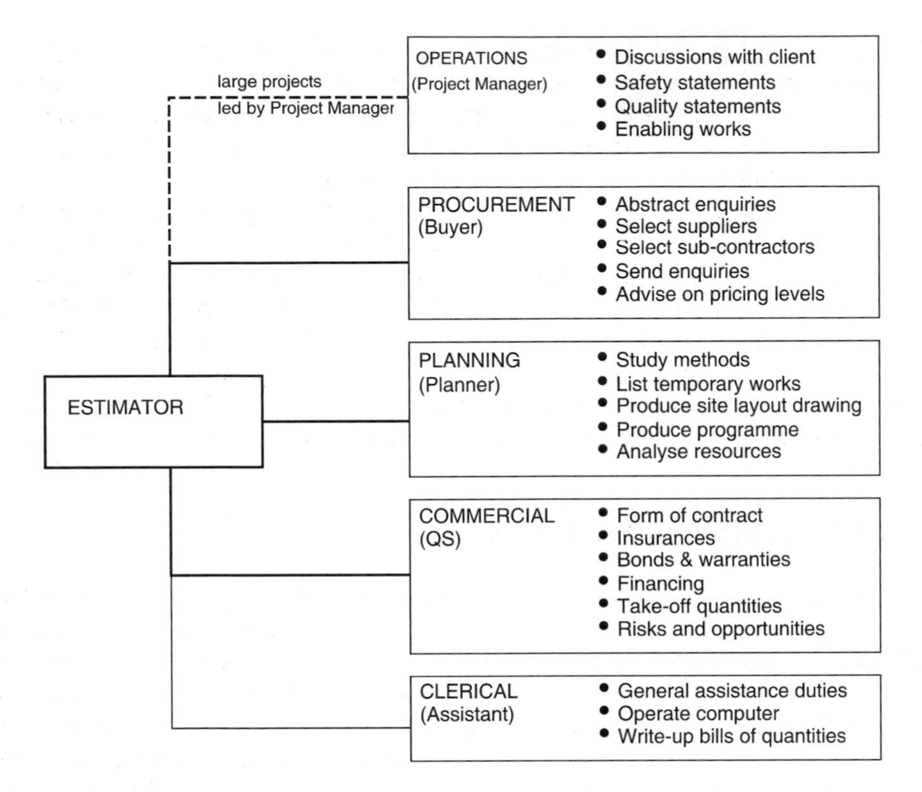

Fig. 7.5 *Coordination of the estimating team*

control of this activity in the estimating section. A compromise would be for the estimator to abstract the materials and sub-contract packages, forming part of the estimate, and ask the buyer to select suitable companies to receive enquiries and coordinate and prepare documents for despatch. This allows the estimator to keep control over what prices are sought, and the buyer can use his experience to get better prices. The commercial manager, or quantity surveyor, should be given the opportunity to comment on the form of contract proposed, insurances, bonds and financing requirements. If there are any onerous or unusual conditions that may cause problems with an unqualified bid, they should be challenged by writing to the client for a ruling. This should produce a more satisfactory result than leaving it for management to decide at the final review meeting when it is too late for an amendment to be made to the tender documents.

The commercial manager needs to develop an overview of the scheme in order to produce a 'risks and opportunities' register. The register will identify a risk (or opportunity), determine its value and the probability of it happening. Many risks can be reduced or eliminated, during tender stage, by changing the design, or passing the risk to others. The best way to gather a wide range of risks is to bring together the main members of the bid team in a short risk workshop. Each person is asked to write down at least ten risks, and categorize them as construction, financial, commercial, programme or client-based. For a large design and build project, risks can be extended to include: design, life cycle and facilities management.

During the tender period, a planning engineer has a great deal to contribute. Instead of looking at the project from a financial point of view he will start by examining the layout of the site, methods for construction, temporary works requirements, distribution patterns, sequence of work, a preliminary construction programme and resource levels. A civil engineering estimator will usually assess the temporary works, and programme the works himself because this is often the preparation needed for operational estimating techniques. A building estimator will tend to rate the items in a bill of quantities and rely on the planner for a more detailed examination of methods.

Visits to consultants and site

An estimator needs to examine the documents carefully before leaving his office. It may be helpful to mark up a site plan to highlight the main elements, such as areas of scaffolding; access routes; existing and proposed services; fencing and external hard landscaping. The estimator must identify the work items which will be priced on site, including demolitions, alterations and repairs. A preliminary assessment can be made to find areas for general facilities such as site accommodation, cranage, storage areas and hard standings. If any relevant information is missing then it might be questioned at the consultant's office or during the site visit.

In a perfect world there should be no need to visit the consultant's office – the tender documents should define the basis of an agreement to construct. There are, however, some benefits, as follows:

1. A critical assessment can be made of the progress made with the construction drawings, which would be important if there is a need for a quick start.
2. The estimator can seek clarification of layouts and details that were not included with the tender documents.
3. The names of adjoining property owners can be used to find sites for disposal of surplus materials and search for sources of fill.
4. The contractor can consider opportunities to offer an alternative bid in terms of time or design.
5. A contractor can explain its track record and interest in future projects.

A site visit must be made before a tender is submitted and should wherever possible be carried out by the estimator in person. The COEP provides a comprehensive standard report form which will cater for most projects.

Site features and existing buildings can be recorded on camera, not only to remind the estimator of site details during the pricing stage but will prove an excellent means of communication at the final review meeting. Occasionally permission will be required to take photographs, such as inside existing buildings, near sensitive production processes and of existing property where security must be preserved. Permission is not needed for photographs taken of any property from outside the site boundary.

In assessing the site, note should be made of the topography – whether it is situated on a hill or in a valley, for example. Some plants could signal that although the ground is dry in summer the same ground may be waterlogged in winter. The general levels of the site need to be related to the information gained from the site investigation report; sometimes the remains of trial holes can be inspected. The contractor must assess the effect of a change of season from summer to winter. Some of the best information is available from residents and landowners; they often have valuable knowledge about flooding and ground conditions met during previous excavation work.

The site location and access routes must be examined. Urban sites need particular consideration. In built-up areas where cranes are to be used permission may be needed to encroach on air space; loading gantries, hoardings, protected walkways and temporary footpaths all need careful consideration when pricing preliminaries. In country districts there could be a problem with taking large loads through narrow lanes. Signboards may be needed to guide traffic from the nearest main road. Will crossovers be needed from public roads to the site, or will they need to be constructed?

There are some questions about existing utility services which must be answered. Who supplies the services, where are the nearest connections and what sizes and capacities are available? Much effort can be saved if site toilets can be connected to a foul drain. Water will be needed early on when the site facilities are set up. Electricity supplies may be present but are they sufficient to meet the heavy demand of running plant? If an electrically powered tower crane is planned, a large-capacity supply may be needed. The electricity company is usually reluctant to bring the new main in at an early stage unless the contractor is prepared to contribute to the capital cost.

Security is an increasing problem which must be checked during the site visit. Existing boundaries need to be related to the site boundary and compared with the fencing measured in the bill or described in the preliminaries. A short discussion with a resident might highlight the risks.

Finally and most important are the items which can only be priced by assessing the extent of the work on site. These include demolitions (usually given as items in the tender documents with overall dimensions), alterations (given as spot items the complexity can be gauged by the estimator and expressed in work-hours), and clearing site vegetation (measured in square metres and varies from site to site). Where the work is predominantly described in spot items, the estimator will make extended and often repeated visits to site, on occasions meeting sub-contractors.

Enquiries to suppliers and sub-contractors

Contractors have developed procedures to ensure that tenders are based on up-to-date prices for materials and specialist services. As the COEP states, 'The contractor's success in obtaining a contract can depend upon the quality of the quotations received for materials, plant and items to be sub-contracted.' To illustrate the point, the following breakdown has been taken from a typical building estimate, and shows that materials and sub-contracts account for 72% of the estimated costs before on-costs and provisional sums are added:

Breakdown of contractor's costs	
Direct work – Labour	23%
Direct work – Plant	5%
Direct work – Materials	28%
Domestic sub-contractors	44%

Enquiries must be sent promptly but not at the expense of accuracy. Wrong or incomplete information will lead to delay in receiving comparable quotations. An orderly presentation will do much to avoid mistakes. The COEP provides some model forms for use in abstracting and suggests ways of selecting suitable suppliers and sub-contractors. These forms are further developed according to each company's procedures. Figure 7.6 shows a form which can be used for abstracting materials or sub-contract packages and later helps to record the receipt of quotations by highlighting the names of firms which have responded.

Most enquiries consist of either photocopies or electronic copies of the relevant pages from the bills of quantities with a letter and any related specifications and drawings. Ideally those bill items, which are not to be priced, should be crossed out to avoid confusion. With computer-aided estimating systems and spreadsheets, a *sub-contractor* bill of quantities can be generated by choosing only those items relating to a trade package. Historically, sub-contractors were sent two copies of the bill pages which were to be returned. This practice has died out largely because sub-contractors prefer to photocopy their priced bills or return their quotations using a spreadsheet file attached to an email.

With the introduction of CPI and SMM7, bill descriptions now have references to the specification clause which apply to the item of work. This helps the estimator ensure that he sends the correct specification clauses with the enquiries. For tenders based on drawings and specifications, enquiries must clearly state the scope of work to be priced for each sub-contract package. For example, a flooring sub-contractor will need to know whether to include latex screeding, skirtings and expansion joint covers in their price.

CB CONSTRUCTION	ENQUIRY ABSTRACT	[X] materials [ ] sub-contractors	Project	Lifeboat Station		
			Ref. No.	T384	Date:	16.6.08

Ref.	Description	Approx. Quants	Bill pages	Spec pages	Drawings	Names	Telephone
M1	AGGREGATES					SWANFIELD STONE CO	
	DOT TYPE 1	580T		2/5,6		LITTLEGREEN QUARRIES	
	DUST	55T				J.P.HEPPLE	
	20MM SINGLE SIZE	110T					
M2	CONCRETE					SWANFIELD STONE CO	
	C15P	25m^3		2/11-15		DRAY CONC SERVICES	
	C20	40m^3				PINTO CONCRETE LTD	
	C35	400m^3					
M3	REINFORCEMENT					BARBEND FABRICATION	
	A393	1240M^2		2/19,20		OAKFORD REINFORCEMENT	
	10MM HY	1.1T				DOWLAIS STEEL	
	12MM HY	3.8T					
	16MM HY	5.6T					
M4	PRECAST CONC					HILBERG CASTINGS	
	LINTELS	27NR	3/22	2/22		TELGAN CONC PRODUCTS	
	KERBS	360M	4/46				
M5	BRICKS/BLOCKS					SIMGROVE BRICK	
	FACINGS	11000NR	3/19-22	2/28-30		BUSH BROS LTD	
	100MM BLOCKS	550M^2				HEPPLE BRICK CO	
	140MM BLOCKS	200M^2					
M6	JOINERY					GOTHIC JOINERY LTD	
	AS BILL PAGES		3/31-37	2/36-40	SK1,2,3	ST.ANNES TIMBER	
					B/27/204	SHIRE MANUFACTURING	

Fig. 7.6 *Abstract form for materials or sub-contract enquiries*

In the past, estimators and buyers kept lists of companies who could supply materials and services in a card index of names, usually in trade order. Since desktop computers have been introduced with user-definable databases, many contractors maintain name and address files which can be searched for trade contractors within travelling distance of the site. The software is usually able to address the letters and envelopes and keep track of previous performance, although feedback from site on a sub-contractor's behaviour is not always consistent or reliable. Computer-aided estimating software can also maintain a supplier database. An estimator can then link bill pages to his list of sub-contractors, from within the software running on his computer, or using web-access to a central database.

When tendering in an unfamiliar location, the estimator must allow time when visiting the site to tour the area, and note the volume of work going on and what stage individual jobs have reached. Site hoardings which list the sub-contractors being used on a site are a valuable source of information about specialists who are acceptable to your competitors. Another useful starting place for obtaining names is the local newspaper; especially when the newspaper has printed a special promotion naming various sub-contractors linked to the completion of a local building. An estimator will often search the internet or look up nationally published directories. If new sub-contractors are to be invited to tender the buyer should telephone first to ascertain their ability and willingness to prepare a quotation.

Enquiries for materials

Many construction organizations have a standard form of enquiry for suppliers; a typical example is given in Fig. 7.7. Enquiries should give the following information:

1. Title and location of the work; some suppliers complain that enquiries from different contractors are received in varying formats, often making it impossible to decide whether they are for the same job.
2. Description of the materials, supported by specifications.
3. Approximate quantities, so that bulk discounts can be quoted.
4. Date by which the quotation is needed; 7 to 10 days would seem reasonable although those with complex fabrication work to price, joinery suppliers for example, may need longer; the most successful approach where time is limited is to say so, and request cooperation by responding as soon as possible.
5. Name of the estimator dealing with the tender.
6. The contract period with a guide to the dates for deliveries; it can only be a guide because the start date is rarely known and the construction programme is not yet available.
7. Whether firm price or fluctuating price.

Our ref : T384/M5 17th June 2008

Simgrove Brick Ltd
Unit 3, Northbridge Industrial Estate
Northbridge
NB3 5MGG

Dear Sirs,

NEW LIFEBOAT STATION
BEACH LANE, STANSFORD

We are tendering for the above project and ask you to submit your best rates for the following items to be delivered to the site. The project is due to start on Friday 29 August 2008. If you have any queries in relation to this enquiry please contact our Regional Buyer, Mr Frank Applecourt.

Please reply by 28th June 2008 and state any known or anticipated price rises likely to affect our tender.

Yours faithfully,

Item	Description	Approx quants	Supporting documents enclosed
1.	Facing bricks	11,000nr	Bill pages 3/19–22 Spec pages 2/28–30
2.	100 mm blocks	550 m^2	
3.	140 mm blocks	200 m^2	

Fig. 7.7 *Typical enquiry for materials*

8. Minimum discount terms required.
9. Any limitations on access to the site.

The supplier should make every effort to meet the specified submission date and tell the contractor if a delay is expected. If the supplier is himself awaiting information from his sources, he should submit a quotation with a clear statement that prices are to follow.

There is not much point in keeping price lists unless they are updated regularly. It is more important for an estimator to compile a library on material characteristics, quality, sizes and performance standards. Knowledge of available materials and products will enable the estimator to consider alternatives which comply with the specification.

Enquiries to sub-contractors

There are three kinds of domestic sub-contractor who will be approached at tender stage: the conventional sub-contractor who provides a complete service; labour-only sub-contractors who are supplied with their materials and plant; and labour and plant sub-contractors who receive their materials from the main contractor. Sub-letting work to specialists is an attractive arrangement for contractors because much of the technical and financial risk is passed to another party, and a profit is almost guaranteed (providing the work goes to plan). On the other hand sub-contractors can benefit from changes to the scope and specification together with possible variations. Most contracts state that the main contractor shall get written permission before sub-letting any of the work, and in some (now rare) instances employers will not allow the use of labour-only sub-contractors.

Domestic sub-contractors will need a lot of information about the site, contract conditions, programme, the specification and extent of work. The example given in Fig. 7.8 shows a standard enquiry letter which can be stored as a word-processor file and tailored for each contract and trade. This example is lacking guidance on the timing of the work, probably because this can be found in the extract from the preliminaries. On larger projects, an outline programme might be available and the sub-contractor will be asked to prepare his own tender programme with information about extended delivery periods which might affect the progress of the works.

The COEP lists the details to be given in a contractor's enquiry letter, as follows:

- Site address and location (with a map if necessary).
- Name of employer, and professional team.
- Relevant details of main and sub-contract.
- Any amendments to the standard conditions, including bonds and insurances.
- A request for daywork rates.
- Date for return of quotations.
- General description of works.
- Details of access, site plant and other facilities available.
- Where full contract details and drawings can be inspected.
- Contract period, programme and any phasing requirements.
- Any discounts to be included.
- Two copies of the relevant sections of the bills of quantities.

17th June 2008

Dear Sirs,

NEW OFFICES FOR MANIFOLD METALS PLC,
NORTH LANE, STANSFORD

We invite you to tender for the PAINTING work for the Manifold Metals project and enclose the following details which describe the quality and quantity of the work:

Preliminaries pages:	1/2–12	
Bill pages:	3/45–48	(2 copies)
Specification pages:	2/39–40	
Drawings:	D/206/1	
Form of Tender		(2 copies)

The names of the parties, general description of the works, and details of the main contract are given in the extract from the preliminaries. Your form of tender, priced bill of quantities and daywork rates must be delivered to this address to arrive by 7th July 2008.
The form of sub-contract will be SBCSub incorporating all relevant published amendments and the following:

Payments	:	Monthly
Discount to main contractor	:	2.5%
Fluctuations	:	Firm Price
Liquidated damages	:	£1200 per week
Basis of daywork	:	current RICS definition
Retention	:	3%
Method of Measurement	:	SMM7
Defect liability period	:	six months

We will provide all sub-contractors with water, lighting and electricity services near the work and common welfare facilities on site. Sub-contractors will be required to provide the following services and facilities:

a. unloading, storing and taking materials to working areas
b. power and fuel charges to temporary site accommodation
c. clearing-up, removing and depositing in designated collection points on site all rubbish or other surplus or packing materials
d. temporary accommodation and telephones
e. day-to-day setting out from main contractor's base lines

If you have any queries about this enquiry please contact the estimator for the project, Mrs Peggy Carter.
Would you please confirm by return your willingness to tender by the date for tender?

Yours faithfully,

Fig. 7.8 *Sample enquiry letter to domestic sub-contractors*

- Copies of drawings and schedules where applicable.
- Services and attendances to be provided by the main contractor.
- A clear statement of how fluctuations will be dealt with.

Labour-only sub-contractors usually quote under different arrangements; in particular they often expect:

1. Setting out by the main contractor.
2. Weekly or fortnightly payments by the main contractor.
3. Modified retention sums to reflect the extent of their work.
4. Materials delivered, unloaded and sometimes taken close to the point of fixing.
5. Major items of plant (which will be used by other sub-contractors) to be provided by the main contractor.

Although most enquiries are in the form of photocopies of bill pages, it is important that the estimator clearly states the portions of the work which the labour-only sub-contractor is expected to carry out. As an example, a concrete specialist might be asked to price placing concrete, fixing reinforcement and labour and materials in fixing formwork. Another firm might be asked to lay concrete in floor slabs and power float the surface.

The Construction Industry Board produced in 1997 a 'Code of Practice for the Selection of Sub-contractors', which recommends a tendering procedure which mirrors that suggested in its parallel publication for main contractors. In other words, sub-contractors should be asked for their willingness to tender, there should be full tender documents, sufficient time must be given for preparing tenders, and sub-contractors should be told about their performance.

For design and build projects, there are additional responsibilities for sub-contractors, not least the development of the concept design and completion of working drawings. Sub-contractors are expected to submit, with their tender, risks that have been identified and priced in their offer. It is important that the main contractor ensures there is no duplication of risk allowances in the tender.

Sir John Egan's report, *Rethinking Construction*, brought immediate changes to the way in which the construction industry procures supplies and services. John Egan used his experience of other industries to highlight the benefits of smarter procurement through integrated supply chains.

There are efficiencies in working with suppliers and sub-contractors who become part of a close working relationship. Through strong supply chain management, vendor lists are kept small, problems can be shared and organizations begin to work better together. Term contracts can be set up with material and plant suppliers whereby prices are fixed for any site in any location for a fixed period of time.

Defence procurement has taken these concepts further. The prime-contracting route, adopted by Defence Estates, relies on strong relationships between (prime) contractors and their sub-contractors. Commitments to guaranteed maximum cost and risk assessments are made at prime contractor and sub-contractor levels and shared with the client.

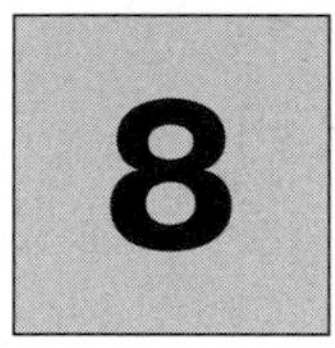

8 Tenders with cost planning

Introduction

This chapter extends estimating into pricing contracts with special terms at the pre-contract stage. This includes building work which is procured through:

- The Private Finance Initiative (PFI);
- Prime contracting (Ministry of Defence);
- Schools in batches such as Academies and Building Schools for the Future (BSF);
- Health projects using 'Procure 21' and 'LIFT' procurement arrangements;
- Private clients who have adopted framework agreements;
- Target-cost contracts.

It is also becoming common for estimators to provide cost plan advice to developers and project sponsors in advance of a competitive tender stage. Furthermore, there is a role for cost planning in the first stage of a two-stage tender.

There is a clear need for tenders with cost planning when schemes are designed up to RIBA Stage C or D, where full construction details and specifications are not yet available. These procedures bridge the gap between cost plans created by clients' advisers and tenders produced by contractors.

'Top down' and 'bottom up' estimating are terms used to describe the combination of cost planning and traditional estimating in a design and build tender. These arrangements are illustrated in the process overview shown in Fig. 8.1.

An understanding of schedules of accommodation and gross internal floor area are important skills throughout these processes. Areas are used to prepare early cost plans for whole buildings. An alternative approach would be to prepare costs for functional areas, such as hospital wards (£2400/m^2) and operating theatres (£5400/m^2). The Department of Health publishes departmental cost guides to assist trusts in building up their business cases using this method. For a school project, a cost plan could provide rates for classrooms, offices, kitchens, corridors and plant space. This departmental pricing method is worthy of further study and development.

Figure 8.2 shows the processes undertaken by an estimator preparing a tender based on early designs and cost plans (as opposed to bills of quantities).

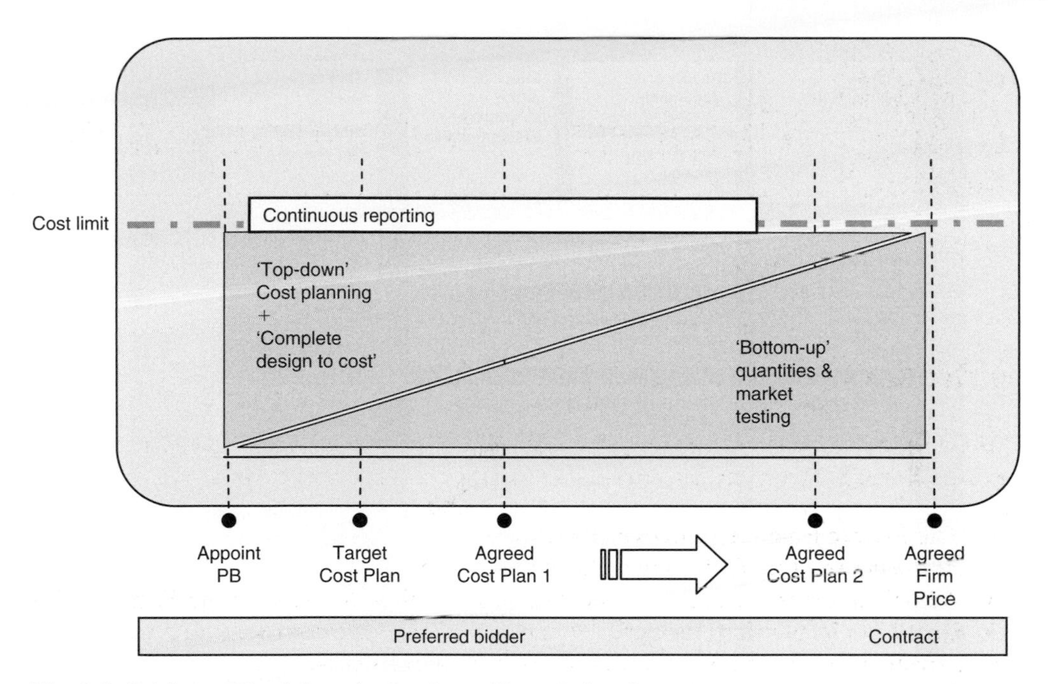

Fig. 8.1 *Pricing methodology for tenders with cost planning*

Terminology

Client's brief (Work Stage C): The client's requirements developed after consideration of any feasibility studies and set out in the strategic brief.

Cost indices: Based on historical records and projected forecasts, they provide a guide to possible cost changes from the date of a cost plan to the tender or award date and for inflation during construction.

Cost limit or budget: Limit of capital expenditure set by viability or funding for the project.

Cost plan: A document showing the estimated cost of all parts of the project and how it is to be spent.

Cost planning: A method used to set, monitor and report on costs during the design process in order to ensure expenditure will be within a client's budget.

Design to cost: As part of a cost planning process, this provides designers with suitable cost statements to allow them to understand the consequences of their decisions.

Estimates: Net estimated cost of carrying out the works for submission to management at the final review meeting.

Functional area method: An approximate estimating method that uses the cost of functional areas from earlier schemes to calculate a rough order of cost at the earliest

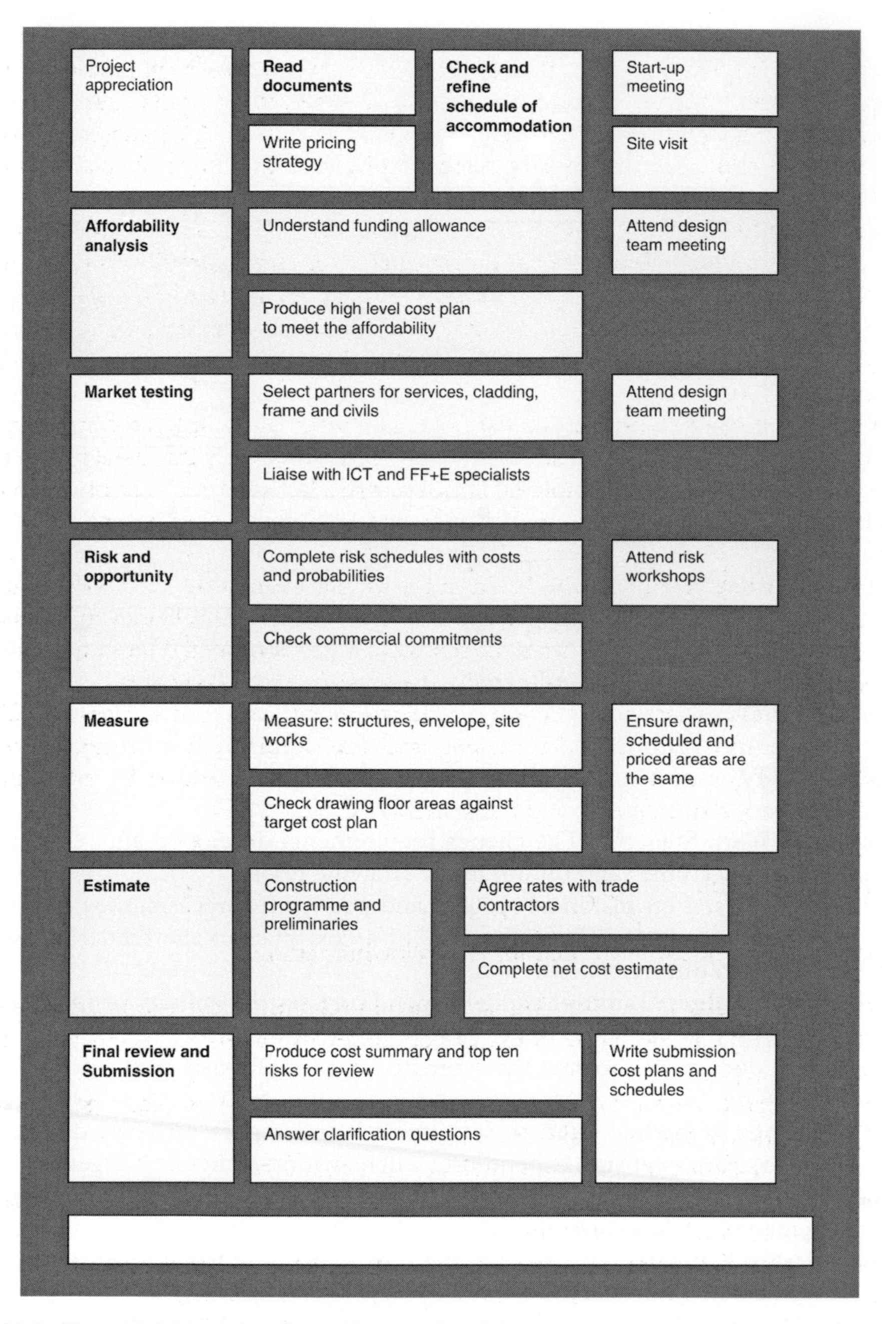

Fig. 8.2 *Flowchart for estimating with cost planning*

stage of a project's development. The unit cost is calculated by analysing previous projects to establish historical data for the cost per square metre for functional areas.

Functional unit method: An approximate estimating method that uses the unit cost from earlier schemes to calculate a rough order of cost at the earliest stage of a project's development. The unit cost is calculated by dividing the total cost of building by the number of functional units.

Gross internal floor area (GIFA): This is measured to the internal face of external walls of all buildings. No deductions are made for internal walls, voids, ducts, stairs, lifts cores, plant rooms and roof top 'pop-ups'. An atrium is measured at its lowest level only. Sums can be added for canopies, balconies, car parking, demolitions and small minor buildings. If connecting corridors have windows and heating, they are measured as GIFA.

Standard Form of Cost Analysis: The Building Cost Information Service (BCIS) Standard Form of Cost Analysis is a standard list of elements used in the UK to provide data which allows comparisons to be made between the cost of achieving various building functions in one project with that of achieving equivalent functions in other projects.

Superficial (floor area) method: An approximate estimating method that uses the cost per square metre of floor space from earlier comparable schemes to calculate a rough order of cost at an early stage of a project's development. The rate is calculated by dividing the total cost of building by the gross internal floor area.

Target cost plan: A schedule of costs that shows an allocation of available funding against elements and abnormals. Elemental rates are obtained from historical data which is re-based for time and location. Abnormal costs are added for site issues, additional statutory requirements and enhancements.

Cost plan aims

Cost planning techniques are needed for the following reasons:

- To assess the validity of the funding allowance in relation to the proposed scheme. This is by looking at the scope of the project in terms of GIFA, specification and abnormals.
- To show a suitable amount of money for each element of the project together with suitable design fees and preliminaries.
- To break down elemental costs into their constituent parts.
- To produce 'design-to-cost' statements for the design team to understand the level of specification which is affordable.
- To ensure there is continuous reporting of costs and to predict the eventual price for the project – sometimes called a 'no surprises' pricing service.
- For some schemes this approach will identify the need for additional funding early in the tender stage.

An early cost plan is a broad guide to the price for a development. In order to improve the accuracy of a cost plan, a pricing strategy is needed for many of the following issues:

1. The extent of market testing needed to secure a robust tender.
2. Catering equipment to new build and retained estate.
3. Existing items of furniture and equipment that can be transferred.
4. FM services area requirements to new build and retained estate.
5. Site supervision with core team for multiple buildings.
6. Design fees need to be set at a competitive level.
7. Minimizing decant costs.
8. Avoiding constructing buildings in phases with phased handovers.
9. Concrete vs steel frame solutions.
10. Understanding the infrastructure, connection charges and services upgrade.
11. Passing ground risk and asbestos risk to the next stage.
12. Agree ICT infrastructure scope with M&E sub-contractors.
13. Achieving the required margin to support the business.
14. Ensuring the cash flow position is at a sustainable commercial level.
15. Producing selling rates using submission pro formas.

A 'top-down' approach to pricing takes the affordability budget and allocates sums for all the elements and on-costs in a cost plan. Some actions are:

1. Understand the importance of the schedule of accommodation for the architect to produce his design. The contractor will produce a challenging schedule of areas and share his proposals with the client team.
2. Identify abnormal items at early design team meeting.
3. Agree target cost plan.
4. Agree a shopping list of finishes and envelope finishes.
5. Agree the quality to be offered for furniture, fixtures and equipment (FF&E).
6. Agree landscaping schedule and get landscape architect to monitor costs.
7. Check structural solutions against the cost plan.
8. Agree an allowance for sustainability – in particular the methods for energy conservation and renewables.
9. Show levels of refurbishment on the drawings as hatched areas and legends explaining the work required.
10. Undertake value-engineering exercises when design solutions are proposed.
11. Agree the target costs for preliminaries with the site management team.
12. Agree design fees and obtain quotations for surveys which meet the target cost plan.
13. Obtain continuous cost advice from the M&E contractor.
14. Check the drawings for any redundant areas (but ensure all areas are measured to the RICS definition).

15. Concentrate the pricing effort on externals works and abnormals. Include temporary works, demolitions and decant accommodation, which can add considerable costs without any long-term benefit to the customer.
16. Submit drawings, accommodation schedules and cost plans with the same areas.

Funding and affordability

Traditionally, estimators received tender documents comprising a set of drawings, bills of quantities and specifications. With the scope of works fixed, the competition was to find the lowest price. The main features of modern procurement practice demand a better understanding of risk, more opportunities for the contractor to develop the design to meet an output requirement and price. Since the funding has been secured prior to the issue of tender documents, the focus is now on best value for a fixed budget.

The contractor receives clients' budgets in a variety of forms. Examples are:

Defence projects	A rough order of cost within ∀40%, or An assessment study which gives a preferred option.
Health projects	Outline business case (OBC) based on a public sector comparator scheme.
School projects	A fixed funding allowance based on a formula for build cost and fittings plus abnormals relating to the site conditions.
Commercial clients	An investment appraisal or cost plan developed by the cost adviser.

In gaining an understanding a client's budget, it is important to interrogate the constituent parts such as:

- Time – when was the budget produced and when was the project due to start?
- VAT – does the figure include VAT?
- Professional fees – design fees are usually included but what about legal and other consultancy costs?
- Specialist equipment – including medical equipment, loose furniture, garage equipment, computers, data links etc.
- Location factor.
- Enabling works – does the budget include advance work such a service diversions – which will be carried out by others?
- Third party income – such as retail units, cafes, evening classes.
- Sale of land – does the business case rely on an income form another source?

Consider employing a financial analyst to check that the client has applied for all available funding. An example for a batch of schools is given in Fig. 8.3. This shows

Stansford High School BsF funding allowances based on stated areas

Description		Funding allowance		Smithfield School	
		1Q03		new build	refurb
Pupils			900		
Construction base cost					
New build		1 080	2 284 m^2	2 466 720	
Refurbishment		£700	2 188 m^2		1 531 600
Minor works		£150	4 609 m^2		691 350
SEN new build		£1 228			
SEN refurb		£800			
ICT infrastructure (containment)		£225	900	202 500	
Sub-total			9 081 m^2		
Inflation adjustment using PUBSEC	3Q07	1.2431		602 301	548 487
Site costs	% + no infl				
Site costs new build		12%		369 641	
Site costs refurb+minor works		8%		–	224 410
Abnormals	% + no infl				
Abnormal costs new build		5%		154 017	
Abnormal costs refurb		9%		–	252 461
FF+E					
New build (fixed) 1000 less 20%	£/pupil	£800	25%	181 090	
Refurbishment (fixed) 500 less 20%	£/pupil	£400	24%		86 739
Minor works	£/pupil	no money	51%		–
SEN new build	£/pupil	£3 500			
SEN refurb	£/pupil	£1 750			
Inflation adjustment using RPIX	rate pa	11.0%		19 920	9 541
Professional fees					
New build	% + no infl	12.5%		450 500	
Refurbishment	% + no infl	15.0%			492 298
TOTALS				**4 255 506**	**3 870 568**
Total new build and refurb					**8 126 074**
Funding gained by Authority					**8 502 500**

Fig. 8.3 *Funding allowances for a secondary school*

that the available funding would be £8,126,074. In practice, the Authority has secured £8,502,500, which is sufficient for the project and would provide additional enhancements such as covered courtyards and flexible break-out spaces.

Benchmarking

Benchmarking involves techniques to compare performance and practices against peer organizations in order to bring about improvements.

Examples of benchmarking are:

- A comparison of costs such as building costs or salaries.
- Comparing key performance indicators.
- Making a statistical appraisal of efficiency, such as speed of construction, profit levels or recycling waste.
- An examination of production techniques used by other organizations.
- Understanding the strategic objectives of organizations.

Benchmarking against leading businesses is a route to improving performance quickly.

Internal benchmarking is comparing performance, processes or costs of departments or sites within an organization. By definition this is inward looking and can lead to complacency. External benchmarking is to compare your organization with a business of similar size and composition. Organizations can also set up benchmarking practices in the form of knowledge databases, i.e. maintaining a database of best practices and lessons learnt.

In addition to cost benchmarks, in each market sector, client bodies produce guidelines for room sizes and additions for circulation and support accommodation.

Cost benchmarking

It is unlikely that a construction company could exchange and compare rates and prices with a competitor. This would put parties in danger of contravening competition legislation. There are, however, a number of techniques which can be used (see Fig. 8.4).

Re-basing cost data

In order to compare costs for similar schemes, it is important to identify any abnormal costs which have been incurred in the source projects. This means three adjustments are needed:

1. Identify and isolate abnormal costs.
2. Adjust the sample schemes to a common start date.
3. Adjust the sample schemes to the same location factor, usually 1.

It would be advantageous to adjust each cost plan for number of storeys and function of buildings. This is rarely possible. The answer is to make the three adjustments listed above and then select a project which is the closest match to the new project.

In this example, on-costs have not been re-based because they have been applied as percentage additions to re-based build costs.

Type	Source	Comments
Internal	Developing a company database	Costs from tenders are analysed and prepared in a standard elemental cost plan format. The database includes details of size, type and specification of buildings
Internal	Collecting and evaluating costs from work in progress	Gathering feedback from projects on site needs careful analysis but has the benefit of being up-to-date
Internal	Selectively collecting cost information from previous tenders – successful and unsuccessful	Ad hoc process which draws on previous tenders to produce a set of cost plans for a particular project
External	Using the Building Cost Information Service knowledge-based system	BCIS provides a standard form of cost analysis for thousands of projects. BCIS has tools to search in various categories
External	Gaining the assistance of clients and consultants	Clients and consultants have access to cost levels from competitors
External	Publications	There are articles in journals giving project studies with indicative cost plans

Fig. 8.4 *Table showing internal and external cost benchmarking*

The next stage is to examine the site in order to add for the actual abnormal costs such as demolitions, piling, sprinkler systems, highway works, sloping site and service diversions.

Design yardsticks are used in a variety of ways, to produce economic designs and limit inefficiencies. When a cost planner sets up a target cost plan, he assumes certain standards, for example number of storeys, shape of buildings, extent of external envelope, amount of glazing and sizes of plant rooms.

In addition, in each market sector, client bodies produce guidelines for room sizes and additions for circulation and support accommodation. The estimator's role is often to establish and monitor all these design parameters (see Fig. 8.6).

Another kind of yardstick is the standard cost produced for certain kinds of accommodation and equipment (see Fig. 8.7).

Target cost plans

The first cost plan is a comparison between the client's funding and a broad look at the scope of work. The following example is for a typical secondary school where the project involves a mix of new build and retained buildings (Fig. 8.8).

Elements	Sample school	Proposed school	Notes
Base date (at start)	2Q06	1Q08	
Inflation index	166	183	
Location	1	0.97	
No of pupils	1 200	1 220	
GIFA (m^2)	10 100	10 220	
Substructure	82.00	87.69	Sample X 183/166 X 0.97
Frame	118.00	126.18	Ditto
Upper floors	40.00	42.77	Ditto
Roof	80.00	85.55	Ditto
Stairs	12.00	12.83	Ditto
External walls	125.00	133.67	Ditto
Windows + extl doors	55.00	58.81	Ditto
Internal walls	65.00	69.51	Ditto
Internal doors	27.00	28.87	Ditto
Wall finishes	24.00	25.66	Ditto
Floor finishes	42.00	44.91	Ditto
External works	128.00	136.88	Ditto
Abnormals	55.00	58.81	Ditto
	1 387.00	1 483.17	Nett total
Preliminaries	189.00	202.01	13.62% of nett
Contingencies	40.00	42.72	2.88% of nett
Design fees + surveys	190.00	203.19	13.70% of nett
Inflation (embedded)	0		
	1 806.00	1 931.09	
OH+P	95.00	101.58	5.26%
Total	1 901.00	2 032.67	

Fig. 8.5 *Example of re-basing a sample scheme cost plan to a proposed project with adjustments for inflation and location*

Since the price is generally fixed before the tender period, the contractor needs to look at a strategy which gives the best value, which often means the maximum amount of new buildings.

It soon becomes clear that most effort needs to be directed to setting and monitoring the area and understanding the extent and cost of abnormals. In the above example, the project appears to be affordable. In fact the position could be improved if the

Control type	Requirement	Comments
Number of storeys	At least 70% of building area to be accommodated in three-storey blocks	Not including double height halls and attria
External wall to floor ratio	Not to exceed 0.50	Wall area to include parapets and gable ends
Glazed areas in external walls	Not to exceed 30% of external wall areas	Including windows and curtain walling
Area of plant rooms	Not to exceed 5% of GIFA	Not including vertical ducts

Fig. 8.6 *Example of estimator's design guide for a school building*

Type	Example	Cost	Unit
Functional units	Student bedroom	32 000	£/nr
Building	Sports Centre	2 250	£/m^2
Elements	Frame	150	£/m^2
Functional areas	Operating theatres	4 000	£/m^2
ICT allowance	For a secondary school	1 450	£/pupil
FF&E allowance	For a secondary school	1 000	£/pupil

Fig. 8.7 *Examples of cost yardsticks for buildings and equipment*

temporary (decant) accommodation could be reduced in size or eliminated by better phasing of the programme. If a surplus is available, then it could be used for:

1. Creating additional circulation space;
2. Upgrading the quality of finishes;
3. Enhancing the external environment;
4. Offering a saving to the client.

The next stage is to build up a full elemental target cost plan (Fig. 8.9).

Design to cost

The 'design to cost' process is needed at the early design stages of a project and again during the detailed design stage when a contract is secured. It is a control process used to:

- translate to target cost plan into design criteria which the design team can understand;
- monitor the design against the cost plan to ensure the available funding is not exceeded.

Funding			Woodgate Academy	notes
Pupils	nr		925	
Published area	m^2		8 900	
Published rate	£/m^2		1 886	Average for new and refurb
Published funding	£		16 785 400	
Non construction costs	£		1 100 000	This is for ICT hardware
CAPEX TARGET	**£**		**15 685 400**	**Amount available for construction**
Scheduled area	m^2		8 900	
Additional area	m^2		525	FM and server rooms reqd
Retained areas	m^2		-1 250	Ddt sports hall being retained
Agreed area	**m^2**		**8 175**	
COST PLAN 1	**m^2**	**£/m^2**	**£**	
New build	4 350	2 200	9 570 000	Standard rate 3Q08
Remodel	1 280	1 800	2 304 000	
Refurb	950	1 050	997 500	
Light touch	1 595	400	638 000	Includes alarms and data
Demolition			450 000	Abnormal cost
Asbestos removal			350 000	Ditto
Semi-basement			425 000	Ditto
Decant accomm			500 000	Ditto
CP 1 Total	**£**		**15 234 500**	

Fig. 8.8 *Example of a high-level cost plan for a secondary school*

There are a number of tools available to the estimator. The examples below (Figs. 8.10–8.12) show a 'traffic lights' system for external walls and more detailed breakdowns which can be used to test the affordability of selected materials.

The estimator must check drawings and specifications for compliance with the target cost plan in terms of floor area, elemental costs and specifications. These checks are carried out at the following stages:

- Prior to drawings presented to the planning authority.
- Before seeking acceptance of the design from a client.
- At design freeze.
- Before documents are prepared for tender submissions.

		Rate £/m^2	Total £	
	GIFA (m^2)	10 100		
	Location	0.97		
	Inflation index 3Q08	188		
0	ALTERATION / RENOVATION	0	0	
1	SUBSTRUCTURE	86.00	868 600	
2	SUPERSTRUCTURE		–	
2A	Frame	133.00	1 343 300	
2B	Upper Floors	46.00	464 600	
2C	Roof	87.00	878 700	
2D	Stairs	14.00	141 400	
2E	External Walls	127.00	1 282 700	
2F	Windows & External Doors	58.00	585 800	
2G	Internal Walls & Partitions	69.00	696 900	
2H	Internal Doors	29.00	292 900	
3	INTERNAL FINISHES		–	
3A	Wall Finishes	29.00	292 900	
3B	Floor Finishes	58.00	585 800	
3C	Ceiling Finishes	25.00	252 500	
4	FITTINGS		–	
4A	Fixtures	8.00	80 800	Incl blinds
4B	Fixed furniture	55.00	555 500	
4C	Loose furniture	65.00	656 500	
5	SERVICES		–	
5A	Sanitary appliances	11.00	111 100	
5B	Services equipment kitchen	11.00	111 100	Catering equip
5C	Disposal installations	4.00	40 400	
5D	Water installations	24.00	242 400	
5E	Heat source		–	
5F	Space heating & air treatment	42.00	424 200	
5G	Ventilation systems	21.00	212 100	
5H	Electrical installations	90.00	909 000	
5I	Gas installations	2.00	20 200	
5J	Lifts & conveyor installations	9.00	90 900	Two lifts × 3 storeys
5K	Protective installations	42.00	424 200	
5L	Communications Installations	78.00	787 800	
5M	Special installations	17.00	171 700	
5N	Builders work in connection	11.00	111 100	
5O	Builders attendance			
	Total cost of buildings		**12 635 100**	
	Demolitions		430 000	Abnormals
	Asbestos removal		250 000	Ditto
	Divert culvert		125 000	Ditto
	Temporary sports hall		200 000	Ditto
	Site works	140.00	1 414 000	
	Drainage	40.00	404 000	
	External services	6.00	60 600	
	Minor buildings	2.00	20 200	
	Nett Total	**1 538.50**	**15 538 900**	
	Prelims at 16%	246.16	2 486 224	
	Risk 3%	46.16	466 167	
	Design fees, surveys etc	161.54	1 631 585	
	10.5%			
	Pre-start costs 1.5%	23.08	233 084	
	Inflation on net cost + prelims 6%	107.08	1 081 507	
			21 437 466	
	OH&P 6% on turnover	135.48	1 368 349	
	Tender total	**2 258.00**	**22 805 815**	

Fig. 8.9 *Example of a target cost plan using standard elemental rates*

Design to cost tools

	Construction method	£/m²	Target %	Actual %
Low cost	Flat panel cladding system; 0.7 mm PVF2 coated steel outer skin, 70 mm insulation and 0.4 mm white steel liner.	140	55% of wall area	
	Cavity wall comprising outer skin of facings PC £300/1000; 140 mm blockwork inner skin; including insulation & wall ties.	165		
	Cavity wall comprising outer skin cedar board rainscreen cladding system; 140 mm blockwork inner skin; including insulation & wall ties.	220		
	Polyester powder coated aluminium flat composite insulated cladding system.	220		
	Cavity wall comprising outer skin of 100 mm blockwork; proprietary external self coloured render system; 140 mm blockwork inner skin; including insulation & wall ties	230		
Medium cost	Wide format natural terracotta rainscreen cladding system on and including 140 mm internal block walling.	300	40% of wall area	
	Polyester powder coated double glazed aluminium **windows**; standard colours.	360		
	Curtain walling; with aluminium framing and including all trims, flashings etc.	500		
High cost	Reconstituted stone wall cladding including all fixings, insulation and backing wall to inner leaf.	550	5% of wall area	
	Structural glazing; with secondary support system and aluminium framing including all trims, flashings etc.	660		
	Natural stone wall cladding including all fixings, insulation and backing wall to inner leaf.	660		

Fig. 8.10 *'Traffic light' system for designing external walls to cost*

Gross Internal Floor Area - Drawn	13 758	m^2		
Total target external wall area	8 255	m^2	8 285	Measured
Wall to Floor Ratio	0.60	Target		

	New Charter Academy	%	Component area	Budget cost	£	£/m^2 GIFA
	Target cost		m^2		2 520 000	183
	External walls, windows, doors & brise-soleil					
1	Cavity walls; facing brick outer skin & block inner skin + insulation	35%	2 889	155	447 795	
2	spare	0%	–	–	–	
3	Cavity walls; block outer & inner skin + insulation + self coloured render system	15%	1 238	210	259 980	
4	Cladding; microrib or similar (incl louvres)	7%	482	175	84 350	
5	Cavity wall insulated render finish	0%	–	250	–	
6	Curtain walling; aluminium framed or similar	5%	413	450	185 850	
7	Aluminium louvres				–	
8	Windows; dg aluminium	35%	2 889	375	1 083 375	
9	High quality/feature cladding/glazing	3%	248	600	148 800	
10	Extra over for doors; glazed aluminium		10	900	9 000	
11	Extra over for doors; plantrooms, stores etc		5	400	2 000	
12	Extra over for entrance automatic doors		1	12 000	12 000	
13		100%	8 255		2 233 150	
14	Secondary components, design development etc	2.50%			–	
15	Secondary steelwork, to brickwork & etc		8 255	10	82 550	
16	Soffits to overhangs		250	150	37 500	
17	Brise-soleil (not recommended for PFI schools)		100	250	25 000	
18	Drywall inner leaf		578	55	31 790	
19	Airtight testing		1	12 000	12 000	
20	Mansafe/Window cleaning systems		1	15 000	15 000	
21	Extra for feature cladding to 'main entrance'		1	12 000	12 000	
22	Entrance canopy		1	50 000	50 000	
			Trade total		2 498 990	182
	Trade price allowance from Cost Plan	13 758	m^2	175.90	2 520 000	183

Fig. 8.11 *Design to cost schedule – external walls*

College cost plan		August 08	Design to cost statements
Number of pupils		1 250	
GIFA (m²)		10 000	
	Rate	Totals	
1 SUBSTRUCTURE	**95**	**945 000**	Based on 'normal' foundations, i.e. pads, strip foundations and solid in-situ ground bearing floors. Some suspended ground floor slabs to reconcile levels. 'Easy float' finish to civil/structural engineers design & specifications. Generally floors to be screeded where under floor heating installed. Screed priced in floor finishes element.
2 SUPERSTRUCTURE			
2A Frame	145	1 449 000	Generally steel-framed buildings. Steelwork stanchions, upper floor beams roof beams and secondary steelwork
2B Upper floors	50	504 000	In-situ RC suspended ground floor slabs, tamped or 'easy float' finish. Upper floors to be in-situ RC on metal permanent formwork. Generally floors to be screeded where under floor heating installed. Screed priced in floor finishes element.
2C Roof	95	945 000	**Pitched Roofs** Aluminium built up standing seam roof system incorporating 25 mm sound insulation and approx. 200 mm mineral fibre thermal insulation to required U-values. System to incorporate either perforated liner tray on cold rolled purlins or perforated structural liner tray. External perimeter eaves line to have cover flashing. **Parapet(s)** to be capped with ppc. aluminium coping profiles, fully mechanically adhered to sub-structure. Single-ply membrane sheet bonded directly onto copings to provide a waterproof seal. **Flat roofs:** Warm roof construction with single-ply membrane roof finish; on 150 mm insulation, on roof support deck. Single-ply membrane sheets mechanically fixed to roof support deck, on structural purlins. Colour of single-ply membrane – dark grey. Walkway system to form maintenance access where required. Roof laid to falls to internal rainwater downpipes or perimeter aluminium gutters and downpipes. **Roof lights** in communal and circulation areas to provide feature lighting and amplify natural light ingress. Suitable access for cleaning externally will be provided for. Rooflights at high elevel in areas such as atria will require the FM contractor to use suitable access equipment **Eaves gutter** – (outside building line), manufactured from pressed aluminium to form uninsulated box gutter, set in-board from eaves line and integrated within roof finishes zone.
2D Stairs	15	151 200	Precast concrete, class 'B' finish staircases, emulsion painted finish to soffits and stringers. Treads, risers and landings finished with vinyl sheet with colour contrasting non-slip nosings. Balustrades to be 50 mm diameter PPC or Painted mild steel handrails on PPC or Painted mild steel stanchions with PPC or Painted mesh panel infills. Handrails/stanchions to be surface or side fixed. 50 mm diameter satin stainless steel handrails to feature staircases. Open tread feature steel staircases where shown (feature areas only).

Fig. 8.12 *Example of an elemental cost plan with design statements*

Controlling areas

A schedule of accommodation is commonly used in the various work sectors to transmit the clients' requirements to the design and build contractor. Figure 8.13 shows the differences between three client groups.

The schedule of accommodation needs to be developed by a specialist in the team. This could be the lead architect; but for large projects, a specialist is employed. In the health sector this is a health care planner and for education, an educationalist advisor would be used.

The role of the specialist would include the following duties:

- Interrogate what types of accommodation and areas are being requested as being suitable for the output specification.
- Look for any redundant rooms/spaces (in comparison with the curriculum in a school for example) that could be better re-deployed or eliminated.
- Check for any areas which could have dual use.

Sector	Output specification	Response from contractor
Defence	Requirements are usually expressed as a list of standard building types. For example: single living accommodation or armoury building. The Schedule of Requirements document sometimes adds particular requirements for the establishment	The bidders use national guidance documents (JSP scales) to build up room data sheets for each building. Room data sheets are required for all main room types
Education	A standard schedule of accommodation is often made available to the bidders using BB98 guidelines. Room Data Sheets are prepared by the authority or school	The authority's schedule is returned with an additional column showing the contractor's net room areas, and comments to explain the differences. Room data sheets are required for all main room types
Health	With a greater number of rooms in a hospital a Trust's schedule can be very detailed. Normally, the bidders are expected to develop the room data sheets and the schedule of accommodation	The Trust's schedule is returned with an additional column showing the contractor's net room areas, and comments to explain the differences. Room data sheets are submitted for the majority of room types.

Fig. 8.13 *Area requirements for various public market sectors*

- Compare the client's schedule with national guidance as the authority may have over-sized some rooms.
- Analyse what area the schedules give for circulation, break out space is sometimes required by a client but not included in a schedule of accommodation. This will identify an early problem.
- Explain the approach to area planning to the client and gain their support for any changes.
- Ensure that the schedules include facilities management (FM) and information and communication technology (ICT) spaces.
- Include any third party spaces on the schedules.
- When the design is ready for printing, near the end of the bid, transfer the drawn areas to the schedule of accommodation and include any notes explaining any significant differences.

Once the schedule has been developed into a working solution, it needs to be divided between areas which can be found in the retained estate and those to be provided as new buildings. Work to retained buildings is classified under the following headings:

- structural remodelling
- remodelling and refurbishment
- refurbishment
- light touch (refreshment of finishes and M&E fittings).

Some clarity will be needed on backlog maintenance.

The target developed area will be set at xxx m^2 with any canopies or value-added areas being line items.

Bidders produce their own schedules of accommodation for the following reasons:

1. To challenge the client's schedule of areas;
2. To ensure the client understands the areas being offered;
3. To set an affordable target for designers.

It is essential that **drawn areas will be measured and priced**. Schedules of accommodation for each new building are to be provided showing net room areas, non-scheduled areas and gross building areas. An example is given in Fig. 8.14.

Gross internal floor areas (GIFAs) will be measured to the internal face of external walls of all buildings. No deductions are made for internal walls, voids, ducts, stairs, lifts cores, plant rooms and roof top 'pop-ups', because they all contribute to the building area. An atrium will be measured at its lowest level only. Canopies without walls, windows and heating will be priced as line items and not included in the GIFA. If connecting corridors have windows and heating, they will be measured as GIFA.

		Authority			Bidder			
Rm nr	Room type	Nr of rooms	Room area (m^2)	Total area (m^2)	Nr of rooms	Room area (m^2)	Total area (m^2)	Comments
S2	Headteacher's office	1	15	15	1	15	1	
S3	Meeting room	1	20	20	1	20	20	
S4	ICT technician's office	1	15	15	1	15	15	
S5	Maths store	2	8	16	1	12	12	Single room
S38	Staff toilets	1	27	27	1	25	25	More efficient use of space
S39	Circulation space	1	1 331	1 331	1	1 450	1 450	
S40	Plant space	1	135	135	1	145	145	
S41	Partitions	1	217	217	1	227	227	
	TOTAL GIFA			6 876			7 125	

Fig. 8.14 *Typical schedule of accommodation for a school*

New build and refurbishment

It is becoming more common for an invitation to tender to ask the contractor to decide between replacing buildings and refurbishing them. Clearly this gives the bidder more flexibility to meet the output specification while meeting the affordability target. The invitation to tender might suggest that certain buildings are replaced with new, not in the requirements document but by reference to an outline business case (OBC).

One benefit of the Private Finance Initiative is the ability to re-structure the finances to expand the scope of construction, beyond the assumptions made in the OBC. So, if a better unitary payment can be achieved, there could be headroom to extend the amount of new build accommodation.

Pricing the refurbishment of buildings

In developing a target cost plan for a scheme which includes the refurbishment of existing buildings, certain assumptions will be needed. The categories given in Fig. 8.15 can be used. This provides a clear definition of work coverage for each heading. This table will need to be modified for each project. **For clarity, each category of work must be shown hatched or coloured on the floor plans.**

Types of refurbishment	Scope of work
Light touch Cost range 15–20% of new build cost (BsF funding allows 14%)	• No work to structure • Small areas of re-pointing • Some minor repairs to int. doors, frames and glazing • Minor making good plastered walls • Replace damaged ceiling tiles • Making good a proportion of floor finishes • Redecoration throughout • No furniture or fittings • Minor repairs to soiled or leaking services, re-lamping and replacing some diffusers • Additional data wiring may be required • Check and maintain alarms • Additional funding will be needed for upgrading work to meet DDA standards and BREEAM.
Refurbishment Cost range 20–55% of new build cost	As above plus: • Forming occasional door openings in structural walls • Construct occasional non-load-bearing partitions. • Making good finishes, with substantial replacement to floor finishes • Replace minor fittings • Move or add a small number of radiators and lights • Consider new alarm systems and extend BMS into these areas
Reconfigure Cost range 55–75% of new build cost (BsF funding allows 65%)	As above plus: • Minor foundations for structural alterations • Breaking through slab for alterations to drainage • Alterations including steel supports, localized strengthening, supports to form larger internal openings • Provision of non-load bearing walls to form new room layouts • Replace approx. 50% of internal doors, frames and glazing • Replace handrails to meet Building Regulations • New tiling to wet areas • New floor finishes throughout • Replacement of ceiling grid and tiles • Replacement of many fittings and all soft furnishings • New kitchen fittings • Substantial alterations and upgrading to engineering services • Containment and mains to remain

Fig. 8.15 *Continued*

Structural remodelling Cost range 75–120% of new build cost	As above plus: • New foundations for structural alterations • Moderate alterations to existing structure with some new beams and columns • Removal of stairs and infill slabs • New beams for removal of load-bearing walls • Floor structures to remain but voids for new lifts and light wells • Some of the new walls may be load-bearing • Majority of door sets and internal glazing to be replaced • Replace all FF&E and soft furnishings, new kitchen fittings • New sanitary ware and services to wet areas • Replace all engineering services but some containment and mains will remain • New voice and data cabling • New lift • Additional funding might be required for sprinkler system

Fig. 8.15 *Cost plan categories for refurbishment of existing buildings*

It is vital to check the client's funding assumptions. In the case of Building Schools for the Future (BsF), for example, there is no funding for FF&E to any of the refurbishment categories of work. It must be assumed therefore that all FF&E can be transferred.

Condition surveys

Where backlog maintenance has been inadequate, or completely absent for a number of years, a building condition survey could be used to define the work needed to bring the building up to a safe and serviceable standard. The survey can be used to price work which is needed immediately and over the next few years. Backlog maintenance work can be priced in the capital cost or the life cycle fund.

Unfortunately, condition surveys carried out by clients prior to the issue of tender documents are often reports on the **visible** state internally and externally, and can be misleading in the following areas:

1. Only listing items over a certain value, for example: only items over £400 are recorded.
2. Unlikely that intrusive asbestos surveys have been carried out.

Description	Quant	Specify	Supply	Fix	Maintain	Replace	Fixed S&F	Loose Fix only
							£	£
Chair	543	Client	Client	CB	Client	Client		500
Notice board	24	CB	CB	CB	FM	FM	3 800	
Cutlery trolley	2	transfer	Client	CB	FM	FM		40
Stage lighting	1	M&E	M&E	M&E	FM	SPV	In M&E cost plan	

Analysis for submission	£	£/m^2
Fixed FF&E	1 789 040	162.64
Loose FF&E	850 300	77.30

Fig. 8.16 *FF&E schedule for a secondary school*

3. Latent defects are by definition not evident. **A latent defect is a fault in the building that could not have been discovered by a reasonably thorough inspection before the project is started.**
4. Some surveys exclude work needed to comply with Building Regulations, Disability Discrimination Act, fire regulations and targets for renewables.

Furniture, equipment and ICT

The estimator must have an understanding of the scope of items included in the FF&E schedule (extract shown in Fig. 8.16). This example also includes responsibilities for specifying the items, supply, fixing, maintaining and replacing them in the future. There are often additional items which need to be included in the cost plan: in BCIS Element 4, Fittings and Furnishing, there are items such as receptions desks, notice boards, window blinds, shower curtains, sanitary accessories, white goods and catering trolleys.

The process to price FF&E is:

1. Produce a Schedule of Accommodation for the contractor's design.
2. Produce room data sheets for the contractor's design.
3. Produce the list of FF&E for the contractor's design.
4. Add responsibilities to the FF&E list.
5. Supplier (or consultant) prices the schedule.
6. Add items priced by the estimator.

Clarity is needed on definitions of 'fixed' and 'loose' FF&E because often the client needs the two categories priced separately. It is also important to understand how many items of furniture and fittings can be transferred from an existing school. At an early stage in the tender process this is usually expressed as a transfer assumption, such as: '35% of the FF&E requirement can be transferred from the existing school'.

Item	Category	Specify	Price	Install	£
Access control (hardware)	Access control	ICT	D+B	ICT	
Access control (software)	Access control	ICT	D+B	ICT	
Door locks	Access control	D+B	D+B	D+B	
Readers	Access control	ICT	D+B	ICT	
Software	Access control	ICT	D+B	ICT	
Servers	Access control	ICT	ICT	ICT	
Cabling	Access control	ICT & D+B	D+B	D+B	
BMS hardware	BMS	D+B	D+B	D+B	
BMS software	BMS	D+B	D+B	D+B	
Network cabling	BMS	ICT/D+B	D+B	D+B	
Link to IP network	BMS	ICT	ICT	ICT	
Cashless shopping	Cashless	ICT	D+B	ICT	
Tills and units	Cashless catering	ICT	D+B	ICT	
Software	Cashless catering	ICT	D+B	ICT	
CCTV external (hardware)	CCTV	ICT	D+B	ICT	
CCTV external (software)	CCTV	ICT	D+B	ICT	
CCTV internal (hardware)	CCTV	ICT	D+B	ICT	
CCTV internal (software)	CCTV	ICT	D+B	ICT	
Cameras	CCTV	ICT	D+B	ICT	
Camera brackets	CCTV	ICT	D+B	ICT	
Cabling	CCTV	ICT	D+B	D+B	
Underground ducting	CCTV	D+B	D+B	D+B	
ID & login cards	Consumable	ICT	ICT	ICT	
Display screen (LCD/plasma)	Display screens	ICT	D+B	ICT	
Mounting bracket	Display screens	ICT	D+B	ICT	
32" LCD	T/L/M – devices	ICT	ICT	ICT	
Etc, etc.					

Fig. 8.17 *ICT interface schedule*

ICT in schools

A schedule of ICT-related items needs to be drawn up at the start of the tender. Responsibilities for specifying equipment, pricing and installation must be clear. All the items are priced by the ICT provider or ICT advisor and split between the ICT and contractor cost plans. This information must be passed on to the engineering services contractor because the contractor items fall with the BCIS element 5L, Communications.

Fixtures, fittings and medical equipment in hospitals

The methods used to produce equipment schedules are better understood in the health sector where specialists are experienced in scheduling and pricing equipment. The process is shown in Fig. 8.18. The activities are:

1. Produce an agreed schedule of accommodation, wherever possible improving the client's schedule with modern practice and area efficiencies.
2. Produce room data sheets for the most common rooms starting with the client's sheets but adding items relating to the contractor's design.
3. Add equipment for the facilities manager such as workshop equipment and cleaning equipment.

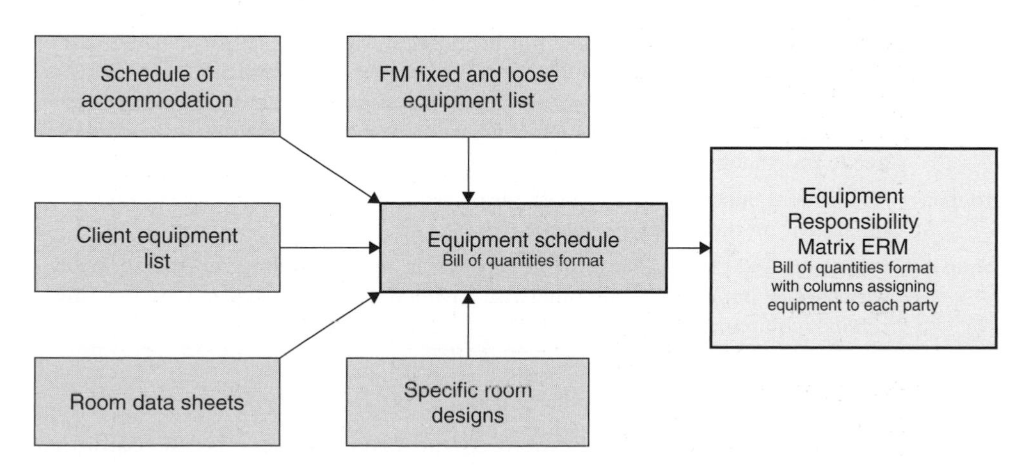

Fig. 8.18 *Inputs to an equipment responsibility matrix*

4. Use the above to list all the equipment in all of the rooms.
5. Add responsibilities for:
   - specification
   - purchase
   - fix
   - maintain
   - replace
   - transfer assumption.

Hospital equipment has for many years been grouped into the following categories:

Group 1	Items (including engineering terminal outlets) supplied and within the terms of the building contract. This means equipment that is normally fixed to the building fabric and/or attached to or forming part of the building services. This may have an impact on space and/or building services to be provided and will normally be supplied and fitted by the contractor (e.g. worktops, autoclaves, sanitary fittings, bedhead units, theatre lights, terminal outlets).
Group 2	Items which have space and/or building construction and/or engineering service requirements and are fixed within the terms of the building contract but supplied under arrangements separate from the building contract. Certain major items of equipment, such as major diagnostic imaging equipment or linear accelerators etc., will require particular attention. In a PFI contract the Trust will need to consider the risk, potential delay, cost and implication for life cycle replacement and maintenance of not including it within the contract. The Trust will also need to consider that this type of equipment has implications in terms of space, access, shape of rooms, environmental requirements, engineering services and commissioning. The inclusion will have to be considered against the clinical needs and changing technology.
Group 3	As Group 2 but supplied and fixed (or placed in position) under arrangements separate from the building contract.
Group 4	Items supplied under arrangements separate from the building contract, possibly with storage implications but otherwise having no effect on space or engineering service requirements.

The extent of private sector involvement in the provision of medical equipment varies between PFI projects. The Health Trust provides information on the assumptions made for the provision of equipment in the Public Sector Comparator (PSC).

This is developed from estate, equipment and capital databases to reflect their expectations for the PFI scheme. It will not be sufficient to rely on NHS Standard equipment groupings alone but should be considered as follows:

Category A	Equipment which is supplied, installed, commissioned, maintained and replaced by the Project Company and is essential and integral to the building.
Category B	Equipment which is supplied by the Trust. It is installed, commissioned, and maintained by the Project Company but replaced by the Trust.
Category C	Equipment which is supplied, installed and replaced by the Trust and maintained by Project Company.
Category D	Equipment which is supplied, installed, commissioned, maintained and replaced by the Trust.

The Trust may have assumed that some categories of equipment will need to be transferred from the existing hospital. This equipment should be identified against the above categories. It should also be carefully assessed as being useable at the anticipated time of transfer and the decommissioning, transfer, re-installation and insurance costs and responsibilities established. Facilities management equipment is classified as Category A unless the services are not being transferred to the project company.

Transfer assumptions need to be clear and need to distinguish between new build and refurbished areas. A suggested format is given in Fig. 8.19.

Building types	Medical	Non-medical Equipment	Non-medical fixtures
New build, refurbish and reconfigure	50% transfer 100% life cycle	50% transfer 100% life cycle	0% transfer 100% life cycle
Light touch and retained areas	100% transfer 100% life cycle	100% transfer 100% life cycle	100% transfer 0% life cycle

Fig. 8.19 *Definition of equipment required for different parts of a hospital*

Aspects of good practice

- Understand the relevance of Groups 1–4 and Categories A–D.
- The term 'medical equipment' can be difficult to define when reading a business case or cost plan. In one document it might mean 'big ticket' items and in another all fittings, fixtures and equipment.

- Ensure that builder's fittings and fixed furniture are included.
- Catering equipment could be added to the ERM. Include regeneration trolleys in the list of catering equipment.
- Check that the ERM includes cold rooms and white goods.
- FM equipment should be specified by the FM services provider and added to the ERM.
- Give the M&E specialist a list of equipment which he needs to include in his price. For example, theatre lights and surgeons' panels.
- Obtain or produce a list of Trust supplied items.
- Develop a clear understanding of the transfer assumptions, including:
  - Who takes down and fixes transferred equipment and fittings?
  - Who maintains and renews the equipment?
  - How much builder's work is needed?
  - Who manages the procurement and logistics for Client supply and transferred equipment?

Adjustments for location

Location tables show pricing levels for various locations relative to the national average. There are a number of geographical location factors available for construction work. The most common are:

- NHS Estates and Facilities Division Quarterly Briefing which has 9 regions.
- BCIS Location Factors which are for UK regions, counties and towns.
- DfES Location Factors based on an earlier version of BCIS figures.

Boundaries are drawn using the named political regions and districts, but this does not mean that the likely costs change when a boundary is crossed. Perhaps an attempt

Area	Factor	Sample
Chester-Le-Street	0.98	12
Durham	1.01	29
Darlington	1.10	19
Chesterfield	0.95	16
Northampton	0.98	50
Bradford	0.96	56
Scarborough	1.04	12
BCIS May07		

Fig. 8.20 *Typical factors for adjusting costs for location*

should be made to draw lines around regions were costs are known to be at a similar level, but this is rarely done by national bodies.

So what makes one area different from another? The likely differences are:

1. Availability and cost of labour;
2. The cost of transport for materials deliveries;
3. Distance from distribution centres;
4. Demand for new buildings and contractors' workload;
5. The cost of local accommodation and housing;
6. Market forces;
7. Types of building work in the area.

Building costs are not necessarily the same within one location and so the factors must be seen as trends. Data can become distorted by large variations within a region caused by how desirable a project is to the local contractors, the complexity of the project and the form of procurement.

For example: A fire station has been built in Durham for £1850/m^2. To make a location adjustment for a fire station in Northampton, using the table below, the calculation is:

$$£1850 \times 0.98/1.01 = £1795/\text{m}^2$$

For major contracts there are different considerations such as bringing specialist labour from all parts of the UK; employing staff who travel from project to project; paying large-scale project rates agreed with trade unions; requirements for sub-contractors to take on onerous contract provisions.

Cost plan format

Generally there are two software options which are considered for cost plans: a proprietary estimating system and a spreadsheet. Here a spreadsheet approach is used because it gives the most flexible solution to creating a cost plan and exchanging data with other people.

'Linear' bill of quantities

The term 'linear bill of quantities' is used to describe a complete bill of quantities which has to reside in one worksheet. In this way, the estimator can interrogate the bill to produce summaries for elements and sub-contract packages. It also allows multiple filtering of costs in order to answer a question such as: 'What is the total

Ref	Bldg	Description	Qu	Unit	Rate	Total	BCIS	Sub	Other sort
	filter +					Sub-total *	Filter +	Filter +	Filter +
1	Block A	Description 1	24	m^3	34.00	816	1	S1	civils
2	Block A	Description 2	22	m^3	36.00	792	1	S1	civils
3	Block B	Description 3	34	nr	124.00	4 216	2h	S12	carp
4	Block B	Description 4	67	t	1560.00	104 520	2a	S2	frame

Fig. 8.21 *Example of a linear bill of quantities*

for internal doors in block B?' It could also be asked to total the number of doors in blocks B and C.

Features are:

- Automatic summaries using the 'sumif' function.
- Items can be entered in any order, avoiding spaces and headings.
- The costs can be interrogated to give answers using single or multiple filters.
- The 'sub-total' (*) box will give the total cost for the selected code when filtered.

Figure 8.21 is an example of a linear bill of quantities.

A preliminaries workbook is produced in a separate standard document in order to produce figures in a well-known format for the review team. For some cost plans, the summary of preliminaries can be transferred as bill items to the above cost plan. In this way, a tender total can be shown, and a submission elemental cost plan is readily produced.

Tender timetable and reviews

The bid manager, responsible for the production of the estimate, must ensure that a timetable is established which highlights the key dates in the production of the estimate and tender (see Fig. 8.22). It will be an essential document for all those associated with the tendering function.

All people associated with the tender must confirm that they are able to provide the necessary data in the format required, in accordance with the agreed timetable.

Activities	Reviews	Dates
Affordability analysis		
High level cost plan		
Pricing strategy		
Procurement strategy		
Challenging schedule of accommodation		
Sustainability strategy		
Life cycle strategy		
FM and financing strategy		
Agree all of the above	**Review 1** Mid-tender	
Design meetings		
Meetings with trade partners		
Meetings with ICT and FF+E advisors		
Design freeze		
Final measurement of area and high value works		
Agree construction programme		
Produce preliminaries workbook		
Produce net cost estimate		
Risk meeting		
Estimate summaries for final review		
Final review meeting	**Review 2** Final	
Check drawn areas match schedules and cost plan		
Check the designs and specifications match cost plan		
Fill in submission schedules		
Submit tender		

Fig. 8.22 *Timetable for the production of a cost-planned tender showing tender reviews*

Final review

The final review of an estimate and its conversion to a tender is the responsibility of management and is a separate commercial function based upon the cost estimate and its supporting reports and documents.

The accountability of the estimator should be limited to the proper preparation of the predictable cost of a project. It must not be considered that the estimator's responsibility is to secure work for the company; that is the responsibility of management.

For large projects, the final review stage could comprise two meetings: the first to review the estimate through a detailed examination of rates, quotations and preliminaries; the second would be for a director to receive a short briefing and consider how well the team has met the affordability targets, check this project against benchmarks for similar projects, understand the commercial responsibilities and risks, and ensure the required margin is achievable. The need for a formal approach to finalizing tenders should be regarded as fundamental to competent tendering.

Final Review Meeting Agenda			
Project:	Regional Medical Centre, Northampton	Date:	18.12.08
Attendees	Graham Moss	Martin Piper	John Williams
	David Stott Apol	Jas Zafar	*Martin Brown Apol*
	Nicholas Bates	Tim Burnett	
	Stuart Colbeck	*Kevin Goodyear Apol*	
	Mark Doyle	Paul Thurlow	

Agenda

Introductions
Tender date: 5th January 2009
Financial close: 15th June 2009

Part one: TECHNICAL 45 mins:

1.1 Client brief and description of the project
1.2 Design concept and rationale
1.3 Extent of surveys
1.4 Construction methods and logistics
1.5 Phasing and programme
1.6 Enabling works
1.7 FF+E and ICT infrastructure

Part two: PRICING 30 mins:

2.1 Cost plan summary
2.2 Benchmarking and links with demonstration/comparable projects
2.3 Scope, areas, quantities and assumptions
2.4 Supply chain input, market conditions and industrial relations
2.5 Equipment and IT pricing
2.6 Preliminaries
2.7 Design fees
2.8 Top ten risks and opportunities
2.9 Tender summaries

Part three: COMMERCIAL and SUBMISSION 45 mins:

3.1 Commercial background, competition and follow-on work
3.2 Client team
3.3 Contract conditions, bonds and insurances, acceptance period
3.4 Submission requirements
3.5 Negotiation process
3.6 Continuous improvement
Please ensure that any information brought to the meeting is in a condensed format to enable adherence to the timings.

Fig. 8.23 *Continued*

Record of Meeting – Action & Decision Notes/Qualifications	
Northampton Medical Centre	
1. Ask client for guidance on submission of FF&E costs – how to split 'fixed' and 'loose'	SC
2. Look at changing cladding sub-contractor from Downer to Epicon	TB
3. Query condensation issues in the pool area with design team	TB
4. Check latest guidance on insurances, may need to be uplifted to 0.9%	JW/SC
5. Check with Telford project how the waste package is being procured	SC
6. Make a statement about the level LADs in the commercial clarifications.	KG
7. Check that the margin has been applied to turn over	SC
8. Send Joe Davies a copy of the preliminaries workbook for second opinion	SC
9. Shopping list needed with the tender to show the remaining value engineering items which we would wish to develop further	KG
10. Ensure that the architect removes named products from the specification. It is too early to select materials for a stage C design.	TB
Prepared By: Nicholas Bates Date: 19.12.08	

Fig. 8.23 *Typical tender finalization meeting agenda*

An agenda for review meetings is indispensable. The standard review agenda given in Fig. 8.23 can be used for most cost-panned tenders. The agenda should be circulated to those required to attend, with times shown for each agenda item. A list of attendees would typically include:

- Technical/pre-contract director
- Operations director
- Commercial director

- Bid manager/director
- Lead architect
- Estimator.

and possibly: FM manager, life cycle surveyor, design manager, planner and project manager.

The mark-up shown on the final summary should be set early in the tender stage so that the cost plan can be developed to inform the design. At the end of the final review meeting, there may be other decisions to be made before proceeding to the submission stage, including clarifications and qualification, how a positive cash flow can be achieved, and the need for an alternative tender.

Value drawdown charts

Aims

A positive cash flow is essential for the health of a project, and is needed to serve the financial security of the business. The estimator's role is two-fold:

1. Predict the amount due in stage payments;
2. Examine the income and expenditure to arrive at a secure cash flow position.

	Group elements	CP2 £k	Duration	at FC £k	Sep	Oct	Nov	Dec
1	Sub-structure	881	4	32	212	212	212	..
2	Superstructure	4 956	12	180		398	398	..
3	Internal finishes	1 168	9	0				..
4	Fittings and furnishings	910	6	25				..
5	Services	3 811	15	120			246	..
6	External works	1 572	15	150	95	95	95	..
7	Abnormals	908	10	45	86	86	86	..
8	Preliminaries	2 408	20	185	111	111	111	..
9	Risk	435	20	40	20	20	20	..
10	Inflation	796	9	60	82	82	82	..
11	Design fees	1 280	9	640	71	71	71	..
12	Pre commencement costs	241	1	240	1			
13	Overheads and profit	1 712	20	184	76	76	76	..
14	Below-the-line items	400	3	0	133	133	133	..
	Column totals	21 478		1 901	888	1 285	1 531	..
	Cumulative drawdown £k	**21 478**			**2 789**	**4 074**	**5 605**	**..**
	% of total value				13%	19%	26%	..

Fig. 8.24 *Typical value drawdown chart for a PFI project*

The cost plan workbook is an ideal vehicle in which to produce value drawdown schedules. This relieves the estimator of another task at the end of the tender stage. When the final programme has been produced, the activity bars can be adjusted accordingly. The example value drawdown chart, given in Fig. 8.24, shows a simple set of capital cost values which can be entered into a financial model for a PFI tender.

Periodic and milestone payments

Generally, a contractor needs monthly payments based on costs incurred for the project. There are some contracts where payments are based on stages of completion, particularly where there are no bills of quantities which can be used to value the works.

If payments are based on stages of completed work, a milestone payment schedule is produced, and completed milestones are valued and paid to the contractor on a monthly basis.

In this example (Fig. 8.25), the estimator has avoided 'all-or-nothing' type milestone arrangements. It is important that payments are valued on the basis on a money-loaded programme. Milestone values can then be calculated but the milestones should be large in number, small in value and should not involve the 100% requirement. This is achieved by using starts, sections of work and substantive completions.

Milestone number	Milestone description	Milestone assessment criteria	Milestone value £	Cumul. value £
M1	Start on site	Occupying site	550 250	550 250
M2	Site set up	Substantially complete site set up	522 000	1 072 250
M3	Pad foundations	Pad founds (Bldg 23) substantially complete to grid A	236 000	1 308 250
M4				
M5				

Fig. 8.25 *Typical layout for a milestone payments schedule*

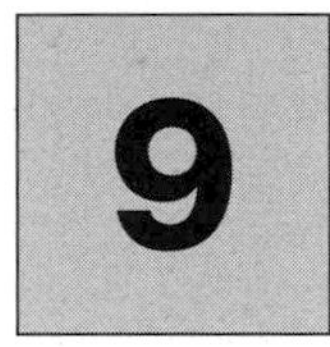

9 Tender planning and method statements

Introduction

The estimating team will consider construction methods and employ planning techniques to:

1. Highlight any critical or unusual activities;
2. Examine alternative sequences and phasing requirements;
3. Calculate optimum durations for temporary works and plant;
4. Reconcile the labour costs in the estimate with a programme showing resources;
5. Determine the general items and facilities priced in the site overheads workbook;
6. Check whether the time for completion is acceptable.

The effort needed will depend on the size and complexity of the project, the proposed use of heavy plant and the design of major temporary works. Estimating for civil engineering work in particular is dependent on an examination of alternative methods and pre-tender programmes. A civil engineering estimator usually produces a resourced programme to price major aspects of the work operationally.

Pre-tender programmes are prepared by either the estimator or planning engineer, or more likely by working together. The choice of staff depends on company policy, size of project, availability and type of work. The planning engineer's contribution can be seen as producing an appraisal of labour and plant resources and general items – in other words the estimator expresses his solutions in terms of cash, the programmer deals with time. The aim is to reconcile one with the other.

The role of the planning engineer

In a competitive market it is important to look for ways to construct the project more economically. Applying planning techniques can have opposite consequences. Increasing the value of the tender when problems are identified and reducing the estimate when methods can be adopted which reduce individual and overall durations. The team must, however, look for a practical solution which will be used to produce a robust cost of construction. The role of the planning engineer is wider than just producing a programme. His input to a tender can also include:

1. Producing site layout drawings, which are used to locate temporary facilitates, such as concrete batching plant, cranage, access routes, restrictions, areas for accommodation and storage, location of services, overhead service, temporary spoil heaps, and areas which will need reinstatement.
2. Examining the most suitable methods in relation to the design and the temporary works required.
3. Preparing method statements not only for pricing purposes but also for submission to clients or consultants when requested.
4. Producing cash flow forecast charts for management and clients who need them.
5. Providing staff structure and resource histograms for general labour, production labour and plant.

The planning engineer will often have a better understanding of current site practice and will be better placed to collect data from monitoring exercises on site. His experience of completed work will be important especially where the overall duration of a project could be reduced. Shorter contract periods can have a substantial effect on the cost of preliminaries where time-related costs (mainly staff, site accommodation, cranage and scaffolding) account for as much as 12–20% of a tender figure.

Method statements

Method statements are written descriptions of how items of work will be carried out. They usually deal with the use of labour and plant in terms of types, gang sizes and expected outputs.

There are many reasons why method statements are prepared during the tender stage. It is unlikely that an estimator will prepare a written method statement for his own use but if any of the following requirements exist then he will commit his thoughts to paper:

1. The client's advisers may ask for a method statement to accompany the tender, to satisfy themselves that the contractor has an understanding of the technical challenges and has considered suitable ways of overcoming them.
2. The quality management scheme adopted by the organization may require method statements for work worth more than a certain value.
3. Management contractors usually ask for method statements where there may be interface problems with other works contractors on the site.

4. In satisfying the need for safe systems of work, an estimator might develop a method statement with a demolition contractor, for example, before agreeing a price to be incorporated in the tender.
5. Large-scale activities needing a combination of items of plant and labour are difficult to price on a unit rate basis and cannot be started without an examination of methods and resources.
6. Where the estimator has investigated an alternative design he will need to assess the effect these changes will have on other elements of the construction.
7. Part of the handover information prepared for successful tenders is a description of the assumptions made by the estimator.

Many contractors are reluctant to submit a detailed method statement at tender stage because their ideas could be used by other parties without any financial return. A preliminary document can be prepared (see Fig. 9.1) based on the broad assumptions made at tender. It is likely to include extracts from the company's manuals for safety and quality management and some development of the client's pre-tender health and safety file. This method statement can also be of benefit to the contractor because it is a suitable vehicle for:

1. Qualifying the tender;
2. Identifying dates when information is required from the client or his advisers;
3. Indicating when instructions are required for dealing with named sub-contractors and provisional sums;
4. Explaining the limitations of temporary works; a contractor might have allowed for earthwork support but not sheet piling, for example.

Logistical planning

Operational staff work on logistical planning during the tender stage but often there are inadequate cost allowances in the cost plan. When the project moves to site, the logistics plan has to be re-worked to meet the budget. Good practice demands:

- An understanding of site specific and core requirements;
- That there is an appropriate balance of costs between the main contractor's preliminaries and trade contractors' scope of works;
- Engage with specialist logistics companies to benefit from their expertise;
- Engage with waste management specialists to explore costs for resources, segregation and disposal techniques;
- That the tender remains competitive.

Outline Method Statement
for
New Offices
for
Fast Transport PLC
North Lane, Stansford

Site location	The project is located in the existing transport yard of Fast Transport PLC, North Lane, Stansford. Access will be through the main entrance gate.
Restrictions/access	Incoming traffic will be directed to use the north access road and will leave the site along the road next to the canteen. The live oil and gas mains will be protected during the contract period and the fibre optics cable will be carefully exposed by hand dig and protected in accordance with the specification prior to piling equipment entering the site.
Sequence of work	Our tender programme T354/P1 shows the preferred sequence of activities. The aim is to start at the east end of the building progressing to the west. The site will be filled with a stone layer on a ground improvement mat immediately after the site is levelled. Concrete floors will be started after the columns have been cast and before the upper floors are constructed. External paving will be carried out in the last quarter of the contract period. The drain connecting manhole 3 to the existing foul sewer will be completed early to provide disposal from temporary facilities.
Design development	Detailed drawings of the roof cladding will be produced by our specialist sub-contractor. These will clarify the scope of the work giving fixing details, sequencing and weathering procedures. Roof flashings will not be made until formal approval has been received by our sub-contractor.
Temporary works	An independent scaffold will be erected to each external face of the building, and a mechanical hoist will be provided near the north-west corner.
Safety	Anyone working on or visiting the site will be required to wear safety helmets and operatives will use other protective clothing depending on the type and location of work. The sides of the drain trench next to the oil tank will be supported with trench sheeting and we will provide barriers next to all excavations where a danger exists. The agent will attend regular meetings with the planning supervisor and cooperate with site regulations to maintain the client's good safety record. The health and safety plan will be developed by our construction team prior to starting any affected works, and sub-contractors will have contributed to any relevant planning for their works. The health and safety file will be prepared as the project progresses. The safety performance of the site is monitored by line management who report to regular safety audit meetings; and external consultants inspect our compliance with current legislation at intervals of no more three weeks.
Supervision	Our management structure for the project is shown in the diagram attached. We will adopt a flexible approach to site supervision, providing sufficient operatives in suitable disciplines to meet our programme requirements.
Quality plan	The site manager will be responsible for drawing up a quality plan for the project with assistance from the area planning engineer. The control and monitoring framework is given in the company's general procedures and QA manual.

Fig. 9.1 *Example of tender method statement for submission to a client*

Tender programmes

The tender programme will fix an overall time for the project, from which the estimator can determine times for sections of work in main stages such as:

1. Design and mobilization;
2. Substructure;
3. Independent structures;
4. Superstructure;
5. Engineering services;
6. Internal trades, finishes and fixtures;
7. External works.

Information about these periods is essential to the estimator, enabling him to calculate times for:

1. Staff requirements;
2. Site accommodation;
3. Mechanical plant and equipment;
4. Temporary works such as falsework and scaffolding;
5. Increased costs for firm-price tenders;
6. Work affected by the seasonal weather changes such as drying out buildings, heating, protection and landscaping.

The overall section durations can be used to check workforce levels and items of plant such as excavators and cranes that often remain on site for continuous periods. There may be times of excessive demand for plant and labour, which will call for a levelling exercise to balance resource needs.

The estimator must be clear about what he wants from the programme so the planner will concentrate on what is important. To illustrate the point, an estimator has brought together all the labour costs amounting to £91,250 for a clear run of brickwork comprising 250,000 facing bricks. He priced the brickwork items with an all-in rate for bricklayers of £12.50 and labourers £9.50 per hour.

Assuming each bricklayer is serviced with half the time of a labourer, each bricklayer's effective rate is:

$$12.50 + (9.50/2) = £17.25$$

The total time included in the rates would be:

$$£91,250/£17.25/\text{hr} = 5,275 \textit{ hours' work}$$

CB CONSTRUCTION LIMITED — Proposed Offices for Fast Transport Limited — Tender Programme No: T354/P1

Activity	Quant	Output hrs/unit	Man wks	Gang size	Duration wks
1 Setting up and setting out					
2 Excavation and filling	1035 m^3	0.13	2.9	JCB+lab	3
3 Foundations formwork	820 m^2	1.20	21.9	4 carp	5
4 Foundations concrete	419 m^3	1.20	11.1	3 lab	4
5 Underslab drainage	178 m	0.50	2.0	2 lab	2
6 Concrete ground floors	137 m^3	1.00	3.0	4 lab	1
7 Columns formwork	279 m^2	1.20	7.5	2 carp	4
8 Columns concrete	21 m^3	3.50	1.6	3 lab	1
9 Floors and beams formwork	962 m^2	1.60	34.2	4 carp	8
10 Floors and beams concrete	203 m^3	2.20	10.0	4 lab	3
11 External walls	1870 m^2	1.50	62.3	8 brklayer	8
12 Roof timbers	2300 m	0.20	10.2	4 carp	3
13 Roof covering	685 m^2	0.50	7.6	4 carp	2
14 Windows	£25 500	sub-contract			4
15 Services 1st fix	£43 000	sub-contract			6
16 Plasterwork and partitions	£43 500	sub-contract			8
17 Joinery	£34 000	sub-contract			6
18 Ceilings	945 m^2	1.00	21.0	4 fixers	5
19 Services 2nd fix	£28 000	sub-contract			5
20 Painting	3500 m^2	0.30	23.3	4 paint	6
21 Floor coverings	£12 300	sub-contract			3
22 External work and drainage	£16 700	labour	52.0	5 lab	10
PRELIMINARIES	Agent	(24 plus 3 mobilization)			27
	Engineer				17
	Foreman				26
	Crane				7
	Forklift				10

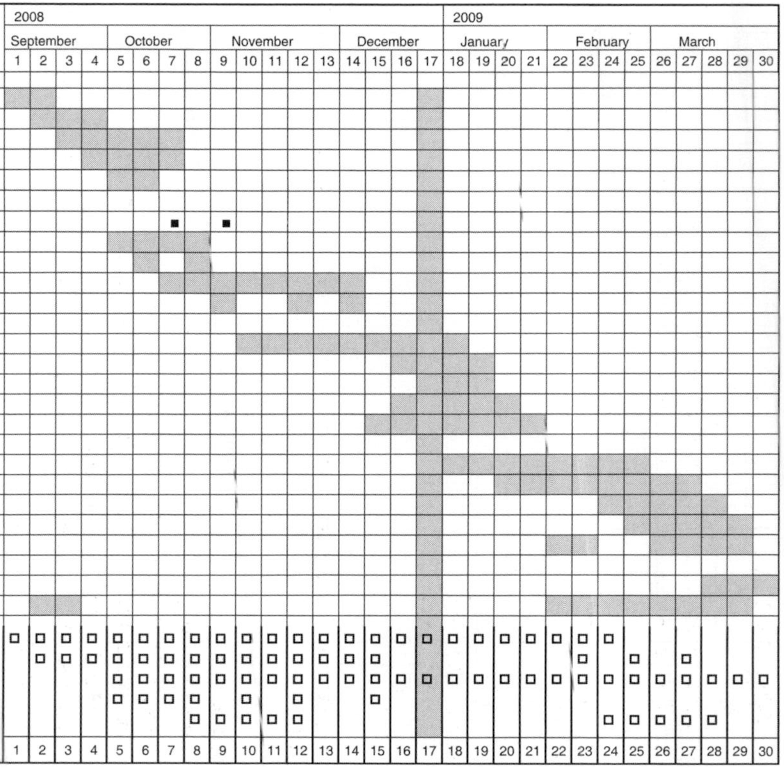

Fig. 9.2 *Example of a tender programme*

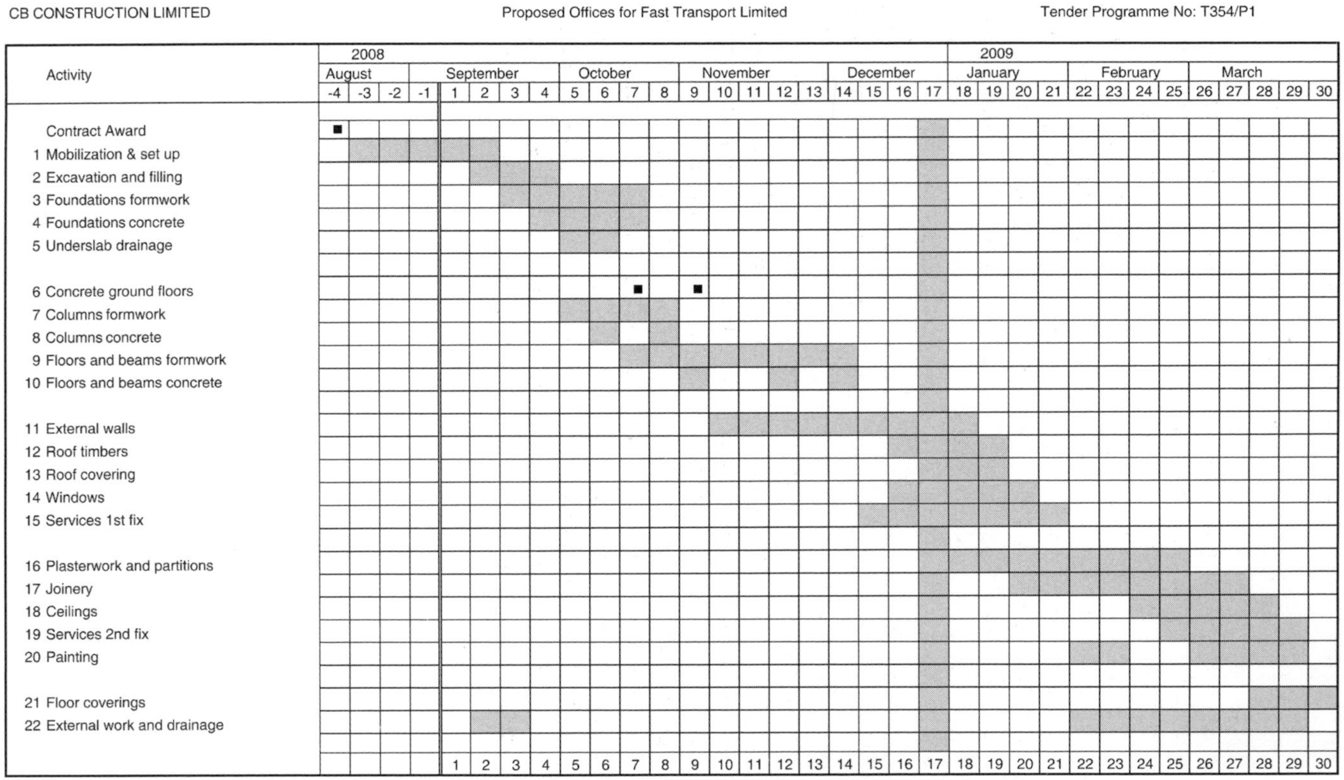

CB CONSTRUCTION LIMITED

Proposed Offices for Fast Transport Limited

Tender Programme No: T354/P1

Activity

2008

2009

August | September | October | November | December | January | February | March

-4 | -3 | -2 | -1 | 1 | 2 | 3 | 4 | 5 | 6 | 7 | 8 | 9 | 10 | 11 | 12 | 13 | 14 | 15 | 16 | 17 | 18 | 19 | 20 | 21 | 22 | 23 | 24 | 25 | 26 | 27 | 28 | 29 | 30

Contract Award
1 Mobilization & set up
2 Excavation and filling
3 Foundations formwork
4 Foundations concrete
5 Underslab drainage

6 Concrete ground floors
7 Columns formwork
8 Columns concrete
9 Floors and beams formwork
10 Floors and beams concrete

11 External walls
12 Roof timbers
13 Roof covering
14 Windows
15 Services 1st fix

16 Plasterwork and partitions
17 Joinery
18 Ceilings
19 Services 2nd fix
20 Painting

21 Floor coverings
22 External work and drainage

1 | 2 | 3 | 4 | 5 | 6 | 7 | 8 | 9 | 10 | 11 | 12 | 13 | 14 | 15 | 16 | 17 | 18 | 19 | 20 | 21 | 22 | 23 | 24 | 25 | 26 | 27 | 28 | 29 | 30

Fig. 9.3 *Example of a programme submitted with a tender*

The planner has decided to use an average output of 50 bricks per hour:

250,000 bricks at 50 bricks/hour = 5000 *hours work*

This is clearly close to the estimate. Now that the number of working hours has been established, a duration is calculated by dividing by the number of productive hours in a week and the number of bricklayers. The programme might dictate the number of gangs required. This will not change the rate but will alter the cost of ancillary facilities such as scaffolding and mixers. Most activity durations can be derived from the product of quantities and standard outputs (see Fig. 9.2), but parts of the tender will be based on offers received from specialist sub-contractors and labour-only sub-contractors. These firms will be asked to provide information about the time they will be on site and the effect of delivery periods on the main contractor's programme.

The tender programme must allow for recognized public/industry holidays, inclement weather and the peak summer holiday period which leads to a slowing of progress which may be reflected in output. Clearly, very little work is carried out on site during the two-week Christmas shutdown. One week is lost at Easter and some planners believe that about two weeks are lost in the summer due to operatives' annual holidays. Scaffolding, fixed cranes and supervision are items that will incur the largest costs during shutdown periods and must be included in the preliminaries schedule.

When a client or his advisers request a programme at tender stage the contractor will submit a preliminary or outline programme, such as the example given in Fig. 9.3. The contractor is often unclear about the role of such a programme in vetting tenders. Sometimes contractors have used the opportunity to offer completion sooner than expected and thereby try to gain an advantage over the competition. The drawback is that if the project is delayed, but still finishes within the original duration, the contractor will have difficulty recovering the costs of delay and disruption to the work.

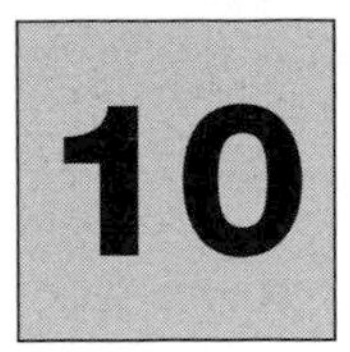

Resource costs – labour, materials and plant

Introduction

There was a time when the unit costs of labour and plant were calculated from first principles, the assumption being that the company employed operatives in sufficient numbers to carry out the work and provided its own plant. A more realistic approach today would be to find the current market rates paid for labour near the site and look at the market prices for plant hire. This information is readily available as feedback from current jobs and plant hire rates can be obtained from plant specialists. Another change has come with computers. The importance of establishing accurate rates for labour, materials and plant, before pricing the bill of quantities, has reduced because programs allow the estimator to change unit rates for resources at any stage of the tender period.

Labour rates

A method for estimating all-in rates for labour is given in the CIOB Code of Estimating Practice (COEP). This has been adopted by many publications, professional bodies and contracting organizations as a reasonable basis for calculating the cost to employ an operative. The example given in the Code uses the formula:

Hourly rate = annual cost of employing an operative/actual hours worked

During the first half of the twentieth century, builders calculated labour rates by looking at weekly costs. This was a little easier to do but lacked the precision of the current method. The main reasons for calculating costs and hours on an annual basis are:

1. To include the effect of annual and public holidays on the number of hours for payment;
2. Overtime working often depends on the proportion of summer and winter working, because longer working hours are available and used in the summer period.

The COEP calculation is clearly a theoretical approach that should be checked periodically against recorded costs. The main variance is commonly the amount paid for 'bonuses', such as attraction money, plus rates for semi-skilled operatives, spot bonuses and locally agreed payments.

The estimator needs to be aware of some of the difficulties associated with calculating labour costs, and should answer the following questions:

1. Are there enough skilled operatives in the area? If not, will they need to be paid increased rates to work on the site or is there a need to import labour from outside the area?
2. How many operatives will be paid travelling expenses and will any key people receive a subsistence allowance?
3. Will there be any local union agreements which affect the wage levels, such as those found in the petrochemical industry?
4. Will bonus payments and enhanced wages be self-financing?

Some organizations, typically those that employ their own regular labour force, build up labour rates for every job. This allows changes to be made for the type of work, time of year and location. It must be said, however, that in recent years the nationally agreed wage rates have not reflected the rates paid in the marketplace. On the one hand where skilled labour is scarce, labour costs rise, and during times of recession labour rates fall. There is an argument that an estimator will price work quicker if a constant labour rate is used for several months. Global adjustments can always take place at the final review stage providing an analytical approach to pricing is used. Where computer databases are used, fine-tuning of the labour element can take place at any time before tender submission.

Figure 10.1 illustrates the all-in rate calculation using a spreadsheet model. Travelling and subsistence costs have been omitted on the assumption that they are better assessed when calculating the preliminaries. Changes can be made to any of the figures and the following are the items that might change from job to job:

1. Time of year – the proportion of work carried out during 'summer' weeks.
2. Number of hours worked each week – the normal working hours are thirty-nine per week throughout the year, but in the summer more working hours can be achieved.
3. The allowance for bad weather – depends on time of year, exposure to the weather and height above sea level.
4. Attraction bonus – is the non-productive element needed to match the going rate for skilled and semi-skilled people?
5. Trade supervision – is rarely included in the all-in rate today because it is better to consider all aspects of site supervision while assessing preliminaries.
6. Extra payments for special skills – the Working Rule Agreement specifies many additional payments (to be added to the labourer's rate) principally for plant operatives.

CB CONSTRUCTION LTD	**Fast Transport Ltd**				**2007/08**
Description		Entry col	Calc col		
SUMMER PERIOD	Number of weeks	30			
	Weekly hours	44			
	Total hours		1320		
	Days annual hols	14			
	Days public hols	5			
	Total hrs for hols		-167		
WINTER PERIOD	Number of weeks	22			
	Weekly hours	39			
	Total hours		858		
	Days annual hols	7			
	Days public hols	3			
	Total hrs for hols		-78		
SICKNESS	Number of days (say winter)	6	-47		
TOTAL HOURS FOR PAYMENT			1886		
% Allowance for bad weather deduct		3	-57		
TOTAL PRODUCTIVE HOURS			1829		
		craftsman	labourer	**craftsman**	**labourer**
ANNUAL EARNINGS	Basic wage	379.08	285.09		
	Attraction bonus (say)	35.00	17.00		
	Total weekly rate	414.08	302.09		
	Hourly rate of pay (39th)	10.62	7.75		
Annual earnings =	1886 x hourly rate			20 029	14 617
	Public holidays 62.4 hours x rate			716	522
ADDITIONAL COSTS	NON-PRODUCTIVE OVERTIME (time+half only)				
	Hours per week ..summer		2.50		
	Hours per week ..winter		0.00		
	Hours per year ..summer		65.50		
	Hours per year ..winter		0.00		
	Cost of non-prod overtime			637	479
	SICK PAY excluded from calculation				
TRADE SUPERVISION					
No. of tradesmen per foreman		7.00			
Plus rate for foreman		2.00			
% of time on supervision		50.00		1 700	1 313
WORKING RULE AGREEMENT					
Skill rate .. per hour		0.40			754
		Sub-total		23 082	17 686
OVERHEADS					
	1 NATIONAL INS	12.80	%	2 363	1 672
	2 HOLIDAYS WITH PAY	226.20	hrs	2 402	1 753
	3 RETIREMENT BENEFIT	10.90	per week	567	567
	4 TRAINING LEVY	0.50	% on wages	127	97
		Sub-total		28 541	21 775
SEVERANCE PAY and SUNDRIES		1.5	%	428	327
		Sub-total		28 970	22 102
EMPL. LIABTY & 3rd PARTY INS		2.5	%	724	553
ANNUAL COST OF OPERATIVE				29 694	22 654
Divide by Total Productive Hours ..		1829			
	COST PER HOUR =			**£16.23**	**£12.38**

Also consider Construction Industry Joint Council pay conditions:

Storage of tools — Maximum liability £600
Subsistance allowance — £28.17 per night
Sick pay — £101.09 per week
Death benefit — £22,000
National insurance is 12.8% above earnings threshold of £100.00 per week

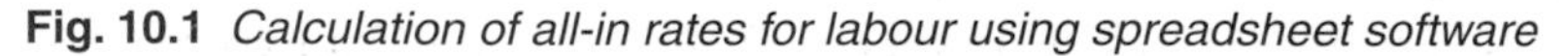

Fig. 10.1 *Calculation of all-in rates for labour using spreadsheet software*

7. Employer's liability insurance – although related to the labour value, may be part of a general assessment of liabilities in the preliminaries schedule.

Spreadsheets are used for these repetitious calculations because various combinations can be tried out, hence the phrase 'what if calculations'. In Fig. 10.2, supervision and insurances have been removed so that they can be considered in pricing preliminaries. A longer working week is envisaged during a 35-week summer period. The overall effect is a reduced hourly rate.

For the analysis of rates throughout this book, the labour rates have been rounded off to £17.00/h for craftsmen and £12.00/h for labourers. These are the labour rates calculated from first principles in this chapter and reflect rates during the period June 2007 and May 2008.

Material rates

Quotations should be obtained for all materials, not only because prices can fluctuate unpredictably but also because the haulage rates to various sites could be different, depending on their distance from the supplier; and the size of loads can dramatically affect the transport costs. The following factors are considered by the estimator in building up the material portion of a unit rate:

1. Check the materials comply with the specification – the estimator may consider the use of an alternative product if it is cheaper and from experience is a satisfactory choice that the contract administrator is likely to accept. A common example is the use of cement replacements and additives in ready-mixed concrete which ironically are readily accepted by the Department of Transport and water industry and sometimes rejected by architects for building work. Many specifications envisage the use of alternatives with the statement 'subject to the approval of the Contract Administrator' for example.
2. The supplier may want payments for the costs of transport or small load charges. Ready-mixed concrete suppliers for example impose extra payments for part loads. The cost can be significant and must be considered where small concrete pours are expected.
3. Some products are manufactured in fixed sizes that are the minimum that can be ordered. An estimator may have received a price of £3.00 per metre for polythene pipe for a job which needs only 15 m. If the minimum coil size is 30 m then the estimator must consider the likelihood of using the pipe on another site that might involve a storage cost. Alternatively it might be more realistic to allow £6.00/m (including waste) in this tender.
4. The quantity required for each unit of work must be considered for each material. Estimators should keep a note of the conversion factors they need for commonly

CB CONSTRUCTION LTD **Fast Transport Ltd** **2007/08**

Description		Entry col	Calc col	craftsman	labourer
SUMMER PERIOD	Number of weeks	35			
	Weekly hours	50			
	Total hours		1750		
	Days annual hols	14			
	Days public hols	5			
	Total hrs for hols		-190		
WINTER PERIOD	Number of weeks	17			
	Weekly hours	39			
	Total hours		663		
	Days annual hols	7			
	Days public hols	3			
	Total hrs for hols		-78		
SICKNESS	Number of days (say winter)	6	-47		
TOTAL HOURS FOR PAYMENT			2098		
% Allowance for bad weather	deduct	3	-63		
TOTAL PRODUCTIVE HOURS			2035		
		craftsman	labourer	**craftsman**	**labourer**
ANNUAL EARNINGS	Basic wage	379.08	285.09		
	Attraction bonus (say)	35.00	17.00		
	Total weekly rate	414.08	302.09		
	Hourly rate of pay (39th)	10.62	7.75		
Annual earnings =	2098.2 x hourly rate			22 283	16 261
	Public holidays 62.4 hours x rate			780	569
ADDITIONAL COSTS	NON-PRODUCTIVE OVERTIME (time+half only)				
	Hours per week ..summer		5.50		
	Hours per week ..winter		0.00		
	Hours per year ..summer		171.60		
	Hours per year ..winter		0.00		
	Cost of non-prod overtime			1 668	1 254
	SICK PAY excluded from calculation				
TRADE SUPERVISION					
No. of tradesmen per foreman		7.00			
Plus rate for foreman		2.00			
% of time on supervision		0.00		0	0
WORKING RULE AGREEMENT					
Skill rate .. per hour		0.40			839
		Sub-total		24 730	18 924
OVERHEADS					
	1 NATIONAL INS	12.80	%	2 574	1 831
	2 HOLIDAYS WITH PAY	226.20	hrs	2 402	1 753
	3 RETIREMENT BENEFIT	10.90	per week	567	567
	4 TRAINING LEVY	0.50	% on wages	136	103
		Sub-total		30 409	23 178
SEVERANCE PAY and SUNDRIES		1.5	%	456	348
		Sub-total		30 865	23 525
EMPL. LIABTY & 3rd PARTY INS		0	%	0	0
ANNUAL COST OF OPERATIVE				30 865	23 525
Divide by Total Productive Hours ..		2035			
	COST PER HOUR =			**£15.17**	**£11.56**

Also consider Construction Industry Joint Council pay conditions:

Storage of tools	Maximum liability £600
Subsistance allowance	£28.17 per night
Sick pay	£101.09 per week
Death benefit	220 00

National insurance is 12.8% above earnings threshold of £100.00 per week

Fig. 10.2 *Calculation of all-in rates for site working 50 hours/week with an extended summer period of 35 weeks and supervision and insurances priced in preliminaries*

used materials. For example, a half-brick wall has 60 bricks per m^2, 2.1 tonnes of stone may be needed for each cubic metre of hardcore, and 0.07 litres of emulsion paint might be the coverage for work to plastered ceilings.

5. Unloading and distributing materials are activities that can be priced in the unit rate calculation or dealt with as a general site facility in the preliminaries. Often a combination of both is needed. With facing bricks, for example, the price for bricks will include the cost of mechanical off-loading, whereas distributing bricks around the site could be catered for by including a forklift and a distribution gang in the project overhead schedule.
6. If the specifications, or preliminaries clauses, call for samples of certain materials the estimator needs to ascertain the cost. Usually a supplier will provide samples without charge. Testing of materials, on the other hand, is usually undertaken by an independent organization, and as such must be specified or preferably included as an item in the bill of quantities. The cost of testing will be assessed when the overheads are calculated.
7. An allowance for *waste* is difficult to estimate. The standard methods of measurement state that work is measured net as fixed in position (SMM7 3.3.1) and the contractor is to allow for any waste and square cutting (SMM7 4.6 e and f) and overlapping of materials (Fabric reinforcement E30.M4 for example). CESMM3 Section 5 states that the quantities shall be calculated net using dimensions from the drawings and that no allowance shall be made for bulking, shrinkage or waste. The questions that the estimator must consider are: is there a selection process needed on site to achieve the quality specified (such as picking facing bricks to produce a specific pattern)? Are the materials likely to suffer damage in the off-loading and handling stages? Is the design going to lead to losses in cutting standard components to fit the site dimensions? Is the site secure from theft and vandalism? Will the finished work be protected from damage by following trades? Has the company had previous experiences with the materials? Will some materials be used for the wrong purpose, such as using facing bricks below ground level to avoid ordering a few cheaper bricks?

An estimator will need help in making these decisions. Guidance can be found in price books or research papers and the company should collect information from previous projects.

Plant rates

The plant supply industry can provide a wide range of equipment throughout the UK. It can offer hire or outright purchase, and in some cases lease and contract rental schemes. The following steps can be taken at tender stage to assess the mechanical plant to be used.

Step 1

Identify specific items of plant needed by looking at quantities and methods. The machine capacities can be found by assessing the rates of production required. Examine the tender programme for overall durations.

Step 2

Obtain prices; the *sources of plant* are:

(a) purchase for the contract;
(b) company-owned plant;
(c) hire from external source.

In practice the *sources of prices* are:

(a) calculate from first principles;
(b) internal plant department rates;
(c) hirers' quotations;
(d) published schedules.

Step 3

Compare plant quotations on equal basis perhaps by using a standard form (the Code of Estimating Practice provides a typical Plant Quotations Register).

Step 4

Calculate the all-in hourly rate for each item of mechanical plant. The main parts of the calculation are:

(a) cost of machine per hour (including depreciation, maintenance, insurances, licences and overheads).
(b) all-in rate for operator (the operator may work longer hours than the plant because of the time needed for minor repairs, oiling and greasing; the National Working Rule Agreement suggests how much time should be added to each eight-hour shift; it also lists extra payments for continuous extra skill or responsibility in driving various items of plant).
(c) Fuel and lubricants (the amounts of fuel consumed will depend on the types and sizes of plant; the average consumption during the plant life is used).

(d) Sundry consumables (where, for example, the plant specialist is unable to accept the risk of tyre replacement on a difficult site or any costs beyond 'fair wear and tear').

The cost of bringing plant to site is usually dealt in assessing preliminaries (project overheads) when the transport of all plant and equipment is considered.

Step 5

Decide where to price plant – either in the unit rate against each item of measured work or in the preliminaries. This decision might be made for the estimator if the company's procedures dictate the pricing method. Plant that serves several trades should be included in the preliminaries, such as cranes, hoists, concrete mixers, and materials-handling equipment. Estimators also price the erection of fixed plant in the preliminaries together with the costs of dismantling plant on completion.

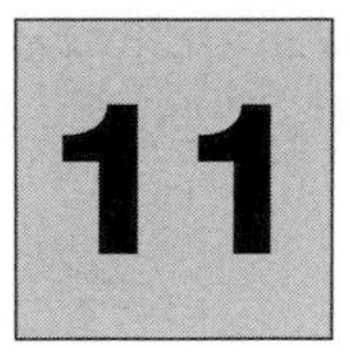

11 Unit rate pricing

Introduction

The estimator must press on with the pricing stage without delay and cannot afford to wait for written quotations from sub-contractors and suppliers.

Once basic rates have been calculated for labour and plant, pricing notes can be written for work which will not be sub-let; such as placing concrete, alterations and brickwork. The pricing form in Fig. 11.1 shows estimator's notes for fixing ironmongery with spaces for the prices from suppliers.

Computer-aided estimating systems allow early pricing to start, using the rates contained in the main library of resource costs. When quotations arrive, the resource costs can be updated in the job library. Estimators can make good progress using this approach but must be careful to check that all the prices are confirmed by suppliers (preferably in writing) before the tender is submitted.

Components of a rate

Unit rates are usually a combination of rates for labour, plant, materials and sub-contractors. *Only the direct site cost is included* because management will develop a better understanding of the pricing level if on-costs are dealt with separately. There is a more extreme view that rates should ignore some or all of the following:

1. General site plant such as cranes and plant for materials distribution such as tractors and trailers, dumpers and forklift trucks.
2. Small plant, tools and safety equipment.
3. General labourers assisting craftsmen, unloading materials, distributing materials and driving mechanical plant.
4. Difficult working conditions such as access, restricted space and exposure to the weather.

The estimator must think about the way in which each operation will be carried out. The following factors must be considered in calculating the cost:

1. Quantity of work to be done.
2. Quality of work and type of finish specified.
3. The degree of repetition.

Many clients assume that unit rates are for all the obligations and risks associated with the work, and in some cases include a statement such as: 'the rates inserted in the bill of quantities are to be fully inclusive of all costs and expenses together with all risks, liabilities, obligations given in the tender documents'. Does this mean that a proportion of preliminaries should be included in all the rates or can the contractor insert rates for all general obligations and overheads in the preliminaries bill? Generally, estimators are doing the latter.

Method of measurement

The classification tables in SMM7 set out the work which is to be included in the unit rate. For example, when working space is measured to excavations, the contractor is to allow for additional earthwork support, disposal, backfilling, work below ground level and breaking out. Clearly the estimator must be aware of the coverage rules before pricing the work. With CESMM3, items for excavation include working space as well as upholding sides of excavation and removal of dead services. In addition, bills of quantities often have a preamble (civil engineering work) or rules for measurement (building), which list the changes to the standard measurement rules. A typical example is the statement 'the contractor shall allow all methods necessary to withhold the sides of excavations including where necessary trench sheeting or sheet piling'. This is a significant change to SMM7 because sheet piling is normally measurable under D32 Steel Piling.

Pricing notes

There are many ways to present pricing notes. Standard forms help the estimator produce clear information, which can be read by others.

The form shown in Fig. 11.1 would allow an estimator to price labour and plant himself and add rates received from a labour-only sub-contractor when they arrive. A direct comparison can then be made. This is similar to the example given in Fig. 12.1 in the following chapter for the comparison of sub-contractors' rates. The

CB CONSTRUCTION LIMITED

Project	Fast Transport Ltd		
Ref. no	T354	Date	5.7.08

Trade	Ironmongery
Page	1

					Total		LABOUR	17.00	PLANT		MATERIALS					LOSC	
Item	Item	Item	Description	Hrs	Quants	Unit	rate	£	rate	£	basic	sund	waste	rate	£	rate	£
3/26A	4/15F		Overhead door closer	1.50	12	nr	25.50	306.00	-	-			2.50			19.00	228.00
3/26B	4/15G		200 mm flush bolt	0.75	5	nr	12.75	63.75	-	-			2.50			8.50	42.50
3/26C	4/15F		Mortice dead lock	0.75	8	nr	12.75	102.00	-	-			2.50			11.25	90.00
								471.75		-					-		360.50

Fig. 11.1 *Estimator's build-up sheet for fixing ironmongery*

form used throughout this chapter for pricing notes was typically used for detailed build-ups. Its use has declined with the growth of computing where pricing notes are stored and recovered in many formats such as price build-ups and resource summaries.

Pricing notes are not always clearly presented by estimators. Where time is short, they sometimes produce their notes in the bill of quantities either in the margin or on the facing page. At the final review stage, management would then examine the rated bill of quantities because summaries for labour, materials and plant will not be available. This example of poor practice is fortunately not common. It is more likely that, if time is short, brief descriptions are put into a spreadsheet together with quantities and rates inserted. This in effect is using the spreadsheet as a calculator which allows some filtering and analysis to take place.

Construction staff need to be aware, however, that any tender notes may be useful to understand the logic used at tender stage but the costs may have been changed by management at the final review meeting. A computer system, on the other hand, will produce an up-to-date report of resources with all changes made after the review stages. There is no doubt that computer reports are quick to produce and can provide comprehensive site budgets and valuations. Very few give reports on the logic used to build up rates, which means that some manual notes or method statements may still be necessary.

Model rate and pricing examples

The way in which unit rates are built up differs from company to company and between trades. Calculations for earthworks, for example, are based on the use of plant, and formwork pricing depends on the making and re-use of shutters. A checklist of items to include in a rate could be used by trainee estimators or anyone pricing an item for the first time. The 'model rate' calculation given in Fig. 11.2 has more components than any one item would need but illustrates the components to think about.

The pricing information sheets given in this chapter (Figs. 11.2.1–11.2.40) contain typical outputs and pricing notes for the categories of work found most often in building and small civil engineering projects.

Most of the data has been expressed in terms of decimal constants which are used for entering resources using computers. Unfortunately, this approach gives some strange results and unfamiliar figures. With excavation of trenches, for example, estimators think in terms of how many cubic metres could be dug in one hour (say $5\,m^3/hr$), and not the reverse (an output such as $0.20\,hr/m^3$).

CB CONSTRUCTION LIMITED PRICING NOTES

Project		Trade	**MODEL RATE**	Date	
Ref. No.			**Unit rate pricing**	Sheet No.	

Typical bill description	Z20 Section of construction material Section of construction material fixing to brickwork ....160 m

item details					analysis				net unit rate
ref:	description	quant	unit	rate	lab	plt	mat	s/c	
Mat	**Unit price from supplier**	1	m	2.82			2.82		
	Delivery and packing charges	1/160	item	30.00			0.13		
	Overlap (usually sheet materials)								
	Penetration (usually aggregates)								
	Nails, plugs, screws, adhesives etc	3	nr	0.16			0.48		
	Mortar (usually bricks, blocks & kerbs)								
	Waste - cutting from larger pieces								
	Waste - breakages before fixing								
	Waste - during fixing	0.05	m	3.43			0.17		
	Waste - residue from large packs								
Lab	**Unload, store and distribute**	0.01	hr	12.00	0.12				
	Craftsman at all-in rate	0.2	hr	17.00	3.40				
	Labourer assistance at all-in rate	0.04	hr	12.00	0.48				
Plt	**Electric drill and masonry bit**	0.2	hr	0.50		0.10			
	(Small tools and equipment usually								
	priced in preliminaries as a small								
	percentage added to all labour costs)								
	Total net rate				4.00	0.10	3.06		7.16
ADD	**a proportion of overheads & profit**								
	(the rest will be shown as items								
	in the preliminaries bill)	10	%		0.40	0.01	0.31		0.72
	TOTAL UNIT RATE	1	m		4.40	0.11	3.37		7.88

Fig. 11.2 *Model rate calculation*

CB CONSTRUCTION LIMITED PRICING NOTES

Project		Trade	**MODEL RATE**	Date	
Ref. No.			**Unit rate pricing**	Sheet No.	

Typical bill description	Z20 Section of construction material Section of construction material fixing to brickwork ....160 m

item details					analysis				net unit rate
ref:	description	quant	unit	rate	lab	plt	mat	s/c	
Mat	**Unit price from supplier**	**1**	**m**	**2.82**			**2.82**		
	Delivery and packing charges	**1/160**	**item**	**30.00**			**0.13**		
	Overlap (usually sheet materials)								
	Penetration (usually aggregates)								
	Nails, plugs, screws, adhesives etc	**3**	**nr**	**0.16**			**0.48**		
	Mortar (usually bricks, blocks & kerbs)								
	Waste - cutting from larger pieces								
	Waste - breakages before fixing								
	Waste - during fixing	**0.05**	**m**	**3.43**			**0.17**		
	Waste - residue from large packs								
Lab	**Unload, store and distribute**	**0.01**	**hr**	**12.00**	**0.12**				
	Craftsman at all-in rate	**0.2**	**hr**	**17.00**	**3.40**				
	Labourer assistance at all-in rate	**0.04**	**hr**	**12.00**	**0.48**				
Plt	**Electric drill and masonry bit**	**0.2**	**hr**	**0.50**		**0.10**			
	(Small tools and equipment usually								
	priced in preliminaries as a small								
	percentage added to all labour costs)								
	Total net rate				**4.00**	**0.10**	**3.06**		**7.16**
ADD	**a proportion of overheads & profit**								
	(the rest will be shown as items								
	in the preliminaries bill)	**10**	**%**		**0.40**	**0.01**	**0.31**		**0.72**
	TOTAL UNIT RATE	**1**	**m**		**4.40**	**0.11**	**3.37**		**7.88**

Fig. 11.2.1 *Model rate pricing sheet*

PRICING INFORMATION	GROUNDWORKS EXCAVATION

SMM7 NOTES Work section D20 Excavating and filling	CESMM3 NOTES CLASS E EARTHWORKS
1 Information given in tender documents:	
a. ground water level	
b. details of trial holes or boreholes	
c. live services and features retained	
2 Working space measured separately	Excavation deemed to include working space
3 Excavating below water measured separately	
4 Earthwork support is measured whether needed or not, except to faces ne 0.25m high and faces next to existing structures	Excavation includes upholding sides
5 Interlocking steel piling must be measured (D32)	Piling for temporary works not measured
6 Excavating foundations around piles identified	Excavating foundations around piles identified
7 Underpinning measured in Section D50	Class E includes excavation for underpinning

EXCAVATION		Average outputs - hr/m³						
		hand dig	small		medium		large	
			JCB 3CX	JCB JS150	JCB 3CX	JCB JS150	JCB JS150	CAT225
Topsoil		2 - 3	0.30	0.20	0.20	0.15	0.10	0.08
Reduce levels	ne 0.25m	2 - 3	0.30	0.12	0.20	0.11	0.09	0.05
& basements	ne 1.00m	2 - 3	0.20	0.10	0.15	0.09	0.07	0.04
	ne 2.00m	3 - 4	0.20	0.10	0.15	0.08	0.06	0.04
	ne 4.00m	4 - 5	0.25	0.12	0.20	0.11	0.09	0.05
	ne 6.00m		0.30*	0.15	0.25*	0.13	0.11	0.07
Trenches/Pits	ne 0.25m	2 - 3	0.35	0.22	0.25	0.17	0.11	0.07
	ne 1.00m	3 - 4	0.25	0.20	0.20	0.14	0.10	0.06
	ne 2.00m	4 - 5	0.25	0.20	0.20	0.12	0.09	0.05
	ne 4.00m	5 - 7	0.30	0.25	0.25	0.20	0.12	0.07
	ne 6.00m		0.35*	0.30	0.30*	0.25	0.15	0.09

* May be beyond range of machine

BREAKING OUT EXISTING MATERIALS hr/m³	ROCK	CONC	R CONC	MASONRY	SURFAC'G	these outputs are for breaking only ADD the following:
Compressor & labourers	3.00	2.00	3.50	1.50	0.75	25% for trench work
JCB 3CX and breaker	0.50	0.40	0.55	0.25	0.20	25% to excavation rate
JCB 812 & breaker	0.35	0.25	0.40	0.15	0.10	25% to loading rate
CAT 225 and breaker	0.30	0.20	0.30	0.10	0.08	25% to removal rate

Excavation outputs are normally expressed as m³/h. These tables use decimal constants for computer applications

Excavation outputs depend on:

Quantities	**small, medium and large in the table is a guide to quantity of excavation**	
Ground conditions	**the data above are based on 'normal' ground conditions (firm clay)**	
Bucket size	**outputs based on:**	**JCB3CX [backhoe/loader] with a bucket capacity of 0.30m³**
		JCB JS150 [backacter] with a bucket capacity of 0.60m³
		CAT225 [backacter] with a bucket capacity of 1.20m³
Location	**outputs assume reasonable access for plant and lorries**	
Disposal	**where lorries have clear access, the above outputs are sufficient to excavate & load**	
Trimming	**the outputs provide for trimming if labour is included in the excavation rate; trimming should be priced separately for large areas and sloping surfaces**	

Fig. 11.2.2 *Groundworks excavation data sheet*

PRICING INFORMATION	GROUNDWORKS DISPOSAL AND FILLING

SMM7 NOTES Work section D20 Excavation and filling	CESMM3 NOTES CLASS E EARTHWORKS
1 Disposal off site is stated	Disposal of excavated material is deemed to be
2 Disposal on site is stated	off site unless otherwise stated
3 Only design-imposed locations are stated	The location for material for disposal on site is given
4 Only design-imposed handling provisions are stated	Double handling measured where expressly required
5 Kind and quality of fill materials are stated	Materials for imported filling are given
6 Compaction of filling is measured separately	Filling is deemed to include compaction
7 The filling quantity is the volume of void filled	
8 Filling measured m³ and compaction m²	Filling to stated thicknesses measured m²
9 Filling thickness is that after compaction	
	Penetration of filling over 75 mm deep is measured

LOADING OF EXCAVATED MATERIAL

outputs for loading lorries/dumpers

	m³/h
JCB3CX	10
JCB JS150	15
CAT 225	20+

REMOVAL OF EXCAVATED MATERIAL

	tip located	
	on site	off site
Average speed to tip	10 mph	15 mph
Average time on tip	3 min	6 min
Average speed to return	15 mph	20 mph

DEPOSITION AND COMPACTION OF FILLING MATERIALS		Output hr/m³					
		JCB 3CX + 2 Labs & roller		JCB JS150 + 2 Labs & roller		CAT 943 + 2 Labs & roller	
quantities		*small*	*medium*	*small*	*medium*	*medium*	*large*
Filling to excavations	ne 0.25 m	*0.25*	*0.17*	*0.12*	*0.10*	*0.08*	*0.05*
	over 0.25 m	*0.17*	*0.12*	*0.10*	*0.08*	*0.07*	*0.04*
Making up levels	ne 0.25 m	*0.25*	*0.17*	*0.12*	*0.07*	*0.05*	*0.03*
	over 0.25 m	*0.17*	*0.12*	*0.10*	*0.05*	*0.04*	*0.03*
Blinding surfaces		*0.33*	*0.20*	*0.25*	*0.20*		
		Output hr/m³					
		1 Lab & roller/rammer		1 lab & tandem roller		CAT 943 & towed roller	
Compacting open excavation/ground		*0.10*	*0.05*	*0.05*	*0.03*	*0.02*	*0.01*
Compacting filling [if not priced above]		*0.20*	*0.10*	*0.06*	*0.04*	*0.02*	*0.01*
Compacting under foundations		*0.20*	*0.10*				

Outputs are normally expressed as m³/h. This table uses decimal constants for computer applications

Material from site spoil heaps will need to be loaded and transported to the filling site

Conversion factors including consolidation t/m³				
	Ashes	1.30	Gabion stone	1.50
	Blast furnace slag	2.10	Crushed limestone	1.95
	Sand	1.75	Scalpings	2.10
	Stone dust	1.75	DOT type 1	2.30

Fig. 11.2.3 *Groundworks disposal and filling data sheet*

				£	p
	D20 EXCAVATING AND FILLING				
	Excavating				
	Topsoil for preservation				
A	275 average depth	380 m²			
	To reduce levels				
B	1 m maximum depth, commencing 275 below existing ground level	246 m³			
	Trenches exceeding 300 wide				
C	1 m maximum depth, commencing 600 below existing ground level	38 m³			
	Extra over excavation irrespective of depth for breaking out				
D	rock (approximate)	26 m³			
	Working space allowance to excavations				
E	trenches, backfilling with selected excavated material	94 m²			
	Earthwork support				
	To faces of excavation				
F	2 m maximum depth, distance between opposing faces not exceeding 2 m	140 m²			
	Disposal				
	Excavated material				
G	off site	237 m³			
	Selected excavated material				
	Filling to excavations				
H	over 250 thick	47 m³			
	Hardcore as D20.M010				
	Filling to make up levels				
I	over 250 thick, obtained off site	252 m³			
		To collection			

Fig. 11.2.4 *Groundworks sample bill of quantities*

CB CONSTRUCTION LIMITED PRICING NOTES

Project		Trade	EXCAVATION	Date	
Ref. No.			Unit rate pricing	Sheet No.	1

item details					analysis				net unit rate
ref:	description	quant	unit	rate	lab	plt	mat	s/c	
	The type of plant to be used should be selected by examining the								
	nature of the ground and quantities for excavation and disposal								
	Page 3/1 of the bill of quantities would be considered in relation to								
	other excavation in the works such as drainage and external works								
	The total excavation on this page is :								
	topsoil 380 x 0.275 =	105	m^3						
	to reduce levels	246	m^3						
	trenches	38	m^3						
		389	m^3						
	The total disposal on this page is :								
	disposal off site	237	m^3						
	filling to excavations	47	m^3						
	topsoil retained on site	105	m^3						
		389	m^3						
	For a machine excavating at 10 m^3/h (on average) there would appear								
	to be at least a week of work . The additional costs of transporting								
	a larger (backacter) machine is justified because it will be needed								
	to break out rock and place filling materials								
A	<u>Topsoil 275 mm deep</u>								
	JCB JS150	0.15	hr	28.00		4.20			
	Banksman	0.15	hr	12.00	1.80				
	Consider lorry or dumper if spoil								
	to be taken away from building area								
		1	m^3		1.80	4.20			6.00
	Topsoil 275 mm dp (x 0.275)	1	m^2		0.50	1.16			1.66
B	<u>Excavating to reduce levels</u>								
	JCB JS150	0.09	hr	28.00		2.52			
	Banksman	0.09	hr	12.00	1.08				
	Excavating to reduce levels	1	m^3		1.08	2.52			3.60

Fig. 11.2.5 *Excavation to reduce levels pricing sheet*

CB CONSTRUCTION LIMITED PRICING NOTES

Project		Trade	EXCAVATION	Date	
Ref. No.			Unit rate pricing	Sheet No.	2

item details					analysis				net unit rate
ref:	description	quant	unit	rate	lab	plt	mat	s/c	
C	Excavating trenches								
	JCB JS150	0.14	hr	28.00		3.92			
	Banksman	0.14	hr	12.00	1.68				
	Labourer trimming	0.14	hr	12.00	1.68				
	Excavating trenches	1.00	m³		3.36	3.92			7.28
D	EXTRA breaking out rock								
	Assuming 20% in red lev and 80% in trenches								
	JCB JS150 and breaker [20% .35 hr]	0.07	hr	40.00		2.80			
	JCB JS150 and breaker [80% .44 hr]	0.35	hr	40.00		14.00			
	Add 25% to excavation rates red lev				0.27	0.61			
	Add 25% to excavation rates trench				0.84	0.98			
	Add 25% to disposal rate	0.25	m³	15.80		3.95			
	EXTRA for breaking out rock	1.00	m³		1.11	22.34			23.45
E	Working space allowance								
	Excavation as for trenches	1.00	m³		3.36	3.92			
	Assume 75% filling and 25% disposal								
	JCB JS150 and roller [75% of 0.08 hr]	0.06	hr	31.00		1.86			
	Labourers (2nr) [75% of 0.16 hr]	0.12	hr	12.00	1.44				
	Additional earthwork support	not priced							
	Additional disposal	0.25	m³	15.80		3.95			
		1.00	m³		4.80	9.73			14.53
	Assuming average thickness is 250 mm								
	Working space allowance	1.00	m²		1.20	2.43			3.63
F	Earthwork support								
	For shallow trenches support may not be required (nil rate)								
	but the trenches may have sloping sides								
	Once an assessment is made of the average over-excavation								
	the working space rate (above) can be used								
	Assuming average thickness is 300 mm								
	Earthwork support	1.00	m²		1.44	2.92			4.36

Fig. 11.2.6 *Excavation trenches pricing sheet*

CB CONSTRUCTION LIMITED PRICING NOTES

Project		Trade	EXCAVATION	Date	
Ref. No.			Unit rate pricing	Sheet No.	3

item details					analysis				net unit rate
ref:	description	quant	unit	rate	lab	plt	mat	s/c	
G	Disposal off site								
	In this case it is assumed that 75% of material								
	can be loaded directly into lorries at the time of excavation								
	This means that 25% is loaded as a separate operation								
	JCB JS150 [15 m³/hr x 25%]	0.017	hr	27.00		0.46			
	The speed of loading is less for material loaded								
	directly at the time of excavation, say 10 m³/hr								
	The average rate of loading is therefore:								
	25% at 15 m³/hr and 75% at 10 m³/hr = 11.25 m³/hr								
	The other assumptions made for the calculation are:								
	Lorry capacity 16T	6.4 m³							
	Distance to tip	5 m							
	Tip charges per load	£27.50							
	Landfill tax (inert material)	£2.50	/T						
	from April 2008								
	Round trip calculation								
	Load 6.4 m³ at 11.25 m³/hr	34	min						
	Haul to tip at 15 m/hr	20	min						
	Time on tip	6	min						
	Return to site at 20 m/hr	15	min						
	Total	75	min						
	So each lorry will achieve 60/75 = 0.80 trips per hr								
	and carry 6.4 x 0.80 = 5.12 m³/hr								
	If the maximum speed of loading is 15 m³/hr,								
	three lorries are needed at £27.00 per hour								
	Lorry cost is therefore 3 x 27.00 = £81.00/hr								
	The average rate of disposal is 11.25 m³/hr								
	Lorries	0.09	hr	81.00		7.29			
	Tip charges £27.50 ÷ 6.4 m³					4.30			
	Landfill tax 1.5T/m³ x £2.50					3.75			
	Disposal off site	1	m³			15.80			15.80
	Note: Standard rate for Landfill Tax is £32/T April 2008								

Fig. 11.2.7 *Excavation disposal off site pricing sheet*

CB CONSTRUCTION LIMITED PRICING NOTES

Project		Trade	EXCAVATION	Date	
Ref. No.			Unit rate pricing	Sheet No.	4

item details					analysis				net unit rate
ref:	description	quant	unit	rate	lab	plt	mat	s/c	
H	Filling with selected excavated material								
	Assuming transport over short distances can								
	be provided by a site dumper, and an allowance								
	has been made for this in the preliminaries:								
	JCB JS150 and roller	0.08	hr	31.00		2.48			
	Labourers (2nr)	0.16	hr	12.00	1.92				
	Filling	1	m^3		1.92	2.48			4.40
I	Hardcore filling								
	Hardcore price from supplier	1.95	T	10.00			19.50		
	Waste	0.19	T	10.00			1.90		
	(Penetration into the ground would be								
	considered for hardcore beds under 250 mm)								
	JCB JS150 and roller	0.05	hr	31.00		1.55			
	Labourers (2nr)	0.1	hr	12.00	1.20				
	Hardcore filling	1	m^3		1.20	1.55	21.40		24.15

Fig. 11.2.8 *Excavation filling pricing sheet*

PRICING INFORMATION	IN SITU CONCRETE

SMM7 NOTES Work section E10 In situ concrete	CESMM3 NOTES CLASS F IN SITU CONCRETE
1 Kind and quality of materials and mixes stated	Concrete mix may be related to BS 5328, or a mix
2 Tests of materials and finished work stated	designed by the contractor, or a mix prescribed
3 Limitations on pouring methods stated	in the specification; with items given separately
4 Methods of compaction and curing stated	for provision and placing of concrete
5 Requirements for beds laid in bays to be given	
6 Concrete assumed to be as struck or tamped finish	Finishes to concrete measured separately
7 Concrete measured net with no deduction for :	Volume of concrete includes that occupied by:
..reinforcement, sections under 0.50 m², voids under	reinforcement, cast-in items ne 0.1 m³, rebates,
.. 0.05 m³	grooves and chamfers ne 0.01 m², large and small
8 Details of concrete sections given on drawings	voids, and joints in in-situ concrete
9 Beds include blinding, plinths and thickenings	Placing concrete in blinding measured separately

	waste %	Output - operative hrs/m³					
		small		medium		large	
		plain	reinf	plain	reinf	plain	reinf
Mass filling	10.0	*1.45*		*1.25*		*1.00*	
Foundations	7.5	*1.65*	*2.00*	*1.40*	*1.65*	*1.10*	*1.30*
Ground beams	5.0	*2.40*	*2.90*	*2.00*	*2.40*	*1.60*	*1.90*
Isolated foundations	7.5	*1.65*	*2.00*	*1.50*	*1.75*	*1.25*	*1.50*
Blinding beds	35.0	*2.75*		*2.40*		*2.00*	
Beds ne 150 mm	10.0	*1.35*	*1.65*	*1.30*	*1.55*	*1.20*	*1.45*
150 – 450 mm	7.5	*1.20*	*1.45*	*1.15*	*1.35*	*1.05*	*1.25*
over 450 mm	5.0	*1.10*	*1.30*	*1.00*	*1.20*	*0.90*	*1.10*
Slabs ne 150 mm	5.0		*4.00*		*3.00*		*2.50*
150 – 450 mm	5.0		*3.00*		*2.50*		*2.25*
over 450 mm	5.0		*2.50*		*2.00*		*1.50*
Troughed slabs	5.0		*3.50*		*3.00*		*2.50*
Walls ne 150 mm	7.5	*3.50*	*4.20*	*2.75*	*3.30*	*2.00*	*2.40*
150 – 450 mm	5.0	*2.80*	*3.35*	*2.30*	*2.75*	*1.80*	*2.15*
over 450 mm	5.0	*2.00*	*2.40*	*1.80*	*2.15*	*1.60*	*1.90*
Filling hollow walls 50 mm th	20.0	*7.00*	*8.00*	*6.00*	*7.00*	*5.00*	*6.00*
Filling hollow walls 75 mm th	15.0	*6.00*	*7.00*	*5.00*	*6.00*	*4.00*	*5.00*
Beams	7.5	*4.50*	*5.40*	*3.50*	*4.20*	*2.50*	*3.00*
Beam casings	10.0	*4.95*	*6.00*	*3.85*	*4.65*	*2.75*	*3.30*
Columns	7.5	*4.50*	*5.40*	*3.50*	*4.20*	*2.50*	*3.00*
Column casings	10.0	*4.95*	*6.00*	*3.85*	*4.65*	*2.75*	*3.30*
Staircases	7.5	*3.60*	*4.30*	*3.00*	*3.60*	*2.40*	*2.90*
Upstands and kerbs	10.0	*6.00*	*7.00*	*5.00*	*6.00*	*4.00*	*5.00*

SLOPING ITEMS	Add 5% to labour for concrete laid up to 15°, and 10% for concrete over 15°
LABOUR RATE	The effective rate for an operative is found by dividing the cost of a concrete gang by the number of operatives in the gang
WASTE	Waste includes losses due to small quantities and irregular levels for beds and blinding
	Part load charges should be considered for very small quantities
CURING	The outputs include labour for protecting and curing fresh concrete
REINFORCEMENT	Consider concrete saving for members reinforced over 5% by volume

Fig. 11.2.9 *In situ concrete data sheet*

CB CONSTRUCTION LIMITED PRICING NOTES

Project		Trade	IN SITU CONCRETE	Date	
Ref. No.			Unit rate pricing	Sheet No.	

item details					analysis				net unit rate
ref:	description	quant	unit	rate	lab	plt	mat	s/c	
	Hourly rate for concrete gang:								
	Working ganger	1	hr	12.75	12.75				
	Labourers (4nr)	4	hr	12.00	48.00				
	Carpenter in attendance	1	hr	17.00	17.00				
	* Poker vibrator (2nr)	2	hr	2.50		5.00			
	* Concrete pump	1	hr	54.00		54.00			
	Rate for concrete gang	1	hr		77.75	59.00			136.75
	Effective rate for one operative (÷5)	1	hr		15.55	11.80			27.35
Mat	Concrete price from supplier	1	m³	82.00			82.00		
	Waste	0.08	m³	82.00			6.56		
Lab	Concrete operative	1.25	hr	15.55	19.44				
Plt	* Vibrator and pump	1.25	hr	11.80		14.75			
	Rate for in situ concrete	1	m³		19.44	14.75	88.56		122.75
*	(plant may be priced in prelims)								

Fig. 11.2.10 *In situ concrete unit rate pricing sheet*

CB CONSTRUCTION LIMITED PRICING NOTES

Project		Trade	IN SITU CONCRETE	Date	
Ref. No.			Operational pricing	Sheet No.	

Typical bill description	E10 IN SITU CONCRETE Reinforced in situ concrete; mix B, 20 mm aggregate; beds thickness 150–450 mm ....160 m³

item details					analysis				net unit rate
ref:	description	quant	unit	rate	lab	plt	mat	s/c	
	For a concrete slab 40 x 20 m								
	Assume 1nr bay 4m wide cast per day								
	Volume cast =								
	40 x 4 m x 0.2 m thick =	32	m³						
Lab	Ganger	9	hr	12.75	114.75				
	Labourers (4 nr)	36	hr	12.00	432.00				
	Carpenter	4.5	hr	17.00	76.50				
Plt	* Poker vibrator (2 nr)	18	hr	2.50		45.00			
	* Concrete pump	9	hr	54.00		486.00			
Mat	Concrete price from supplier	32	m³	82.00			2624.00		
	Waste 7.5%	2.4	m³	82.00			196.80		
	Rate for one bay	32	m³		623.25	531.00	2820.80		3975.05
	Rate for in situ concrete (÷32)	1	m³		19.48	16.59	88.15		124.22
*	(plant may be priced in prelims)								

Fig. 11.2.11 *In situ concrete operational pricing sheet*

PRICING INFORMATION	FORMWORK

SMM7 NOTES Work section E20 Formwork for in situ concrete	CESMM3 NOTES CLASS G CONCRETE ANCILLARIES [Formwork]
1 Basic finish is given	Finishes described as rough, fair or other
.... where not at the discretion of the contractor	
2 Plain formwork measured separately	Formwork deemed to be plane areas
.... where no steps, rebates, pockets etc.	> 1.22 m wide
3 Rules distinguish between left in and permanent	Formwork left in shall be so described
4 Formwork is measured where temporary	Formwork is measured where temporary
support to fresh concrete is necessary	support to fresh concrete is necessary
5 Tender documents give sizes and positions of	
members, and loads in relation to casting times	
6 Radii stated for curved formwork	Radii stated for curved formwork
7 Formwork to members of constant cross section	Formwork to members of constant cross section
measured square metres	measured in linear metres

LABOUR OUTPUTS	Output - carpenter hr/m²					
	small		medium		large	
	make	F & S	make	F & S	make	F & S
Foundations ne 250 mm	1.10	*1.80*	0.90	*1.60*	0.80	*1.50*
250–500 mm	1.00	*1.70*	0.80	*1.50*	0.70	*1.40*
500 mm–1.00 m	0.90	*1.60*	0.70	*1.40*	0.60	*1.30*
over 1.00 m	0.80	*1.50*	0.60	*1.30*	0.50	*1.20*
Edges of beds ne 250 mm	1.20	*1.90*	1.00	*1.70*	0.90	*1.60*
250–500 mm	1.10	*1.80*	0.90	*1.60*	0.80	*1.50*
Edges of susp slabs ne 250 mm	1.30	*2.40*	1.10	*2.20*	1.00	*2.00*
250–500 mm	1.20	*2.30*	1.00	*2.10*	0.90	*1.90*
Sides of upstands ne 250 mm	1.20	*2.20*	1.00	*2.00*	0.90	*1.90*
250–500 mm	1.10	*2.10*	0.90	*1.90*	0.80	*1.80*
Soffits of slabs horizontal		*1.20**		*1.00**		*0.90**
sloping ne 15°		*1.30**		*1.10**		*1.00**
sloping over 15°		*1.40**		*1.20**		*1.10**
Soffits of troughed slabs		*1.20**		*1.00**		*0.90**
Walls	1.20	*1.70*	1.00	*1.50*	0.90	*1.40*
Walls over 3.0 m	1.20	*1.80*	1.00	*1.60*	0.90	*1.50*
Beams isolated regular shape	1.30	*2.20*	1.10	*2.00*	1.00	*1.80*
irregular shape	1.40	*2.40*	1.20	*2.20*	1.10	*2.00*
Beams attached regular shape	1.40	*2.30*	1.20	*2.10*	1.10	*1.90*
irregular shape	1.60	*2.50*	1.40	*2.30*	1.30	*2.10*
Columns isolated regular shape	1.00	*1.50*	0.80	*1.30*	0.70	*1.10*
irregular shape	1.20	*1.60*	1.00	*1.40*	0.90	*1.20*
Columns attached regular shape	1.20	*1.80*	1.00	*1.60*	0.90	*1.40*
irregular shape	1.40	*1.70*	1.20	*1.70*	1.10	*1.50*

The carpenter rate should include part of a labourer's time for handling materials

Items marked * need a carpenter rate plus a full labourer's rate to erect falsework

'Make' applies to timber shutters. Reduce for hired equipment or proprietary systems

ADD 0.15 hr/m² for fix and strike to walls with formwork one side

A small reduction for formwork LEFT IN is balanced by the labour costs in making

Fig. 11.2.12 *Formwork data sheet*

PRICING INFORMATION		FORMWORK Materials and equipment

Timber formwork		units	Founds	Edges	Soffits	Walls	Beams	Columns
Sheet material	Plywood	m^2/m^2	1	1	1	1	1	1
Timber		m^3/m^2	0.04	0.06	0.05	0.05	0.05	0.05
ADD 10% waste to plywood and timber								
Bolts and nails		kg/m^2	1.0	1.0		1.0	1.5	1.5
Surface preparation	(consider varnish to plywood)							
Number of uses *			high	medium	med/low	medium	medium	high
Falsework and equipment	(see examples)							
Consumables	Mould oil	l/m^2	0.3	0.3	0.3	0.3	0.3	0.3
	Buried fixings	nr/m^2				1.0		
	Nails	kg/m^2	1.0	1.0	1.5	1.0	1.5	0.5

* Number of uses

1 Examine drawings and programme to determine the degree of repetition
2 Consider the standard of surface finish required
3 Consult programme to find time constraints
4 Investigate better quality shutters to increase uses
5 Will there be a salvage value after the work is finished

Key :
high = over 6 uses
medium = 3 to 6 uses
low = under 3 uses

Standard timber shutters can be used up to six times without too much repair and maintenance
At tender stage it can be assumed that the saving from additional uses is countered by the additional costs of repairs.

On the other hand, more than six uses can be achieved with higher quality shutter materials or applied protective coatings

Proprietary formwork systems

The following equipment can be hired for concrete work with little repetition, such as a large machine base or a retaining wall which must be cast in one pour.
- Steel or ply-faced panels (pans)
- Angles
- Soldiers, walings and push/pull props
- Tie rods and accessories
- Radius panels and curved waling tubes

Equipment for foundations

Ground beams and machine bases may need telescopic props
Road forms are a useful alternative for strip footings, blinding and edges of beds

Equipment for soffits

A proprietary falsework system is usually used to support soffit formwork. A weekly rate can be obtained as a rate per m^2 of soffit depending on height and load to be carried

Standard profile GRP trough moulds can be hired
Non-standard profiles are purchased
The choice is dictated by the design

Expanded polystyrene trough moulds are cheaper than GRP but deteriorate quicker

Equipment for walls

Walings, soldiers, push/pull props and shutter ties
Lifting equipment - beam, chains and shackles

Equipment for columns

For a column 0.40 x 0.40 x 3.30 m high
Timber formwork will usually need:

Clamps at ave 500 mm centres	2 sets/m	1.2 sets/m^2
Telescopic props No.3	4 nr/col	0.76 nr/m^2

For a column 400 mm dia there are three options:
1. Cardboard tube for single use
2. GRP shutters for multiple uses
3. Steel shutters for hundreds of uses

Fig. 11.2.13 *Formwork materials and equipment data sheet*

CB CONSTRUCTION LIMITED PRICING NOTES

Project		Trade	FORMWORK	Date	
Ref. No.			Foundations	Sheet No.	

Typical bill description

E20 Formwork for in situ concrete

Formwork with basic finish specification type A to sides of foundations; 250 to 500 mm high .... 60 m

item details					analysis				net unit rate
ref:	description	quant	unit	rate	lab	plt	mat	s/c	
	First calculate cost of 1m² of shutter								
Mat	Plywood	1	m²	12.00			12.00		
	Waste 10%	0.1	m³	12.00			1.20		
	Timber	0.04	m³	255.00			10.20		
	Waste 10%	0.004	m³	255.00			1.02		
	Bolts and nails	1	kg	1.50			1.50		
	Surface preparation	nil							
Lab	Carpenter	0.8	hr	17.00	13.60				
	Labourer (one for four carpenters)	0.25	hr	12.00	3.00				
	Shutter cost	1	m²		16.60		25.92		42.52
Mat	Cost per use assuming 6 uses	÷6			2.77		4.32		7.09
	Equipment (say 4 nr props/m²)	4	nr	0.25		1.00			
	Consumables - mould oil	0.3	l	1.00			0.30		
	Consumables - nails	1	kg	1.25			1.25		
	Consumables - buried fixings	nil							
Lab	Labour fix and strike								
	Carpenter	1.5	hr	17.00	25.50				
	Labourer (one for four carpenters)	0.375	hr	12.00	4.50				
	Formwork rate for one m²	1	m²		32.77	1.00	5.87		39.64
	The average height of formwork will be found from an examination of the drawings;								
	on the other hand a shutter may be made to suit the maximum height which								
	is 500 mm								
	Rate for 500 mm high shutter	1	m		16.38	0.50	2.94		19.82

Fig. 11.2.14 *Formwork to sides of foundations pricing sheet*

CB CONSTRUCTION LIMITED PRICING NOTES

Project		Trade	FORMWORK	Date	
Ref. No.			Soffits of troughed floors	Sheet No.	

Typical bill description

E20 Formwork for in situ concrete

Formwork with basic finish type A to soffits of troughed slabs; profile as detail 1 on drawing D338 550 mm thick; ribs at 900 mm crs; 3.0 to 4.5 m high to soffit.... 660 m²

item details					analysis				net
ref:	description	quant	unit	rate	lab	plt	mat	s/c	unit rate
*	Price from supplier for expanded polystyrene core moulds								
	is £50.00/m delivered to site with a maximum of 4 uses								
Mat	Rate for moulds at 900 crs	1.11	m	50.00			55.50		
	Plywood for continuous deck	1	m²	12.00			12.00		
	Waste 5%	0.05	m²	12.00			0.60		
	Timber packing and sole plates	0.04	m³	255.00			10.20		
	Waste 5%	0.002	m³	255.00			0.51		
	Nails	nil							
	Surface preparation	nil							
	Purchase cost of materials	1	m²				78.81		
Mat	Cost per use assuming 4 uses						19.70		
Plt	Falsework from specialist	6	wks	2.20		13.20			
Lab	Labourer erect falsework	0.7	hr	12.00	8.40				
	Carpenter f & s deck and troughs	1	hr	17.00	17.00				
	Labourer (one for four carpenters)	0.25	hr	12.00	3.00				
	Labourer dismantle falsework	0.3	hr	12.00	3.60				
	Rate for troughed formwork	1	m²		32.00	13.20	19.70		64.90
*	This price was calculated by adding the costs of the moulds								
	needed to make a typical pour, including end pieces.								
	The contractor's programme will show that sufficient moulds								
	will be needed to prepare the next bay while the first pour is curing,								
	to allow continuity of work								

Fig. 11.2.15 *Formwork to soffits pricing sheet*

CB CONSTRUCTION LIMITED PRICING NOTES

Project		Trade	FORMWORK	Date	
Ref. No.			Columns (operational pricing)	Sheet No.	

Typical bill description	E20 Formwork for in situ concrete Formwork with smooth finish type B to isolated columns; circular 340 mm diameter; 3 to 4.50 m high to soffit; in 74 nr .... 274 m²

item details					analysis				net
ref:	description	quant	unit	rate	lab	plt	mat	s/c	unit rate
	Quotation from supplier for circular column shutters:								
	£930 per shutter 340 mm dia 3.50 m long including delivery to site								
	£130 per kicker shutter 150 mm high								
Mat	The programme requires three column shutters								
	Column shutters	3	nr	930.00			2790		
	The programme calls for six column kicker shutters						0		
	Kicker shutters	6	nr	130.00			780		
	Equipment - props (4 x 3 cols)	15	wk	3.25		49			
	Consumables - mould oil	274	m²	0.75			206		
Lab	Carpenter fixes 3 columns per								
	day (74 nr/3) 8.5 hrs per day	25	days	144.50	3613				
	Labourer (one for two								
	carpenters)	12.5	days	102.00	1275				
	No credit value taken								
	Total for 74 columns (274 m²)				4888	49	3776		8713
	Rate for circular columns	÷ 274	1	m²	17.84	0.18	13.78		31.80

Fig. 11.2.16 *Formwork to columns (operational pricing) pricing sheet*

PRICING INFORMATION		BAR REINFORCEMENT

SMM7 NOTES **Work section E30** **Reinforcement for in-situ concrete**	CESMM3 NOTES **CLASS G** **CONCRETE ANCILLARIES [Reinforcement]**
1 Nominal size of bars is stated	Nominal size of bars is stated
2 Bars classified as straight, bent or curved	Bar shapes not given
3 Links are measured separately	Links not separately identified
4 Lengths over 12m to be given in 3m stages	Lengths over 12m to be given in 3m stages
5 Bar weights exclude rolling margin	Mass of steel assumed to be 7.85t/m³
6 Bar weights inc tying wire, spacers & chairs	Reinf items deemed to include supports to bars
.... only when at the discretion of the contractor	
7 Spacers & chairs measured in tonnes	Mass of reinforcement to include mass of chairs
.... where not at the discretion of the contractor	
8 Location of bars not given in description	Location of bars not given in description
9 Details of conc members given on drawings	

Bar size mm	MASS kg/m	WASTE %	WIRE kg/t	SPACERS nr/t	UNLOAD BARS hrs/t	FIXING TIMES (total hrs/t) FOUNDS	SLABS & BEAMS	LINKS	SITE CUT AND BEND
6	0.222	4.0	14	60	3.7	40	42	56	28
8	0.395	3.5	13	55	3.4	33	35	48	23
10	0.616	3.0	11	50	3.2	28	29	40	20
12	0.888	2.5	9	45	3.0	24	25	36	17
16	1.579	2.5	8	40	2.8	22	24	32	15
20	2.466	2.5	7	35	2.5	20	22		14
25	3.854	2.5	5	30	2.3	18	21		13
32	6.313	2.0	4	25	2.2	16	20		11
40	9.864	2.0	3	20	2.0	14	19		10
50	15.413	2.0	3	15	1.8	14	18		10

STRAIGHT BARS	Suppliers will quote lower prices for straight bars
	The fixing times for straight bars can be reduced by approximately 10–15%
DELIVERY COSTS	Basic prices normally include delivery costs
	Small loads can attract a delivery charge typically £25 for loads under 8 tonnes
SPECIAL SHAPES	Preferred shapes to BS 4466 normally included in bar prices
LONG LENGTHS	Additional charges are made for lengths over 12 m
SITE CUTTING	Bars are rarely cut and bent on site except In the case of late design information
	An additional cutting waste would be needed for site cut bars
NETT WEIGHTS	Will steel be charged at calculated weight or weight delivered?

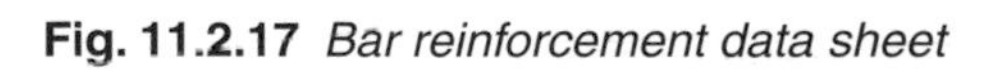

Fig. 11.2.17 *Bar reinforcement data sheet*

CB CONSTRUCTION LIMITED PRICING NOTES

Project		Trade	BAR REINFORCEMENT	Date	
Ref. No.				Sheet No.	

Typical bill description	E30 REINFORCEMENT FOR IN SITU CONCRETE Reinforcement bars grade 460 to BS 4449 Bars, 16mm nominal size, bent.....7.82 t

item details					analysis				net
ref:	description	quant	unit	rate	lab	plt	mat	s/c	unit rate
Mat	Steel per tonne as quotation	1	t	470.00			470.00		
	Waste	0.025	t	470.00			11.75		
	Wire	8	kg	1.20			9.60		
	Spacers	40	nr	0.36			14.40		
Lab	Unload & distribute (plant in prelims)	2.5	hr	12.00	30.00				
	Skilled fixers	22	hr	17.00	374.00				
	Labourer to assist (non-productive)	5.5	hr	12.00	66.00				
	Rate for bar reinforcement	1	t		470.00		505.75		975.75

Fig. 11.2.18 *Bar reinforcement pricing sheet*

PRICING INFORMATION		FABRIC REINFORCEMENT

SMM7 NOTES **Work section E30** **Reinforcement for in-situ concrete**	CESMM3 NOTES **CLASS G** **CONCRETE ANCILLARIES [Reinforcement]**
1 Fabric reference and weight/m² is stated	Fabric ref and weight/m² stated in 2kg/m² bands
2 Laps are not measured	Laps are not measured
3 Laps between sheets are stated	Laps between sheets not stated
4 Fabric inc tying wire, cutting, bending, spacers & chairs when at the discretion of the contractor	Fabric items deemed to include supports Supports to top rabric is measured
5 Location of fabric not given in description	Location of fabric not given in description

BS	MASS	UNLOAD	FIXING TIMES (total hr/m²)					FIXING (hr/m)	
Ref.	kg/m²	FABRIC hr/m²	LARGE BEDS	SMALL BEDS	SLABS	WALLS	BEAMS & COLS	RAKING CUTTING	CIRC CUTTING
A393	6.16	0.02	0.06	0.11	0.14	0.20	0.67	0.40	0.67
A252	3.95	0.01	0.05	0.09	0.12	0.17	0.55	0.32	0.53
A193	3.02	0.01	0.04	0.08	0.10	0.13	0.40	0.25	0.40
A142	2.22	0.01	0.03	0.06	0.08	0.11	0.32	0.22	0.32
A98	1.54	0.01	0.03	0.05	0.07	0.08	0.25	0.17	0.25
B1131	10.90	0.04	0.11	0.17	0.25	0.35	0.90	0.70	1.00
B785	8.14	0.03	0.08	0.14	0.18	0.26	0.80	0.50	0.70
B503	5.93	0.02	0.07	0.11	0.14	0.20	0.67	0.40	0.67
B385	4.53	0.02	0.06	0.10	0.13	0.18	0.60	0.36	0.60
B283	3.73	0.01	0.05	0.08	0.10	0.13	0.40	0.25	0.40
B196	3.05	0.01	0.04	0.06	0.00	0.10	0.30	0.20	0.30
C785	6.72	0.02	0.07	0.12	0.15	0.22	0.80	0.45	0.80
C636	5.55	0.02	0.06	0.11	0.14	0.20	0.70	0.41	0.70
C503	4.34	0.02	0.06	0.10	0.13	0.18	0.60	0.36	0.60
C385	3.41	0.01	0.05	0.08	0.10	0.13	0.45	0.30	0.45
C283	2.61	0.01	0.04	0.06	0.08	0.11	0.32	0.22	0.32
D98	1.54	0.01	0.03	0.05	0.07	0.10	0.30	0.18	0.25
D49	0.77	0.01	0.02	0.04	0.05	0.08	0.20	0.14	0.20

ALLOWANCE FOR WASTE	Large areas	2.50%	For a high proportion of cut sheets
	Small areas	5.00%	the waste must be calculated
ALLOWANCE FOR LAPS	150 laps	10%	
	225 laps	16%	
	300 laps	22%	
	400 laps	31%	
WIRE AND SPACERS	Large areas	2.50%	
	Small areas	5.00%	
CHAIRS	Typically	0.3–0.5 kg/m²	
OPERATIONAL CHECK	Laying fabric reinforcement in some large beds and slabs should be reconciled with labour for laying concrete		

Fig. 11.2.19 *Fabric reinforcement data sheet*

CB CONSTRUCTION LIMITED PRICING NOTES

Project		Trade	FABRIC REINFORCEMENT	Date	
Ref. No.				Sheet No.	

Typical bill description	E30 REINFORCEMENT FOR IN SITU CONCRETE Fabric reinforcement Ref A252; 3.95 kg/m²; 225 minimum laps....556 m²

item details					analysis				net
ref:	description	quant	unit	rate	lab	plt	mat	s/c	unit rate
Mat	Price from supplier	1	m²	2.45			2.45		
	Laps	0.16	m²	2.45			0.39		
	Waste	0.025	m²	2.45			0.06		
	Wire and spacers	0.025	m²	2.45			0.06		
	Chairs	0.5	kg	0.45			0.23		
Lab	Unload & distribute	0.01	hr	12.00	0.12				
	Labourer (see rate below)	0.05	hr	13.00	0.65				
	Rate for fabric reinforcement	1	m²		0.77		3.19		3.96
	Fixing gang for fabric in beds may be made up of:								
	4 labourers @ £12.00/hr	48.00							
	1 skilled fixer @ £17.00/hr	17.00							
	Effective rate (divide by 5)	65.00	£	13.00	/hr				

Fig. 11.2.20 *Fabric reinforcement pricing sheet*

PRICING INFORMATION	BRICKWORK

SMM7 NOTES **Work section F10** **Brick/Block walling**	CESMM3 NOTES **CLASS U** **BRICKWORK, BLOCKWORK AND MASONRY**
1 Type & nominal size brick stated	Type & nominal size of brick stated
2 Thickness of construction stated	Thickness of construction stated
3 Type of mortar stated	Type of mortar stated
4 Type of bond stated	Type of bond stated
5 Facework and number of sides stated	Surface finish and fair facing stated
6 Type of pointing stated	Type of pointing stated
7 Bonding to existing wall stated	Bonds to existing measured separately
8 Building overhand is stated	Building overhand not stated
9 Deemed to include all rough and fair cutting	
... where at discretion of contractor	
10 Deemed to include all mortices and chases	Rebates and chases measured separately
11 Deemed to include raking out joints to form key	
12 Deemed to include returns, ends and angles	
13 Deemed to include centering	
14 Walls include skins of hollow walls	Cavity or composite walls stated

LAYING TIMES for bricks 215 x 102 x 65 mm		COMMONS	SEMI-ENG	FACINGS	MORTAR (m³/m²) nett exc waste	no frog 15% waste	frog 15% waste
half brick	hr/m²	1.10	1.15	1.30	0.018	0.021	0.023
half brick fair one side	hr/m²	1.30	1.35	1.50	0.020	0.023	0.026
half brick fair both sides	hr/m²	1.50	1.55	1.70	0.022	0.025	0.029
one brick	hr/m²	2.00	2.10	2.40	0.046	0.053	0.060
one brick fair one side	hr/m²	2.20	2.30	2.60	0.048	0.055	0.062
one brick fair both sides	hr/m²	2.40	2.50	2.80	0.050	0.058	0.065
1½ brick	hr/m²	2.70	2.90	3.20	0.074	0.085	0.096
1½ brick fair one side	hr/m²	2.90	3.10	3.40	0.076	0.087	0.099
1½ brick fair both sides	hr/m²	3.10	3.30	3.60	0.078	0.090	0.101
Isolated piers	bricks/hr	30	25	20			
Projections	bricks/hr	45	40	35			
Arches	bricks/hr	25	20	15			
Specials	bricks/hr	25	20	15			
Brick coping	bricks/hr	40	35	30			
Unload by hand (lab) *	hr/thou	1.00	1.25	1.50			
Distribute (lab) **	hr/thou	1.00	1.25	1.50			

1 The above outputs are for a bricklayer in a gang of two bricklayers and one labourer (ratio 1:½)
2 Average waste allowance 5%
3 Waste can be more with facework, but reject bricks can be used elsewhere
4 Bucket handle, weather struck and raked pointing can take longer (add 0.15 hr/m²)
* Unloading can be omitted since most deliveries include mechanical off-loading
** Distribution of materials may be priced in the preliminaries

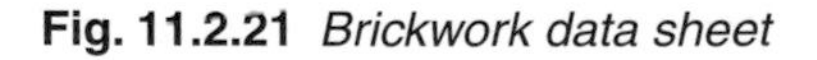

Fig. 11.2.21 *Brickwork data sheet*

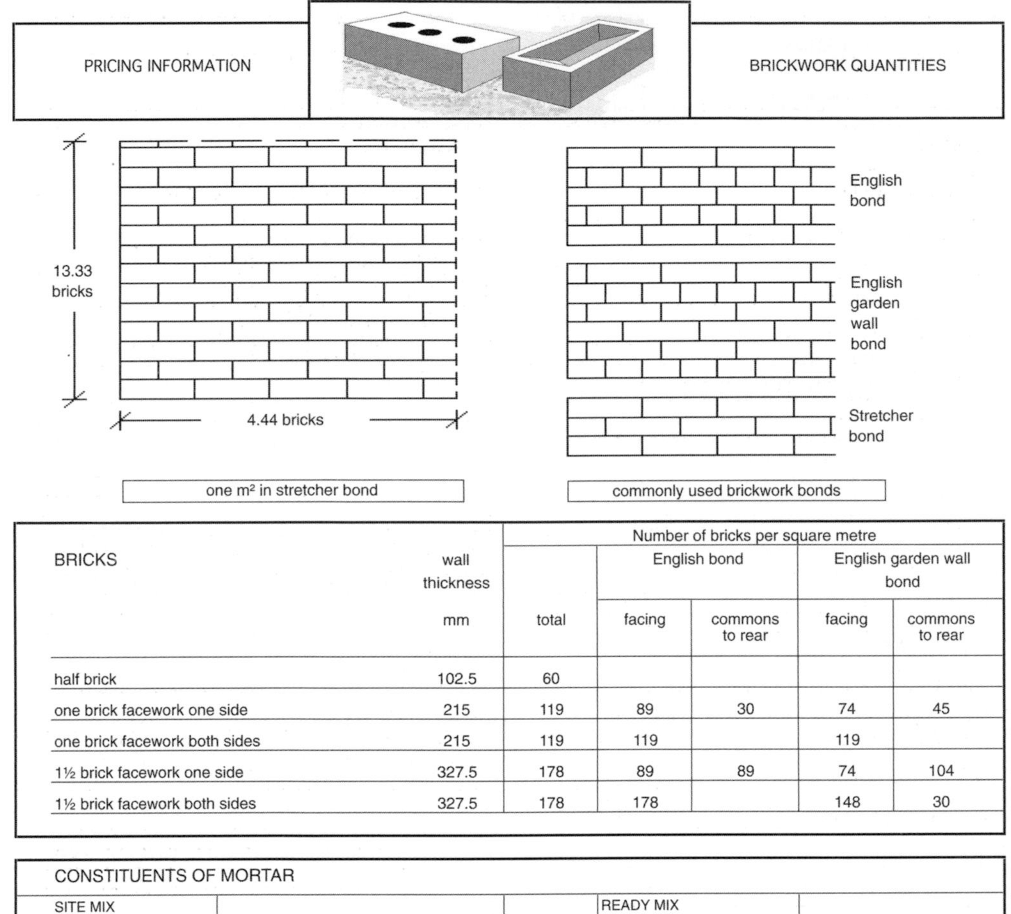

BRICKS	wall thickness mm	Number of bricks per square metre: total	English bond: facing	English bond: commons to rear	English garden wall bond: facing	English garden wall bond: commons to rear
half brick	102.5	60				
one brick facework one side	215	119	89	30	74	45
one brick facework both sides	215	119	119		119	
1½ brick facework one side	327.5	178	89	89	74	104
1½ brick facework both sides	327.5	178	178		148	30

CONSTITUENTS OF MORTAR

SITE MIX CEMENT/LIME/SAND MORTARS BY VOLUME	cement t/m^3	lime t/m^3	sand t/m^3	READY MIX LIME/SAND MORTARS BY VOLUME	cement t/m^3	LSM t/m^3
Bulk density	1.12-1.60	0.50-0.85	1.35-1.60			
Typical dry density	1.44	0.72	1.60	Typical dry density	1.44	1.85
1:3	0.49		1.62			
1:4	0.39		1.74			
1:5	0.32		1.83			
1:6	0.28		1.88			
1:1:6	0.26	0.13	1.71	1:6	0.24	1.75
1:2:9	0.18	0.18	1.75	1:9	0.17	1.85
1:3:12	0.12	0.18	1.75	1:12	0.13	1.85

Fig. 11.2.22 *Brickwork quantities data sheet*

PRICING INFORMATION	BRICKWORK SUNDRIES

	SMM7 NOTES **Work section F30** **Accessories/Sundry items**	CESMM3 NOTES **CLASS U** **BRICKWORK, BLOCKWORK AND MASONRY**
1	Closing cavities; width of cavities stated	Closing cavities; width of cavities stated
2	Bonding to existing; thickness stated	Bonds to existing work measured 'square'
3	Forming cavities; width and ties stated	Cavity construction stated
4	Damp proof courses measured 'square'	Damp proof courses measured 'linear'
5	Joint reinforcement; width stated	Joint reinforcement; width stated
6	Laps in DPC and joint reinf not measured	Laps in DPC and joint reinf not measured
7	Joints in walls measured where designed	Joints in walls measured where designed
8	Proprietary items	Fixings and ties
9	Pointing flashings incl cutting grooves	

Brickwork labours	
Forming cavities	0.03 hr/m²
Closing cavities vert	2.00 hr/m²
Closing cavities horiz	2.00 hr/m²
Bonding to existing	3.50 hr/m²
Prepare wall for raising	1.00 hr/m²
Wedging and pinning	1.25 hr/m²

Designed joints	
Fibreboard to joint ne 200	0.15 hr/m
Fibreboard to joint > 200	0.50 hr/m²
One-part mastic 10x10mm	0.20 hr/m
One-part mastic 20x20mm	0.40 hr/m
Two-part sealant 10x10mm	0.30 hr/m
Two-part sealant 20x20mm	0.60 hr/m

Hoist and bed lintels	
Small precast concrete	0.15 hr/m
Large precast concrete	0.30 hr/m
Small steel	0.20 hr/m
Large steel	0.25 hr/m

Accessories	
Build in butterfly tie	0.01 hr each
Build in twisted tie	0.02 hr each
Build in joist hanger	0.10 hr each
Build in joint reinforcement	0.40 hr/m²

Insulation	
Fix 25mm cavity bats (inc clips)	0.20 hr/m²
Fix 50mm cavity bats (inc clips)	0.25 hr/m²
Fix 75mm cavity bats (inc clips)	0.30 hr/m²

Damp-courses	ne 250	>250
	hr/m	hr/m²
Vertical	0.05	0.30
Raking	0.05	0.30
Horizontal	0.04	0.20
Stepped	0.08	0.40
Laps and waste	15%	10%
Pointing in flashings	0.40 hr/m	

Fig. 11.2.23 *Brickwork sundries data sheet*

CB CONSTRUCTION LIMITED PRICING NOTES

Project		Trade	BRICKWORK	Date	
Ref. No.				Sheet No.	

Typical bill description

F10 BRICK/BLOCK WALLING

Facing brickwork (PC £320 per thousand delivered and off-loaded); cement,lime,sand mortar 1:1:6. Vertical walls one brick thick; English bond; facework both sides with weathered joint as work proceeds .... 334 m²

item details					analysis				net unit rate
ref:	description	quant	unit	rate	lab	plt	mat	s/c	
Mat	For 1m³ of mortar 1:1:6								
	Cement	0.26	t	120.00			31.20		
	Lime	0.13	t	150.00			19.50		
	Sand	1.71	t	10.00			17.10		
Lab	Unloading by general gang	see prelims							
	Mixing by bricklayer's labourer								
Plt	Mixer and dumper	see prelims							
	Rate for 1:1:6 mortar	1	m³				67.80		67.80
Mat	Price from supplier (£340/th)	119	br	0.34			40.46		
	Waste	5.95	br	0.34			2.02		
	Mortar inc 15% waste	0.058	m³	67.80			3.93		
Lab	Distribute bricks (dumper in prelims)	0.17	hr	12.00	2.04				
*	Bricklayers 1 nr	2.80	hr	17.00	47.60				
*	Labourer 0.5 nr	1.40	hr	12.00	16.80				
	Extra for weathered joint both sides	0.3	hr	17.00	5.10				
	Rate for facing brickwork	1	m²		71.54		46.41		117.95
*	Often an effective rate is calculated as follows:								
	one bricklayer plus half labourer = 17.00+6.00 = £23.00/hr								
	in this example the rate for laying bricks would be								
	2.80 hrs @ £23.00 = £64.40/m² plus pointing								

Fig. 11.2.24 *Brickwork pricing sheet*

11.2.25/book4/data blockwork

PRICING INFORMATION		BLOCKWORK

SMM7 NOTES **Work section F10** **Brick/Block walling**	CESMM3 NOTES **CLASS U** **BRICKWORK, BLOCKWORK AND MASONRY**
1 Type & nominal size of block stated	Type & nominal size/thickness of block stated
2 Thickness of construction stated	
3 Type of mortar stated	Type of mortar stated
4 Type of bond stated	Type of bond stated
5 Facework and number of sides stated	Surface finish and fair facing stated
6 Type of pointing stated	Type of pointing stated
7 Bonding to existing wall stated	Bonds to existing measured separately
8 Building overhand is stated	
9 Deemed to include all rough and fair cutting	
... where at discretion of contractor	
10 Deemed to include all mortices and chases	Rebates and chases measured separately
11 Deemed to include raking out joints to form key	
12 Deemed to include returns, ends and angles	
13 Deemed to include centering	
14 Walls include skins of hallow walls	Cavity or composite walls stated

FIXING TIMES for blocks 415 x 215 mm			75	100	150	190	200	215
solid	Lightweight blocks	hrs/m²	0.45	0.50	0.60	0.65	0.70	0.80
	Dense concrete blocks	hrs/m²	0.60	0.65	0.80	0.90	1.00	1.15
	Masonry blocks	hrs/m²	0.75	0.80	1.00	1.20	1.25	1.40
hollow	Lightweight blocks	hrs/m²	0.42	0.46	0.55	0.60	0.62	0.65
	Dense concrete blocks	hrs/m²	0.58	0.61	0.67	0.75	0.80	0.85
	Masonry blocks	hrs/m²	0.50	0.58	0.61	0.65	0.67	0.75
	Unload and distibute (lab)	hrs/m²	0.05	0.07	0.10	0.13	0.13	0.14
	Extra for fairface one side	hrs/m²	0.13	0.13	0.13	0.13	0.13	0.13
	Extra for fairface both sides	hrs/m²	0.20	0.20	0.20	0.20	0.20	0.20
	Mortar quants (exc waste)	m³/m²	0.0050	0.0067	0.0100	0.0127	0.0133	0.0144
	Mortar quants (15% waste)	m³/m²	0.0058	0.0077	0.0115	0.0146	0.0153	0.0166

1 The above outputs are for a gang of two bricklayers and one labourer (ratio 1:½)

2 For heavy blocks marked with [shaded box] allow a labourer with every bricklayer (ratio 1:1)

3 Waste allowance 5%–7½% (except 10% for fairfaced blockwork and small areas)

4 Add to labour for high walls (30%) dwarf walls (35%) casings (60%) filling openings (70%)

Fig. 11.2.25 *Blockwork data sheet*

CB CONSTRUCTION LIMITED

PRICING NOTES

Project		Trade	BLOCKWORK	Date	
Ref. No.				Sheet No.	

Typical bill description

F10 BRICK/BLOCK WALLING

Dense concrete blocks 7N; cement and sand mortar 1:3; stretcher bond; vertical walls; facework one side; flush pointing; 140 mm thick .... 541 m²

item details					analysis				net
ref:	description	quant	unit	rate	lab	plt	mat	s/c	unit rate
Mat	For 1 m³ of mortar mix 1:3								
	Cement	0.49	t	120.00			58.80		
	Sand	1.62	t	10.00			16.20		
Lab	Unloading by general gang	see prelims							
	Mixing by bricklayer's labourer								
Plt	Mixer and dumper	see prelims							
	Rate for 1:3 mortar	1	m³				75.00		
Mat	Price of blocks from supplier	1	m²	10.00			10.00		
	Waste	0.05	m²	10.00			0.50		
	Mortar inc 15% waste	0.0115	m³	100.00			1.15		
Lab	Unload & distribute	0.1	hr	12.00	1.20				
	Bricklayers	0.8	hr	17.00	13.60				
	Labourer	0.4	hr	12.00	4.80				
	Bricklayer pointing one side	0.13	hr	17.00	2.21				
	Rate for blockwork	1	m²		21.81		11.65		33.46

Fig. 11.2.26 *Blockwork pricing sheet*

PRICING INFORMATION	STRUCTURAL TIMBER

SMM7 NOTES	CESMM3 NOTES
Work section G20 **Carpentry/Timber framing/First fixing**	**CLASS O & Z** **TIMBER**
1 Kind, quality and treatment of timber stated	Grade or species and treatment stated
2 Sawn or wrot timber stated	Sawn or wrot timber stated
3 All sizes are nominal	Nominal gross cross-sectional areas given
.... unless stated as finished sizes	Thickness of timber stated
4 Method of fixing and jointing given	Method of fixing and jointing not given
.... where not at the discretion of the contractor	
5 Labours on timbers not measured	Boring and cutting not measured
6 Lengths given if over 6 m	Length of timber given in one of 7 bands Class O

size of member		sectional area m²	plates floor/roof members	wall or partition members	pitched roof members	gutters fascias eaves soffit	supports		herringbone and block strutting	
width	depth						butted	framed		
							grounds, battens, firrings, fillets, upstands, drips etc			
			hr/m	hr/m	hr/m	hr/m	hr/m	hr/m	depth of joists mm	hr/m measured over joists
38	38	0.001	0.09	0.14	0.12	0.18	0.14	0.28		
	50	0.002	0.11	0.16	0.14	0.11	0.16	0.36		
	75	0.003	0.12	0.20	0.17	0.24	0.20	0.40	75	0.22
	100	0.004	0.13	0.22	0.18	0.26	0.22	0.44	100	0.25
	125	0.005	0.14	0.24	0.19	0.28	0.24	0.48	125	0.26
	150	0.006	0.15	0.26	0.22	0.30	0.26	0.52	150	0.27
	175	0.007	0.16	0.28	0.24	0.32	0.28	0.56	175	0.28
	200	0.008	0.17	0.30	0.25	0.34	0.30	0.60	200	0.30
50	50	0.003	0.12	0.20	0.17	0.24	0.20	0.40	225	0.32
	75	0.004	0.13	0.22	0.18	0.26	0.22	0.44	250	0.35
	100	0.005	0.14	0.24	0.19	0.28	0.24	0.48	300	0.40
	125	0.006	0.15	0.26	0.22	0.30	0.26	0.52		
	150	0.008	0.17	0.30	0.25	0.34	0.30	0.60		
	175	0.009	0.19	0.34	0.28	0.38	0.34	0.68		
	200	0.010	0.21	0.40	0.33	0.42	0.40	0.80		
	225	0.011	0.23			0.46				
	250	0.013	0.25			0.50				
	300	0.015	0.30			0.60				
75	100	0.008	0.17	0.29	0.25					
	125	0.009	0.19	0.34	0.28					
	150	0.011	0.23	0.50	0.40					
	175	0.013	0.25	0.55	0.46					
	200	0.015	0.30							
	250	0.019	0.37							
	300	0.023	0.40							
100	100	0.010	0.21	0.40	0.33					
	150	0.015	0.30	0.60	0.50					
	200	0.020	0.37							
	250	0.025	0.42							
	300	0.030	0.45							
150	150	0.023	0.40							
	200	0.030	0.45							
	300	0.045	0.50							

Nails can be priced in the preliminaries with a sum for sundry fixings. Alternatively, allow 2 kg/m³ of timber. For example: if nails cost £1.20/kg the rate for timber 50x100mm would be 2 x 1.2 x 0.5 x 1.0 = 1 p/m

The above outputs are for each carpenter

Labour assisting carpenters can be added to the all-in rate **or** priced in the preliminaries

Average waste for structural timbers generally 7.5%

Outputs will vary significantly depending on the complexity of the work

Fig. 11.2.27 *Structural timber data sheet*

PRICING INFORMATION	JOINERY

SMM7 NOTES

Work section N10 **General fixtures ..** **Work section P20** **Unframed isolated trims ..**		**Class Z** **SIMPLE BUILDING WORKS ..** **Carpentry and joinery**
P20	All timber sizes are nominal sizes unless stated	All timber sizes are nominal sizes unless
	as finished sizes	otherwise stated
	The work is deemed to include ends, angles,	Boring, cutting and jointing not measured
	mitres, intersections except hardwood items	
	over 0.003 m² sectional area	
	Kind, quality and treatment of timber stated	Grade or species and treatment of timber stated
	Sawn or wrot timber stated	Sawn or wrot timber stated
	Method of fixing given .. if not at the	Method of fixing not given
	discretion of the contractor	

Fixing softwood skirtings, architraves, trims, window boards etc Hr/m

size of member		Nailed	Screwed	Plugged &screwed	Screwed &pelleted	size of member		Nailed	Screwed	Plugged &screwed	Screwed &pelleted
19	**19**	0.10	0.13	0.18	0.23	**38**	**19**	0.12	0.15	0.22	0.25
	25	0.10	0.13	0.18	0.23		**25**	0.12	0.15	0.22	0.25
	32	0.10	0.13	0.18	0.23		**32**	0.12	0.15	0.22	0.25
	38	0.12	0.15	0.20	0.25		**38**	0.15	0.19	0.26	0.29
	44	0.12	0.15	0.20	0.25		**44**	0.15	0.19	0.26	0.29
	50	0.12	0.15	0.20	0.25		**50**	0.17	0.21	0.28	0.31
	63	0.12	0.15	0.20	0.25		**63**	0.19	0.24	0.31	0.34
	75	0.15	0.19	0.24	0.29		**75**	0.23	0.29	0.36	0.39
	100	0.17	0.21	0.26	0.31		**100**	0.29	0.36	0.43	0.46
	125	0.19	0.24	0.29	0.34		**125**	0.31	0.39	0.46	0.49
25	**19**	0.10	0.13	0.19	0.23	**50**	**19**	0.12	0.15	0.23	0.25
	25	0.10	0.13	0.19	0.23		**25**	0.12	0.15	0.23	0.25
	32	0.12	0.15	0.21	0.25		**32**	0.15	0.19	0.27	0.29
	38	0.12	0.15	0.21	0.25		**38**	0.17	0.21	0.29	0.31
	44	0.12	0.15	0.21	0.25		**44**	0.19	0.24	0.32	0.34
	50	0.12	0.15	0.21	0.25		**50**	0.19	0.24	0.32	0.34
	63	0.15	0.19	0.25	0.29		**63**	0.24	0.30	0.38	0.40
	75	0.17	0.21	0.27	0.31		**75**	0.29	0.36	0.44	0.46
	100	0.19	0.24	0.30	0.34		**100**	0.31	0.39	0.47	0.49
	125	0.23	0.29	0.35	0.39		**125**	0.4	0.50	0.58	0.60

The above outputs are for each carpenter

ADD 30% to the outputs for fixing hardwood

Average waste allowance is 7.5%. Varies depending on number of short lengths and mitres

ADD for cost of screws

WC CUBICLES	fix
Each door	1.25 hr
Each partition	2.00 hr
Each fascia panel	1.50 hr

KITCHEN UNITS		assemble	fix
Base unit	each	0.65 hr	0.50hr
Wall unit	each	0.55 hr	0.75hr
Worktop	metre		0.50hr

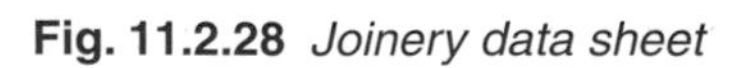

Fig. 11.2.28 *Joinery data sheet*

CB CONSTRUCTION LIMITED PRICING NOTES

Project		Trade	STRUCTURAL TIMBER	Date	
Ref. No.				Sheet No.	

Typical bill description	G20 Carpentry/Timber framing/First fixing Sawn stress graded (SS) timber, pressure impregnated with preservative; pitched roof members; 50 x 175 mm .... 174 m

item details					analysis				net
ref:	description	quant	unit	rate	lab	plt	mat	s/c	unit rate
Mat	Timber price from supplier	1	m	2.35			2.35		
	Nails	0.02	kg	1.50			0.03		
	Waste	0.075	m	2.35			0.18		
Lab	Unload and distribute	see prelims							
	Carpenter	0.28	hr	17.00	4.76				
	One labourer assisting four carpenters	0.07	hr	12.00	0.84				
	Unit rate for timber	1	m		5.60		2.56		8.16

Fig. 11.2.29 *Structural timber pricing sheet*

PRICING INFORMATION	WINDOWS AND DOORS

SMM7 NOTES **Work section L** **Windows/Doors/Stairs**	CESMM3 NOTES **CLASS Z** **SIMPLE BUILDING WORKS ..**
1 Kind and quality of materials given	Shape, size and limits of work given
2 Details given for treatment, tolerances, jointing, and fixing to vulnerable materials	Items are deemed to include fixings and drilling
3 Bedding and pointing frames measured	
4 Ironmongery, trims, surrounds, glazing and fixings deemed to be included where supplied with the component	Glazing is measured separately Ironmongery and frames may be included in items for doors and windows where clearly stated
5 Each leaf of multiple doors counted as a door	
6 Approximate weight is given for metal doors	
7 For glass supplied separately see L40	
8 Ironmongery (P21) includes matching screws	Materials stated for ironmongery
9 Nature of base for ironmongery is given	

FIXING TIMBER WINDOWS	Output - carpenter hr/unit (or perimeter length)					
	casement		box sash		roof windows	
	nr	*m*	nr	*m*	nr	*m*
Windows ne 2 m girth		*0.35*		*0.45*		*0.50*
2 - 4 m girth	1.00		1.40		1.55	
4 - 6 m girth	1.40		1.90		2.20	
over 6 m girth		*0.26*		*0.35*		*0.40*
Bedding frames		*0.10*				
Pointing frames		*0.15*				

FIXING TIMBER DOORS AND FRAMES		Output - carpenter hr/unit including hinges					
		standard		1 hr fire		ledged & braced	
		door	door set	door	door set	door	door set
Doors small *	*m²*	*1.10*	*1.70*	*1.35*	*2.10*	*1.00*	
762 x 1981	nr	1.50	2.60	2.00	3.40	1.40	
838 x 1981	nr	1.60	2.70	2.20	3.60	1.50	
large	*m²*	*0.90*	*1.50*	*1.30*	*2.00*	*0.85*	

* minimum 0.75 hr/door or 1.25 hr/door set

Frames (Jambs, heads and sills)		Output - carpenter hr/m		
		lining & stops	frame & stops	1 hr frame
38 mm thick	m	0.22	0.24	0.26
50 mm thick	m	0.25	0.26	0.30
63 mm thick	m	0.28	0.30	0.35
75 mm thick	m	0.30	0.34	0.40

SPECIFICATION	Specifiers commonly use manufacturers' references to provide technical requirements
LABOUR RATE	The effective rate for a carpenter is found by dividing the cost of a 'carpentry' gang by the number of carpenters in the gang; typically one labourer can service five carpenters
WASTE	Waste mainly due to damage to manufactured components, typically 2½%
	For lengths of doors stops and linings allow 7½%

Fig. 11.2.30 *Windows and doors data sheet*

PRICING INFORMATION	IRONMONGERY

* Hinges	
75 mm butts	0.13 hr/pr
100 mm butts	0.20 hr/pr
125 mm butts	0.25 hr/pr
Rising butts	0.25 hr/pr
300 mm T hinges	0.30 hr/pr
350 mm T hinges	0.35 hr/pr
Double action spring hinge	1.00 hr each

Door closers	
Perko	1.00 hr each
Overhead door closer	1.50 hr each
Single action floor spring	1.50 hr each
Double action floor spring	2.00 hr each
Door selector stay	0.75 hr each

Bolts	
100 mm barrel bolt	0.25 hr each
150 mm barrel bolt	0.33 hr each
200 mm barrel bolt	0.40 hr each
300 mm barrel bolt	0.50 hr each
100 mm flush bolt	0.55 hr each
150 mm flush bolt	0.65 hr each
200 mm flush bolt	0.75 hr each
300 mm flush bolt	1.00 hr each
single panic bolt	1.50 hr each
double panic bolt	2.00 hr each

Door accessories	
Door security chain	0.20 hr each
Door security viewer	0.35 hr each
Lever handles	0.40 hr/pr
Escutcheon	0.15 hr each
Letter plate (and slot)	1.25 hr each
100 mm cabin hook	0.20 hr each
Numerals	0.10 hr each

Wall fittings		
Shelf bracket	150 mm	0.15 hr each
	200 mm	0.17 hr each
	250 mm	0.20 hr each
Handrail bracket		0.17 hr each
Toilet roll holder		0.17 hr each
Bell push		0.20 hr each
Soap dispenser		0.20 hr each
Hat and coat hooks		0.13 hr each

Locks and latches	
Rim latch	0.75 hr each
Rim dead lock	0.75 hr each
Mortice latch	0.75 hr each
Mortice dead lock	0.75 hr each
Mortice deadlock & latch	1.00 hr each
EXTRA for rebated forends	0.50 hr each
Cylinder/night latch	0.75 hr each
Cabinet lock	0.75 hr each
Padlock, hasp and staple	0.40 hr each
WC/bathroom indicator bolt	1.00 hr each

Door handles and plates	
150 mm pull handle	0.13 hr each
225 mm pull handle	0.17 hr each
300 mm pull handle	0.25 hr each
150 mm flush handle	0.35 hr each
225 mm flush handle	0.60 hr each
300 mm flush handle	0.80 hr each
200 mm finger plate	0.20 hr each
300 mm finger plate	0.25 hr each
740 x 225 mm kicking plate	0.55 hr each
810 x 225 mm kicking plate	0.70 hr each

Window accessories	
Casement fastener	0.25 hr each
Casement stay	0.25 hr each
Mortice casement fastener	0.75 hr each
Sash fastener	0.40 hr each
Spiral sash balance	0.75 hr each
Sash pulley	0.50 hr each
Fanlight catch	0.25 hr each

Furniture accessories	
Cupboard catch	0.25 hr each
Magnetic catch	0.20 hr each
Cupboard knob	0.17 hr each
Cabinet handles	0.20 hr each
Curtain track	0.75 hr/m
Window blind	0.75 hr/m
Mirror 400 x 600 mm	0.60 hr each

Floor fittings	
Easyclean socket in concrete	0.40 hr each
Rubber door stop to timber	0.17 hr each
Rubber door stop to conc	0.30 hr each

WASTE	Allow 2½% for replacement of damaged ironmongery
FIXINGS	Check that ironmongery is supplied with matching screws
	Allow for sundry items such as cavity fixings for hollow backgrounds
* FIXING DOORS	The outputs for fixing doors include the fixing of hinges

Fig. 11.2.31 *Ironmongery data sheet*

CB CONSTRUCTION LIMITED PRICING NOTES

Project		Trade	DOORS & IRONMONGERY	Date	
Ref. No.				Sheet No.	

Typical bill description

L20 Timber doors/shutters/hatches

Edward Stockley type KL flush doors; plywood faced for painting; 44 x 762 x 1981 mm .. 27 nr

item details					analysis				net unit rate
ref:	description	quant	unit	rate	lab	plt	mat	s/c	
Mat	Door price from supplier	1	nr	90.00			90.00		
	Waste	0.025	nr	90.00			2.25		
Lab	Unload and distribute	see prelims							
	Carpenter	1.50	hr	17.00	25.50				
	Labourer assistance	0.20	hr	12.00	2.40				
	Unit rate for door	1	nr		27.90		92.25		120.15

Typical bill description

P21 IRONMONGERY

P & M Winfox Ltd Metric Gold range; single action overhead door closers; Ref: 55006 - to softwood .... 27 nr

ref:	description	quant	unit	rate	lab	plt	mat	s/c	net unit rate
Mat	Unit price from supplier inc screws	1	nr	43.25			43.25		
	Waste	0.025	nr	43.25			1.08		
Lab	Unload and distribute	see prelims							
	Carpenter	1.50	hr	17.00	25.50				
	Unit rate for overhead door closer	1	nr		25.50		44.33		69.83

Fig. 11.2.32 *Doors and ironmongery pricing sheet*

PRICING INFORMATION	PAINTING

SMM7 NOTES **Work section M60** **Painting/Clear finishing**	CESMM3 NOTES **CLASS V** **PAINTING**
1 Kind and quality of materials stated	Material to be used is stated
2 Nature of base & preparatory work given	Preparation normally deemed to be included
3 Priming, sealing & undercoats enumerated	Number of coats or film thickness given
4 Method of application given	
5 Rubbing down deemed to be included	
6 Work is internal unless otherwise stated	
7 Work in staircase areas and plantrooms stated	
8 Work to ceilings over 3.50 m stated (except stairs)	Height of work not given
9 Primary classification is the member painted	Primary classification is the type of paint
10 Secondary classification is the surface features	Secondary classification is the background material
11 No deduction for voids ne 0.50 m²	No deduction for voids ne 0.50 m²

	Output - operative hr/unit					
	girth > 300 mm (m²)		isolated ne 300 mm (m)		isolated ne 0.50 m² (nr)	
nr of coats	one	three	one	three	one	three
General surfaces (ADD 10% for ceilings) — Emulsion to plaster	*0.09*	*0.24*	*0.03*	*0.08*	*0.08*	*0.22*
to smooth conc	*0.10*	*0.28*	*0.03*	*0.09*	*0.09*	*0.25*
to board	*0.10*	*0.26*	*0.03*	*0.08*	*0.09*	*0.23*
to tex'd paper	*0.10*	*0.26*	*0.03*	*0.08*	*0.09*	*0.23*
to fair blockwork	*0.11*	*0.30*	*0.04*	*0.10*	*0.10*	*0.27*
Prepare and prime	*0.17*		*0.02*		*0.17*	
Undercoat	*0.15*		*0.02*		*0.15*	
Finishing	*0.17*		*0.03*		*0.17*	
1 pr, 1 uc, 1 fin		*0.45*		*0.14*		*0.41*
Glazed units — panes ne 0.10 m²	*0.24*	*0.68*	*0.04*	*0.22*	*0.21*	*0.61*
panes 0.10 - 0.50 m²	*0.20*	*0.56*	*0.03*	*0.18*	*0.18*	*0.50*
panes 0.50 - 1.00 m²	*0.18*	*0.50*	*0.03*	*0.16*	*0.16*	*0.45*
panes over 1.00 m²	*0.16*	*0.45*	*0.03*	*0.14*	*0.15*	*0.41*
Structural metalwork	*0.23*	*0.65*	*0.05*	*0.21*	*0.21*	*0.59*
Trusses and girders	*0.18*	*0.50*	*0.06*	*0.16*	*0.16*	*0.45*
Radiators	*0.19*	*0.55*	*0.03*	*0.17*	*0.18*	*0.50*
Fencing — Plain open	*0.13*	*0.35*	*0.02*	*0.11*	*0.12*	*0.32*
Close	*0.16*	*0.45*	*0.03*	*0.14*	*0.15*	*0.41*
Ornamental	*0.23*	*0.65*	*0.05*	*0.21*	*0.21*	*0.59*
Gutters	*0.19*	*0.55*	*0.03*	*0.17*	*0.18*	*0.50*
Services (eg. pipes and ducts)	*0.23*	*0.65*	*0.10*	*0.21*	*0.21*	*0.59*

PREPARATION	Washing down, rubbing down, and filling holes included in priming coat
LABOUR RATE	The effective rate for an operative is
HEIGHT ALLOWANCE	ADD 25% for working from ladders
	ADD 15% for working from scaffolding or staging
COVERAGE	Check with paint manufacturer particularly for porous surfaces
	On average 0.07 - 0.08 litres of emulsion or gloss required per m²
	but this could double with surfaces such as blockwork or soft boarding
'OLD' WORK	ADD 25% for painting previously decorated surfaces

Fig. 11.2.33 *Painting data sheet*

CB CONSTRUCTION LIMITED PRICING NOTES

Project		Trade	PAINTING	Date	
Ref. No.				Sheet No.	

Typical bill description

M60 PAINTING AND CLEAR FINISHING

Painting concrete - one mist coat and two full coats emulsion paint - as spec M60; general surfaces over 300 girth.. 270 m²

item details					analysis				net
ref:	description	quant	unit	rate	lab	plt	mat	s/c	unit rate
Mat	Emulsion paint	0.2	l	2.75			0.55		
	Brushes and sundries	1	item	0.10			0.10		
Lab	Painter	0.25	hr	17.00	4.25				
	Unit rate for emulsion painting	1	m²		4.25		0.65		4.90

Typical bill description

M60 PAINTING AND CLEAR FINISHING

Painting wood - one coat primer, one undercoat and one finishing coat - as spec M300; general surfaces; not exceeding 300 girth.. 64 m

ref:	description	quant	unit	rate	lab	plt	mat	s/c	net unit rate
Mat	Paints	0.24	l	3.75			0.90		
	Brushes and sundries	1	item	0.04			0.04		
Lab	Painter - prepare and prime	0.02	hr	17.00	0.34				
	Painter - undercoat	0.02	hr	17.00	0.34				
	Painter - finishing coat	0.03	hr	17.00	0.51				
	Unit rate for oil painting woodwork	1	m		1.19		0.94		2.13

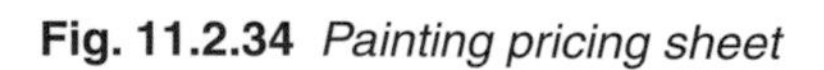

Fig. 11.2.34 *Painting pricing sheet*

PRICING INFORMATION	DRAINAGE PIPEWORK

SMM7 NOTES **Work section R12** **Drainage below ground**	CESMM3 NOTES **CLASS I** **PIPEWORK - PIPES**
1 Kind and quality and nominal size of pipes stated	Kind and quality and nominal size of pipes stated
2 Method of jointing pipes stated	Method of jointing pipes stated
3 Excavating trenches includes earthwork support,	Excavating trenches includes earthwork support,
.. consolidating trench bottoms, backfilling, and	.. surface preparation, pipework, backfilling, and
.. disposal of surplus excavated materials	.. disposal of surplus excavated materials
4 Backfilling with imported materials stated	Backfilling measured in CLASSES K and L
5 Location for disposal stated	Disposal normally at discretion of contractor
6 Average depth of trench given in 250 mm stages	Average depth to invert given in 500 mm bands
7 Difficult conditions or locations stated	Pipe locations identified in descriptions
8 Breaking out hard materials & reinstatement given	Crossings and reinstatement given in CLASS K
9 Dimensions of bed and surround given	Hard dig and beds/surrounds given in CLASS L
10 Pipe fittings measured extra over pipework	Pipe fittings and valves measured in CLASS J.

LAYING PIPEWORK		Labour and plant in laying and jointing pipes in trenches hr/m (pipes) hr/nr (fittings)							
		PVC push fit		CLAY push fit or sleeve		CONCRETE push fit		CHANNEL on mortar bed	
		lab	plt	lab	plt	lab	plt	lab	plt
100	pipe	0.10		0.25		0.25		0.40	
	bend	0.14		0.20		0.20		0.50	
	branch	0.28		0.40		0.40		0.55	
150	pipe	0.15		0.30		0.30		0.50	
	bend	0.18		0.25		0.25		0.65	
	branch	0.32		0.45		0.45		0.70	
225	pipe	0.25		0.40		0.40	0.07	0.50	
	bend	0.25		0.35		0.35	0.07	0.90	
	branch	0.39		0.55		0.55	0.10	1.00	
300	pipe	0.35		0.55	0.10	0.55	0.10	0.90	
	bend	0.39		0.55	0.10	0.55	0.10	1.40	
	branch	0.46		0.65	0.15	0.65	0.15	1.55	
375-450	pipe			0.70	0.25	0.70	0.25	1.05	0.20
	bend			0.80	0.25	0.80	0.25	2.10	0.20
	branch			1.05	0.35	1.05	0.35	2.30	0.30
525-600	pipe			0.90	0.35	0.90	0.35	1.45	0.30
	bend			0.90	0.35	0.90	0.35	2.35	0.30
	branch			1.30	0.45	1.30	0.45	2.60	0.40

LABOUR	Outputs are for drainlayer hours per metre of pipe
PLANT	Outputs are for the machine used to excavate the trench
ADDITIONS	ADD to outputs for filled mortar joints (30%); short lengths (30-50%); deep trenches (30-50%)
DEDUCTIONS	DEDUCT for long lengths (20%), shallow trenches (10%)
JOINTS	For joints which are NOT 'push fit', the laying and jointing calculations should be separated.
WASTE	WASTE is normally 5% on pipes and fittings, and 7.5% on short lengths and channels

Fig. 11.2.35 *Drainage pipework data sheet*

CB CONSTRUCTION LIMITED PRICING NOTES

Project		Trade	DRAINAGE EXCAVATION	Date	
Ref. No.			Unit rate pricing	Sheet No.	

Typical bill description

R12 Drainage below ground

Excavate trench for pipe not exceeding 200 mm nominal size; average depth of trench 750 mm; backfilling with excavated material; disposal of surplus off site ....174 m

item details					analysis				net
ref:	description	quant	unit	rate	lab	plt	mat	s/c	unit rate
	First calculate the rate for excavating and filling trenches								
	Backacter to excavate at 10 m³/hr	0.10	hr	28.00		2.80			
	Banksman	0.10	hr	12.00	1.20				
	Backacter to backfill at 8 m³/hr	0.12	hr	28.00		3.36			
	Two labourers to backfill	0.24	hr	12.00	2.88				
	Plate compactor to backfill	0.12	hr	1.25		0.15			
	EXTRA for taking 25% to tip								
	15.80-(2.88+0.15)	0.25	m³	12.77		3.19			
	cart away less backfill								
	Total rate	1	m³		4.08	9.50			13.58
	For a trench 600 x 750 mm deep, assume NO earthwork support								
	but a small over-dig of say 300 mm x 750 mm deep								
	so volume of excavation = 0.90 x 0.75 m = 0.67 m³								
	Excavate trench	0.67	m³	13.58	2.73	6.37			
	Trim and compact bottom								
	Labourer 10 m²/hr x 0.60 m wide	0.06	hr	12.00	0.72				
	Plate compactor to compact btm	0.12	hr	1.25		0.15			
	Additional cart away for the								
	volume of bed and surround								
	should be added here								
	0.75 x 0.45 for example	0.338	m³	15.80		5.33			
	Rate for drainage excavation	1	m		3.45	11.85			15.30

Fig. 11.2.36 *Drainage excavation pricing sheet*

CB CONSTRUCTION LIMITED PRICING NOTES

Project		Trade	DRAINAGE PIPEWORK	Date	
Ref. No.			Unit rate pricing	Sheet No.	

Typical bill description

R12 Drainage below ground

A. Granular material type A; bed and surround; to 150 mm pipe; 450 wide x 400 deep .... 174 m
B. Clay pipework with flexible joints; in trenches; 150 mm nominal size .... 174 m

item details					analysis				net
ref:	description	quant	unit	rate	lab	plt	mat	s/c	unit rate
A.	Bed and surround								
	First calculate the rate per m³								
Mat	Type A aggregate as quote	1.65	t	10.00			16.50		
	ADD 20% for consolidation and								
	penetration	0.33	t	10.00			3.30		
Lab	Labour previously priced in backfilling trenches								
Plt	Plant previously priced in backfilling trenches								
	EXTRA for disposal of volume occupied by								
	imported filling:	1	m³	15.80		15.80			
	Total	1	m³			15.80	19.80		35.60
	For a machine-excavated trench the minimum width is usually 600 mm								
	So the gross volume per metre is 600x400 = 0.24 m³/m								
	[larger pipe diameters merit a reduction for the volume occupied by the pipe itself]								
	Rate for bed and surround	0.24	m³			3.79	4.75		8.54
B.	150mm pipe in trench								
Mat	Pipe from price list with 15% discount	1	m	9.00			9.00		
	Waste 5%	0.05	m	9.00			0.45		
Lab	Drainlayer	0.3	hr	12.00	3.60				
Plt	This method assumes that the excavator								
	can be employed elsewhere when the								
	pipes are laid								
	Rate for 150 mm pipe	1	m		3.60		9.45		13.05

Fig. 11.2.37 *Drainage pipework pricing sheet*

PRICING INFORMATION	DRAINAGE MANHOLES

SMM7 NOTES	CESMM3 NOTES
Work section R12 **Drainage below ground**	**CLASS K** **PIPEWORK - MANHOLES AND PIPEWORK ANCILLARIES**
Excavation, concrete, formwork, reinforcement, brickwork and rendered coatings measured in accordance with the rules of the relevant work section of SMM7	Items for manholes shall be deemed to include: excavation, disposal, backfilling, upholding sides, concrete work, reinforcement, formwork, brickwork, metalwork, pipework inc backdrops
	Manholes enumerated depending on form of construction
Covers, step irons, channels, benching and building in pipes are enumerated	.... and depths which are given in 1.5 m stages
	Depths measured from tops of covers to tops of base slabs
	Types and loading duties of covers are given
	Hand dig is identified in items
	Hard dig given in CLASS L
	Pipe valves measured in CLASS J.

Typical spreadsheet approach to pricing manholes for a specific (civils) project

			diam	1200		1350		1500	
			depth	ne 1.50 m		ne 1.50 m		1.50-2.00 m	
			cover	A		A		B	
Description	unit	mat rate	lab/plt rate	Mat	Lab/Plt	Mat	Lab/Plt	Mat	Lab/Plt
Excavation	m^3		12.00		72.60		82.84		131.25
Disposal	m^3		15.80		28.44		35.99		62.21
Backfilling	m^3		3.50		14.88		16.19		24.50
Side support	m^2	1.00	1.00	24.20	24.20	27.61	27.61	43.75	43.75
Surface prep	m^2		0.80		1.15		1.46		1.80
Blinding	m^3	81.00	27.00	8.16	2.72	10.33	3.44	12.76	4.25
Base slab	m^3	82.00	22.50	23.62	6.48	29.89	8.20	36.90	10.13
Surround	m^3	81.00	18.50	77.76	17.76	87.48	19.98	97.20	22.20
Formwork	m^2		15.00		72.00		81.00		90.00
Chamber rings	nr			90.00	45.00	140.00	60.00	150.00	90.00
Shaft rings	nr								
Reducer slabs	nr								
Cover slabs	nr			75.00	25.00	100.00	31.00	140.00	41.00
Cover	nr			106.00	22.00	106.00	22.00	56.00	19.00
Benching	m^3	81.00	50.00	23.33	14.40	29.52	18.23	36.45	22.50
Channels	item			30.00	20.00	30.00	20.00	40.00	24.00
Brickwork	m^2	45.00	45.00	20.70	20.70	20.70	20.70	20.70	20.70
Sundries	item								
		totals	£	478.77	387.33	581.54	448.64	633.76	607.29

Fig. 11.2.38 *Drainage manholes pricing schedules*

PRICING INFORMATION	DRAINAGE TRENCHES

Width of trench (m)		Pipe sizes (mm)								
		100	150	225	300	375	450	525	600	750
depth	ne 1.50 m	0.60	0.60	0.70	0.70	0.90	1.10	1.20	1.30	1.45
	1.75-2.75 m	0.70	0.70	0.80	0.80	1.00	1.20	1.30	1.40	1.60
	3.00-4.00 m	0.80	0.80	0.90	0.90	1.10	1.30	1.40	1.50	1.70

Note : These are average widths, the actual widths of drainage trenches will depend on :-

1. The nature of the ground
2. The method of support to the sides of trenches
3. The width of bucket if dug by machine

Outputs for drainage gang excavating and laying drains incl backfill		Pipe sizes (mm)								
		100	150	225	300	375	450	525	600	750
Outputs for a drainage gang of one machine, two drainlayers & one labourer (m/day)										
depth (m)	0.50	35	34	33	30	26	25	23		
	0.75	34	33	32	29	25	23	21	18	
	1.00	32	31	30	28	23	21	19	16	13
	1.25	30	29	27	24	20	18	17	14	12
	1.50	26	25	24	20	16	15	14	12	11
	1.75	21	20	20	17	14	13	12	11	10
	2.00	18	17	16	15	12	11	10	9	8
	2.25	16	15	14	13	11	10	9	9	7
	2.50	15	14	13	12	10	9	8	8	7
	2.75	13	12	11	11	9	8	7	7	6
	3.00	12	11	10	10	8	7	6	6	6
	3.25	10	10	8	8	6	5	5	5	5
	3.50	9	9	8	8	6	5	5	5	5
	3.75	8	8	7	7	5	5	5	5	4
	4.00	7	7	6	6	5	4	4	4	4

Note : These are average production rates, the actual outputs will depend on:

1. The nature of the ground
2. The method of support to the sides of trenches
3. The length of drainage runs and location

If trenches need to be supported, the cost of hiring trench sheets can be added to the drainage gang rate.

For 20 m of trench with both sides supported there would be (20x2)/.33 = 122 sheets (330 mm wide) required

With a typical hire rate of £0.60 per week for a 2400 mm long sheet, the daily rate would be : 122x0.60/5 = £14.65 per day

The use of trench supports would lead to a reduced daily output by the drainage gang.

Fig. 11.2.39 *Drainage trenches data sheet*

CB CONSTRUCTION LIMITED PRICING NOTES

Project		Trade	DRAINAGE EXCAVATION	Date	
Ref. No.			Operational pricing	Sheet No.	

Typical bill description

CLASS I PIPEWORK - PIPES

I112 Clay pipes 150 mm dia in trenches, across farmland runs S2-S12; depth not exceeding 1.5 m .... 302 m
L331 150 mm bed and surround with 14 mm single sized granular material to 150 mm dia pipes .... 302 m

item details					analysis				net
ref:	description	quant	unit	rate	lab	plt	mat	s/c	unit rate
	For excavating a drain run including pipe, bed and surround								
	assume 25 m can be completed in one day by the following gang :								
Plt	Backacter	8.50	hr	28.00		238.00			
	Road tipper	2.00	hr	24.00		48.00			
	Plate compacter	8.50	hr	1.25		10.63			
	Trench sheets (see below)	1.00	day	24.32		24.32			
Lab	Labourers (3 nr)	25.50	hr	12.00	306.00				
Mat	Pipe from price list with 15% discount	25	m	7.70			192.50		
	Waste 5%	1.25	m	7.70			9.63		
	14mm stone						0.00		
	25 m x 0.60 m x 0.45 m x 2.10 t/m³	14.18	t	10.00			141.75		
	Rate for one day's work	25	m		306.00	320.95	343.88		970.82
	Rate for drainage (÷25)	1	m		12.24	12.84	13.76		38.83
	Trench sheets for 25 m = (25x2)/0.33 = 152								
	At £0.80 per week to hire								
	Daily hire rate would be : 152x0.80÷5 = £24.32								

Fig. 11.2.40 *Drainage excavation (operational pricing) pricing sheet*

For clarification the following points should also be kept in mind:

1. Most of the examples are for work measured using the rules of SMM7.
2. Each construction organization should decide how to deal with labour and plant in off-loading lorries and distributing materials on site; either in unit rates or preliminaries.
3. The pricing notes do not bring out the concept of gang sizes. The composition of a brickwork gang may be two bricklayers assisted by one labourer, in other words a 2:1 gang. This is written as the time for a bricklayer and half the amount of time for a labourer.
4. The headings SMALL, MEDIUM and LARGE refer to the quantity or size of an operation.
5. The labour rates used are £17.00 for skilled and £12.00 for unskilled operatives. These rates were realistic between June 2007 and May 2008.
6. Outputs for labour and plant represent average times.

12 Sub-contractors and market testing

Introduction

Sub-contractors can be classified in three main categories:

- Domestic sub-contractors who compete for each package at tender stage
- Domestic sub-contractors who are named by the client or his advisors
- Domestic sub-contractors who have a partnering agreement with the main contractor.

By the end of the 1990s the use of the formal nomination procedure disappeared. It has been replaced by lists of approved sub-contractors given in the tender documents, named sub-contractors where the Intermediate Form is used, and the novation of specialists for design and build schemes. (The word 'novation' means the substitution of a new obligation for an old obligation by the mutual consent of the parties. In construction procurement the term is used where the client has already completed negotiations with a sub-contractor or consultant and invites the contractor to enter the agreement.)

Domestic sub-contractors

The procedures for despatching enquiries to obtain quotations from sub-contractors were given earlier. Great care must be taken when estimating on the basis of sub-contract quotations because the contractor takes responsibility for all the work. It is therefore for all the parties to ensure that quotations are based on accurate and complete information.

All sub-contract quotations should be checked for arithmetical errors and totalled. To compare them on a like-for-like basis the following checks are carried out by the estimator:

1. All the items for that trade should be priced. If there is enough time, the sub-contractor should be asked to provide missing rates, otherwise the estimator needs to insert his own estimate of their value.

2. The rates should be realistic. If a patent error is detected then the sub-contractor should be advised to amend his quotation and tell all the main contractors who have received it.
3. It is sometimes argued (mainly by quantity surveyors) that rates should be consistent throughout the bill of quantities – like items should be priced at similar rates to avoid possible difficulties when valuing variations. This is not reasonable! Anyone vetting a tender must realize that the cost of similar items may vary depending on quantities, location, timing and so on.
4. The sub-contractor should accept the contract conditions without amendment. This will enable the estimator to make fair comparisons between quotations, and avoid any misunderstandings brought about by qualified bids. In practice, quotations are sent with many printed and specific conditions which may conflict with the enquiry documents. These details are often resolved at the negotiation stage.
5. The quotation should be based on the documents that form the main contract. The estimator should not accept a lump sum quotation for work, which will be valued on the basis of an approximate bill of quantities or accept a schedule of rates for a plan and specification project. If a sub-contractor has altered the tender documents, in the bill of quantities for example, there may be a mistake, which should be brought to the attention of the client so that all the contractors will correct the bill before the tenders are submitted.
6. There is a growing tendency for sub-contractors and suppliers to re-type the bills of quantity usually to accord with their interpretation or individual product range. The estimator must be sure that any changes do not represent a significant change to the contract requirements.

Quotations from specialists often need careful comparison using a standard form. The example sheet shown in Fig. 12.1 can be used to compare 'supply and fix' sub-contractors with labour-only contractors; the difference between the two is usually the cost of materials.

A computer can be a great help in comparing sub-contractors' quotations. Spreadsheet software is particularly useful for listing, and comparing rates, and provides a mathematical check. The spreadsheet method also allows rates to be adjusted before they are put in the estimate. Computer-aided estimating packages offer more powerful facilities, in particular:

1. The software will prompt the estimator by showing items that the sub-contractor should have priced.
2. Average rates can be inserted automatically when one of the sub-contractors fails to price an item. This facility can be very misleading, however. In some cases one sub-contractor may price an item at 'nil' because his costs have been allocated elsewhere. Figure 12.2 shows a computer comparison system where an average

CB CONSTRUCTION LIMITED		Project	LIFEBOAT STATION			Trade	STONEWORK
SUB-CONTRACT COMPARISON SHEET		Ref.no.	T384	Date	6.08	Page	1/1

					sub 1		sub 2		sub 3		LOSC		materials				
					PRELUDE STONE		DEANSTONE				CASTLE		RENARD QUARRY				
page	item	description	quant	unit	rate	£	rate	£	rate	£	rate	£	basic	sund	waste	rate	£
															2.5%		
4/13	C	Coping 550x175 mm	23	m	291.63	6 707	331.25	7 619			45.05	1 036	229.13	1.50	5.77	236.39	5 437
	D	Mitred angle	8	nr	35.00	280	inc				inc		22.50		0.56	23.06	185
4/10	E	Jamb 120x160x1114 mm	186	nr	133.75	24 878	156.88	29 179			90.13	16 763	70.63	1.25	1.80	73.67	13 703
	F	Jamb 135x160x1114 mm	38	nr	137.50	5 225	161.75	6 147			91.50	3 477	71.63	1.25	1.82	74.70	2 838
	G	Cill 225x175x839 mm	2	nr	206.38	413	217.00	434			112.63	225	127.78	2.25	3.25	133.28	267
	H	Cill 225x175x1014 mm	94	nr	235.73	22 158	244.13	22 948			112.63	10 587	157.13	2.25	3.98	163.36	15 356
	J	Head 230x225x1014 mm	15	nr	222.45	3 337	293.75	4 406			180.20	2 703	143.85	1.50	3.63	148.98	2 235
	K	Mullion	3	nr	232.81	698	244.13	732			180.20	541	137.50	1.88	3.48	142.86	429
4/11	A	Stainless steel dowels	333	nr	6.75	2 248	6.06	2 019			4.00	1 332	2.25	1.00	0.08	3.33	1 109
					£	65 944	£	73 484				36 664				£	41 559
												41 559					
											£	78 223	Total for labour and materials				

Fig. 12.1 *Example of a sub-contract comparison sheet*

CB CONSTRUCTION LIMITED **Comparison report**

Trade : Structural steelwork									
				Index structures		Flake & Mill		Jones Fabrication	
Item ref	Description	Qty	Unit	Rate	Total	Rate	Total	Rate	Total
1.3.27.a	Preliminaries	1	sum	4 000	4 000	3000.00	3 000	3325.00	3 325
1.3.27.b	Stanchions	33	t	1 295	42 735	1255.00	41 415	1325.00	43 725
1.3.27.c	Roof beams	31	t	1 295	40 145	1255.00	38 905	1320.00	40 920
1.3.27.d	Purlins	210	m	14	2 835	11.00	2 310	15.00	3 150
1.3.27.e	Holding-down bolts	96	nr	2	192	1.50	144	2.00	192
1.3.27.f	Erect steel on site	64	t	380	24 320	280.00	17 920	150.00	9 600
1.3.27.g	Prime steel at works	497	m^2	inc	0	8.00	3 976	4.00	1 988
	Totals				114 227		107 670		102 900

Assuming Jones did not price item g. an average rate has been inserted by computer - must be checked by estimator with sub-contractor

Fig. 12.2 *Computer comparison system showing problem of average rate inserted automatically*

rate has been inserted by the software but the result is a mistake which may lead someone to choose the wrong sub-contractor.

3. The chosen trade rates can be incorporated in the estimate at the touch of a button.

This assumes that the estimator wants to insert the rates as they stand. In some cases a lump sum may be added to certain rates or a percentage may be applied to others. The main reasons for changing the sub-contractor's rates are:

1. The sub-contractor might need specific builder's work not measured elsewhere, drilling holes through the building fabric being the most common example.
2. An estimator might decide to add certain attendances to the measured rate, such as scaffolding for an industrial door installer.
3. There may be specific trade requirements, which are customarily provided by the contractor. For example, a piling firm may ask for surplus soil to be removed by the main contractor, and a plasterer often expects the free use of a mechanical mixer.
4. Site overheads may be given as a separate item by a sub-contractor. In this case the estimator might wish to spread this sum across the net rates.
5. A margin for overheads and profit could be added to all or some of the rates, either because the contractor wishes to spread his overheads through the bill, or for tactical reasons such as work which appears to be undermeasured.

There is a slight danger that adding attendances and margin to rates may confuse site surveyors or buyers. This should not be a problem if staff understands the distinction between internal (net) allowances and the rates given in the client's (gross) bill. The former are target rates for buying materials and services, the latter being the value the contractor will be paid for his services. To keep matters simple, it is customary to deal with attendances and overheads in the assessment of a main contractor's site overheads.

Figure 12.3 shows the attendances to be provided by the main contractor, without charge to the sub-contractor. These attendances are defined in the form of sub-contract, and are normally priced in the preliminaries. There will be some sub-contractors, of course, who will need more than others. A cladding contractor, for example, will need a considerable amount of safety and access equipment whereas a plasterer may only need a small mixer and a supply of clean water.

Whenever a particular sub-contractor is used in the tender, an entry should be made on the summary of domestic sub-contracts (see Fig. 12.4). If a lower quotation is received later in the tender period, an adjustment can be made on this form and carried to the tender analysis reports presented to management at the final review meeting. For some contractors using computer-aided estimating systems, forms are

Example of a sub-contractor attendances checklist

Attendances	Main contractor	Sub-contractor
Site accommodation - hardstanding for all offices	yes	
Site accommodation - sub-contractor's office		yes
Site accommodation - welfare, safety and canteen	yes	
Site accommodation - stores		yes
Temporary roads and hardstandings	yes	
Temporary lighting and power to working areas	yes	
Labour - unloading materials amd plant		yes
Labour - transporting materials to workplace		yes
Labour - hoist driver	yes	
Labour - protecting finished work		yes
Labour - banksman for tower crane	yes	
Labour - banksman for mobile cranes		yes
Labour - clearing/cleaning		yes
Scaffolding - external elevations		yes
Scaffolding - local platform scaffolds	yes	
Scaffolding - internal scaffolding such as birdcage to atrium	yes	
Scaffold towers		yes
Scaffold for commissioning		yes
Scaffold - hoist towers and hoisting plant	yes	
Cranage - unloading lorries		yes
Cranage - tower cranes	yes	
Cranage - mobile cranes		yes
Rough terrain forklift		yes
Waste disposal - take waste to designated container on site		yes
Waste disposal - removing waste from site	yes	

Fig. 12.3 *Sub-contract types and attendances*

not filled in during the tender period because all data can be manipulated at any stage.

Nominated sub-contractors

The nomination procedure suffered from an elaborate set of conditions in the JCT98 contracts which had the effect of turning people away from the practice of nominating specialist sub-contractors and suppliers.

CB CONSTRUCTION LIMITED DOMESTIC SUB-CONTRACTORS SUMMARY

Project	Fast Transport		
Ref.No.	T354	Date	27.6.08

Ref.	Trade	Company	Quotation	Discount offered (%)	Net amount	Firm price Allowance	Alternative/late quotations Company	Net amount	Saving
S1	Roof covering	Beaufort Roofing	17 672	2.5	17 230	nil			
		Brett Technologies	19 560						
S2	Windows	Valley Fabrications	30 641	2.5	29 875	1 225	Archiglass	27 550	2 325
S3	Plumbing	Consort	4 550	nil	4 550	nil			
		Fisher Ltd	5 890						
S4	Plastering & partitions	Swift Services	57 990	nil	57 990	nil	Oscar Finishes	46 760	11 230
S5	Joinery	Projoin Site Services	41 900	nil	41 900	1 935	L.P.Monk	38 450	3 450
S6	Suspended ceilings	Hill Systems	19 882	2.5	19 385	nil			
		Busby Site Services	21 320						
S7	Painting	Tudor Decorations	12 659	2.5	12 343	nil			
S8	Floor coverings	Freedom Finishes	12 615	2.5	12 300	nil			
		Baker Group	14 870						
S9	Electrical installation	Comech Engineering	35 887	2.5	34 990	nil	Beta Technologies	22 860	12 130
		Shelton	36 500						
S10	Mechanical installation	Comech Engineering	25 667	2.5	25 025	nil			
		Shelton	27 410						
S11	Surfacing	W. Smith Contracting	11 800	nil	12 450	nil			
S12	Scaffolding	CCG Scaffolding	see prelims						
		Totals		£	268 037	3 160		£	29 135

Fig. 12.4 *A typical summary form for domestic sub-contractors in a tender*

The use of nominating sub-contractors has declined to the point of extinction because although the main contractor is contractually responsible for all the works there is a reduced liability for the work sub-let under the nomination system. The JCT98 Form of Contract makes the following provisions:

1. Delay by a nominated sub-contractor is a relevant event that can lead to an extension of time under clause 25.
2. Breach by the nominated sub-contractor imposes a duty on the architect to nominate a new sub-contractor if the first is incapable of performance.
3. Failure of design by a nominated sub-contractor under clause 35.
4. Delay caused by a nominated sub-contractor who gives late information.

Market testing rates for cost planned tenders

There are limitations to the amount of market testing that can be carried out with designs developed to RIBA Stages C/D. Nevertheless, a strategy is needed for engaging the supply chain and obtaining some price certainty. Some aspects to consider are:

- Earthworks, demolitions and asbestos removal are normally priced locally
- The supply chain manager should advise how an aggregate spend with other projects can be implemented to promote economies of scale
- Agree where responsibility for design is best placed and ensure it's not priced twice
- Consider design appraisal options. For example: steel vs pre cast vs *in situ* frames.
- Supply chains need to provide data on likely inflation in their rates
- Obtain prices for catering equipment from a specialist with design capability
- Get prices for preliminary items such as temporary accommodation and temporary electrics
- Speak to utilities about likely connection and upgrading costs.

The table of package values shown in Fig. 12.5 shows how a strategy can be set up to achieve market testing of 80% of trade cost.

	Package	£	Comments
	Total estimated cost of packages	**11 100 000**	
1	ME	2 770 000	Partner M+E specialist to provide cost plan advice and cost plan breakdown for use in tender.
2	Frame	1 250 000	Approx. quantities for concrete/steel frames and market tested rates
3	Civil works	1 500 000	Approx. costing from site plan; rates from supply chain or similar recent schemes
4	Envelope and windows	1 300 000	Approx. quantities and discussions needed with cladding specialists
5	FF&E	900 000	Quantities and rates from FF&E advisor
6	Catering equipment	125 000	Obtain quotation from specialist
7	Internal walls	500 000	Cost planning data and advice from market
8	Internal doors	210 000	Cost planning data, feedback data and approx quantities if details available
9	Flooring	400 000	Cost planning data, feedback data and approx quantities if details available
10	Lifts	65 000	Supply chain advice
	Market tested	9 020 000	Represents 81% market tested

Fig. 12.5 *Table of package values targeting 80% of net cost for a school project*

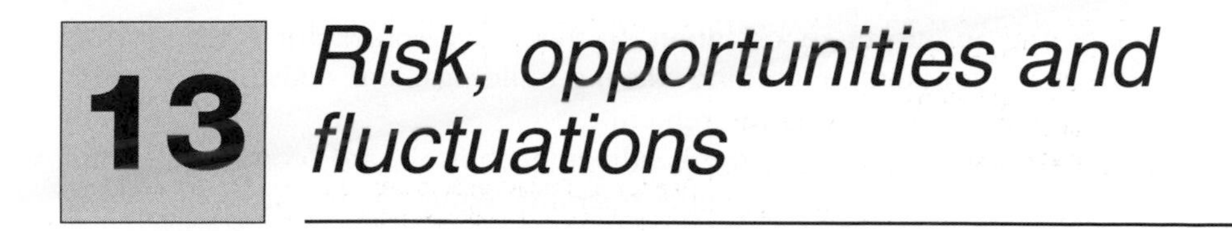

13 Risk, opportunities and fluctuations

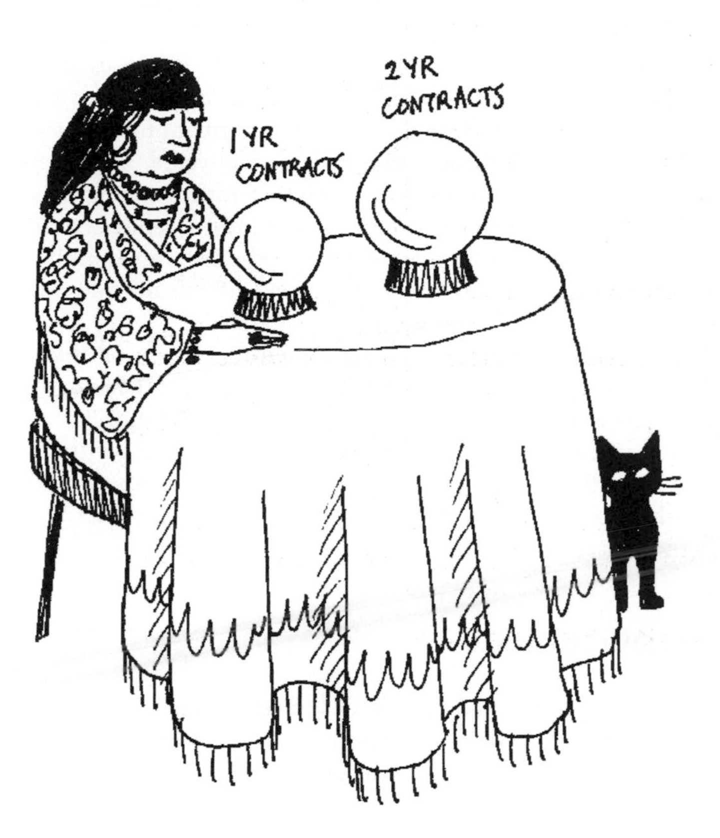

The inflation calculation

Introduction

An estimate is based on a level of information which can be incomplete or uncertain. Increasingly, contractors are engaging in the design and build process earlier. For example, many PFI and framework projects come to a contractor as a site plan, a project boundary and an output specification. In this way a contractor is responsible for dealing with site conditions, upgrading existing buildings, and delivering a building which achieves the performance requirements.

Risks were often considered under the headings 'technical' and 'commercial' but now include design, client interfaces, facilities operator requirements, life cycle, statutory requirements, energy targets and community issues.

For collaborative projects, there is a partnership between the contracting parties. This partnership relies heavily on the sharing of risks, which means there is a joint risk register where the party best placed to manage the risk retains responsibility for the risk and holds the fund.

In the majority of cases – all but the largest of projects where increases are recoverable – contractors must allow for changes in costs that occur during the construction phase. The amount which the estimator adds to the estimate for inflation is a guess, calculated after an examination of price trends over the previous few years, and discussions with suppliers and sub-contractors, in an attempt to predict future trends. If a contract is likely to last for several years then the employer will request a tender based on current prices and any changes will be reimbursed using the methods defined in the contract.

If there is a possibility that inflation will run ahead of the normal allowances, an item could be added to the risk register for 'super-inflation'.

The standard forms of contract have terms for either a *firm* or *fluctuating* price. A firm price is one which will not be varied for changes in the cost of resources, although labour tax fluctuations are usually reimbursed by the employer. In a fluctuating price tender the price is agreed before the job starts but the contract sum can be adjusted for changes in the costs of resources. An estimator needs to understand the clauses dealing with fluctuations so that he can tell the sub-contractors of the risks and calculate his own forecast of increased costs.

Risks and opportunities

For the early stages of a cost planned tender, designed to meet the client's affordability, the estimator will insert an average percentage for risk, which has been derived from feedback from previous tenders and projects on site.

Project	**Estimate package value**	**Tender risk allowance**	**%**	**Actual risk expended**	**%**
A	32 020 000	1 200 000	3.75	925 450	2.89
B	23 562 150	753 900	3.20	980 800	4.16
C	5 251 000	150 000	2.86	250 000	4.76
Totals	60 833 150	2 103 900	3.45	2 156 250	3.54

In this example, a sample of three projects shows the estimated risk allowances averaged at 3.45% and the out-turn requirement was 3.54%. For a new cost plan, the estimator might decide to use an allowance of 3.5% on net cost. Contractor's risk is often in the range 3–4% of trade cost, but varies with the individual site risks, the procurement route, which risks are retained by the client, and the stage of design development.

Risk management

Risk management is the process associated with identifying, analysing, planning, tracking and controlling project risks. In some modern forms of procurement, clients ask for risk schedules to be submitted with tender documents. The aim is to consider which risks are best managed by each party to the main contract. In framework agreements, for example, where contractors are reimbursed actual building costs (with a pain-and-gain share arrangement) the client might wish to manage the technical risk fund and the contractor is expected to manage commercial risks. This is because individual sites are chosen by the client (leading to variable technical risks) and contractual links in the supply chain are under the direct control of the contractor.

In any tender, risk management starts when the tender documents are received, when possible risks are identified and responsibilities are allocated to team members for managing risks and looking for opportunities. As an example, it might be found that there is a risk that skilled labour will not be available in an area, and a large project is being planned on an adjacent site. The risk can be reduced by making an allowance for transporting labour from outside the area. Alternatively, the design could be modified to accommodate off site manufactured elements of the building. Another solution would be to provide training in advance of the programmed activities. In this example the risk has been reduced (mitigated) but a residual risk will need to be managed.

A mitigation plan will decrease risk by lowering the probability of an occurrence. The residual risk could lead to a sum of money and a probability of its occurrence. So, if for example, the labour problem is not fully eliminated, a sum of money will be added to the risk register and a probability applied.

An opportunity is a future event that, should it occur, would lead to a favourable impact upon the project. As with risks there is an uncertainty with the possible occurrence of the event.

With both risks and opportunities, it is important to structure the tender submission in such a way that the risk mitigation and opportunities may be secured. As an example, if a contractor has assumed that a gas tank can be sited at ground level in a car park, he will make this statement as a condition of his offer. The risk of burying the tank is thus transferred to the client.

The submission document is also an opportunity to transfer a risk to the client. This is achieved through a carefully worded qualification or 'pricing assumption' in the tender documents.

Collaborative projects

In many instances today, educated clients do not want contractors to carry all the project risks. The rule is for the best-placed party to own and manage the risk. The following table shows how risks have been shared between the two parties to a contract.

Typical contractor's risks	Typical client's risks
Protecting buildings being refurbished and providing safe routes on site	Access restrictions impose by the client after the contract is signed
Digging out soft spots or providing additional fill than design expected	Unidentified asbestos and contamination beyond the scope of surveys
Bills of quantities do not fully represent the extent of the works	Suitability of existing buildings and incoming services to support the project
Extent of refurbishment might be greater than priced in estimate	Backlog maintenance and hidden defects beyond the contractor's responsibility
Increase in GIFA above current yardsticks is possible through 'design creep'	Changes to requirements leading to increased area
Foreseeable changes to legislation and practice guides	Unforeseeable changes to legislation and practice guides
Inflation	Super-inflation
Delay to planning consent could delay the start and additional staff costs could accrue	Planning consent might be delayed leading to delay beyond period for acceptance
Demolitions and site reparation take longer than planned	Client unable to give access to whole site at start of project

Risk workshops

During the tender period, a number of risk workshops are needed, depending on the project size, complexity and time for producing the tender. The minimum requirement is one meeting near the start and one before the final tender review (settlement) meeting.

The agendas are:

First risk workshop	
Identify and describe risks	90 min
Allocate roles and ownership of risks	15 min
Agree how risks are going to be managed	15 min
Second risk workshop	
Check for any new risks	
Consider programme implications	
Consider probability	4 hours
Consider minimum, most likely and maximum costs	
Set action plan to deal with the risks	
Monte Carlo risk analysis (during or after meeting)	

At each workshop, a cross-section of disciplines is needed. They represent the legal, estimating, planning, purchasing, design and commercial aspects of the bid. This attendance ensures that all aspects of the project are considered.

For larger design and build projects, attendees should include a representative for: area planning, architecture, structure, engineering services, ICT, FF&E, facilities management, asset renewal and financial modelling.

A **client risk** may be derived from contractor's risk; for example, where a contractor's risk may have limited liability, a client risk may capture the residual exposure. Clients do not get good value if contractors carry unlimited risk. So a full assessment is needed to see which risks apply and who should own them.

A **discrete risk** is a risk which may or may not happen, positive or negative, and has probability of less than 100%. If an event is likely to have a probability of 100% then it should be priced in the bill of quantities or included as a cost plan item.

Standard risks are often produced as checklists which can help identify typical risks under various headings. Unfortunately, this can be restrictive to the way people contribute, especially in the first workshop. It is therefore advisable to let people write short descriptions of risks on Post-it notes and these can be categorized by placing on wall charts.

Pricing risks

Pricing a risk should be seen as a method of last resort as it often has an adverse impact on competitiveness. A Monte Carlo simulation determines the combined effect of priced opportunities and risks. Used appropriately it allows the variable element of the bid price to be estimated and at the bid adjudication a risk allowance can be added to the base estimate.

Range of possible outcomes

The distributions which can be selected for each risk are determined by the project team first deciding the range of possible outcomes and then the likelihood of outcomes between the two extremes of the range. Experience has shown that although there are twenty basic shapes which can be modelled, the triangular is the most popular because it is most readily understood. In practice the normal distribution is the most likely distribution to occur in real life situations.

The most common distributions are:

- **Normal distribution** – used under circumstances where there is good knowledge of the out-turn costs, and where there is 95% confidence that the cost will fall within two standard deviations of the mean. If the risk may be affected by a number of unknown factors this is the best distribution to use.

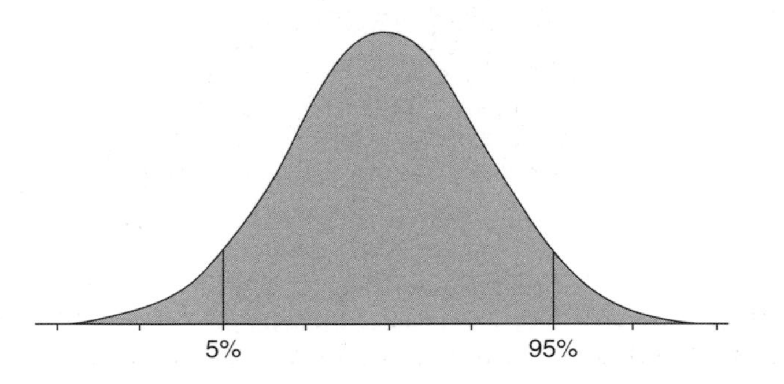

- **Skewed distribution** – where there may be an initial low cost which increases substantially with time. You can specify upper and lower limits for the risk, and a most likely value (which need not be in the middle of the range). See the triangular distribution.

- **Uniform distribution** – the risk cost does not vary. Every value between the absolute limits is equally likely. If the cost is known exactly, the uniform distribution is used.

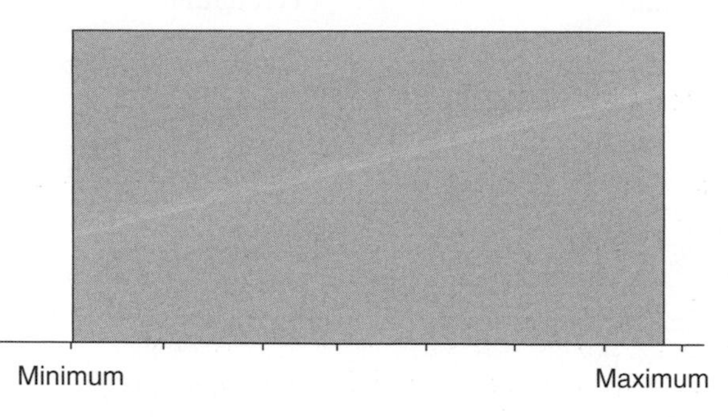

- **Triangular distribution** – this is easy to use, and relies on an assessment of best, most likely and worst outcomes.

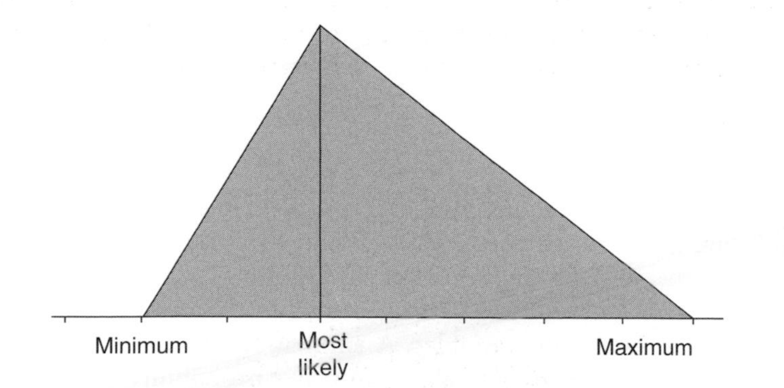

The easiest way to determine the correct distribution to use is to plot out the likely costs associated with different circumstances for the risk being considered.

It is important that the most appropriate distribution is used and that separate careful attention is given to the likelihood of the risk occurring (i.e. the probability). For example, if the project programme is tight there may be a 1 in 5 chance (20% probability) of the services installation being completed late. In going through the risk evaluation process consideration is given to the range of outcomes that could result (best, most likely and worst in the case of a triangular distribution). So for the example quoted it might be that the best outcome is to finish on time (with say two weeks' float) and the worst outcome is 3 months late. The most likely might be judged to be one month late. This equates (at a liquidated damages rate of £50,000 per month) to zero for the best case, £50,000 for the most likely and £150,000 for the worst case.

Comparison of common distributions

	Normal	Triangular	Uniform
Low	There is only a 5% chance that the value will be below this	The value cannot be less than this	The value cannot be less than this
Most likely	Calculated as half-way between low and high	The most likely value	Not applicable
High	There is only a 5% chance that the value will be above this	The value cannot be greater than this	The value cannot be greater than this
Low cut-off	The value cannot be less than this	Not applicable	Not applicable
High cut-off	The value cannot be greater than this	Not applicable	Not applicable

The value at risk and the probability are considered separately.

Monte Carlo simulation

Monte Carlo simulation (named after Monte Carlo, Monaco) is an analytical method used to examine a real-life situation, producing a range of outcomes, when normal mathematical modelling is too complex to produce. Without this method, people tend to arrive at a single outcome using the average position (probability × most likely cost).

Using a spreadsheet macro, the Monte Carlo simulation randomly generates hundreds of values from uncertain variables. A simulation calculates multiple scenarios of a model by repeatedly sampling values from the probability distributions for the uncertain variables and using those values for the cell. The two most common proprietary software packages are Crystal Ball and @RISK.

The following example (Fig. 13.1) shows a risks register which has been put through a Monte Carlo simulation resulting in three confidence levels: P10, P50 and P90. The P50 level gives a value of risk which has 50% certainty. If the project is to be virtually risk-free, then a P90 figure could be used, giving 90% certainty. It is unlikely that a company would choose a total risk figure with less than 50% probability.

Standard fluctuations clauses

The JCT Standard Building Contract lists three options under Schedule 7 for dealing with fluctuations. The choice is for one of three Parts: 1, 2, and 3. The estimator

Risk ID	Description	Identification and impact on programme	Prob.	Min.	Most likely	Max.	Distribution	Action/mitigation	Owner
01	Site conditions	Ground conditions not fully known	50%	£50k	£75k	£120k	Triangular	Qualify and dig trial holes at next stage of bid	TF
02	Existing buildings	Full extent of refurb works not known	60%	£50k	£200k	£350k	Triangular	Qualify extent of work based on condition surveys	HT
	Outputs	Confidence levels	P10	P50	P90				
				£210k	£430k	£760k			

Fig. 13.1 *Typical risks register*

should find this information in the preliminaries section of the bill of quantities. If a method has not been chosen then Part 1 shall apply (firm price).

There are, in fact, four possibilities for dealing with fluctuations:

1	No clause	Not recognized by the JCT contracts but would produce a firm price regardless of statutory or other changes. This is a very popular choice.
2	Part 1	This is the firm price alternative which allows some statutory changes, namely: contributions, levy and tax fluctuations
3	Part 2	Is the full fluctuation option allowing variations in prices of labour and materials?
4	Part 3	Is the 'formula method' where changes in costs are calculated by applying average indices prepared by national bodies published in monthly bulletins?

Part 2 is the traditional method for the recovery of increases in the costs of employing labour and buying materials, but is seldom used today. It is an attempt to calculate the actual increases or decreases in costs incurred by the contractor and his sub-contractors. This method leads to a great deal of work for quantity surveyors because the actual costs of construction must be compared with those at the date of tender. No increases are allowed for the contractor's profit.

Part 3 (the formula adjustment option) is not based on the actual cost changes as in Part 2; instead it uses the changes in indices published monthly by the DOE. There is therefore no need for a basic list of material prices and the administrative work is reduced. There are 49 work categories covering general building work and many of the common specialist activities. The PQS should assign the bill items to work categories so the contractor can assess the way in which increases will be dealt with when the work in valued. Fluctuations are not calculated for the following:

1. Credit for demolition materials.
2. Unfixed materials on site.
3. Plant paid for on a daywork basis (labour is reimbursed at the rate current when the work is carried out).
4. Claims (normally calculated at full value).
5. A non-adjustable element which is deducted from the increased costs payable by local authorities. The deduction which is normally 10% is made because it could be argued that contractors should not receive an addition on overheads and profit.

Works contractors engaged by management contractors under the terms of the JCT Works Contract conditions are reimbursed using one of three methods, which are similar to the JCT 80 clauses 38, 39 and 40. The JCT Management Contract itself does not have provisions for fluctuations, presumably because the management contractor is paid the prime cost of the work.

The ICE Conditions of Contract include a supplementary clause for contract price fluctuations, attached in loose-leaf form. Again the most common arrangement is for a firm price contract without fluctuations but including (under clause 69) provision for statutory labour taxes, levies or contributions. The Central Government form GC/Works/1 uses a similar arrangement for labour tax matters and supplementary conditions for fluctuations using the NEDO formulae method for recovery.

Calculation of non-recoverable increases

In broad terms the calculation shown below is necessary on every contract, including those with fluctuations clauses, because there is usually a shortfall in recovery of increased costs:

A. Forecast increases in costs of resources
B. Forecast amount recoverable
C. *Add* non-recoverable element (A–B) to tender.

Now that analytical estimating is widely used by estimators, fluctuations calculations can be dealt with after the bill of quantities has been rated – all the resources can be examined separately and a forecast of changes can be made. On the other hand, if the total labour element is not known, a wage increase adjustment can be made to the all-in rate before pricing begins.

The tender programme is an invaluable aid in forecasting cost increases, not only for the construction phase but also to calculate the effect of the time between date of tender and start on site. The period for which the tender is to remain open for acceptance should be as short as possible if the employer wants to receive an economical price for the work. Figure 13.2 shows the use of a tender programme in assessing fluctuations. No labour increase is expected in this example until the following June, which is after the project is completed. If the project started in March (Fig. 13.3), there would be a June increase in labour rates but no more costs for staff who have their salaries increased in January.

It can be difficult to forecast changes in costs accurately. A reasonable estimate can be made, however, if individual resources are treated separately, as follows.

Labour

The forecast of labour cost increases is the most predictable part of an estimate because wages change on the same date each year and the increase follows political and economic trends. Historical data can be plotted on a graph if a longer-term view is needed. For labour-only sub-contractors changes are difficult to anticipate and, at times of increased activity, change more quickly.

There are three ways to calculate expected labour costs throughout the currency of the contract:

1. Adjust the all-in rate before pricing the bill so all labour is priced on an average rate. This can be done as follows:

May–June	8 weeks at £16.00/hr	£128.00
July–September	12 weeks at £17.00/hr	£204.00
Total		£332.00
Average rate	£332.00/20	£16.60/hr

2. Increase the total labour costs by a percentage based on the approximate amount of work to be carried out before and after a wage increase.
3. Increase only the trades which are working after the wage increase by examining the tender programme.

Staff

The cost of staff can be split between the amounts before and after the annual review. A simple percentage can then be added to the total salaries after the review.

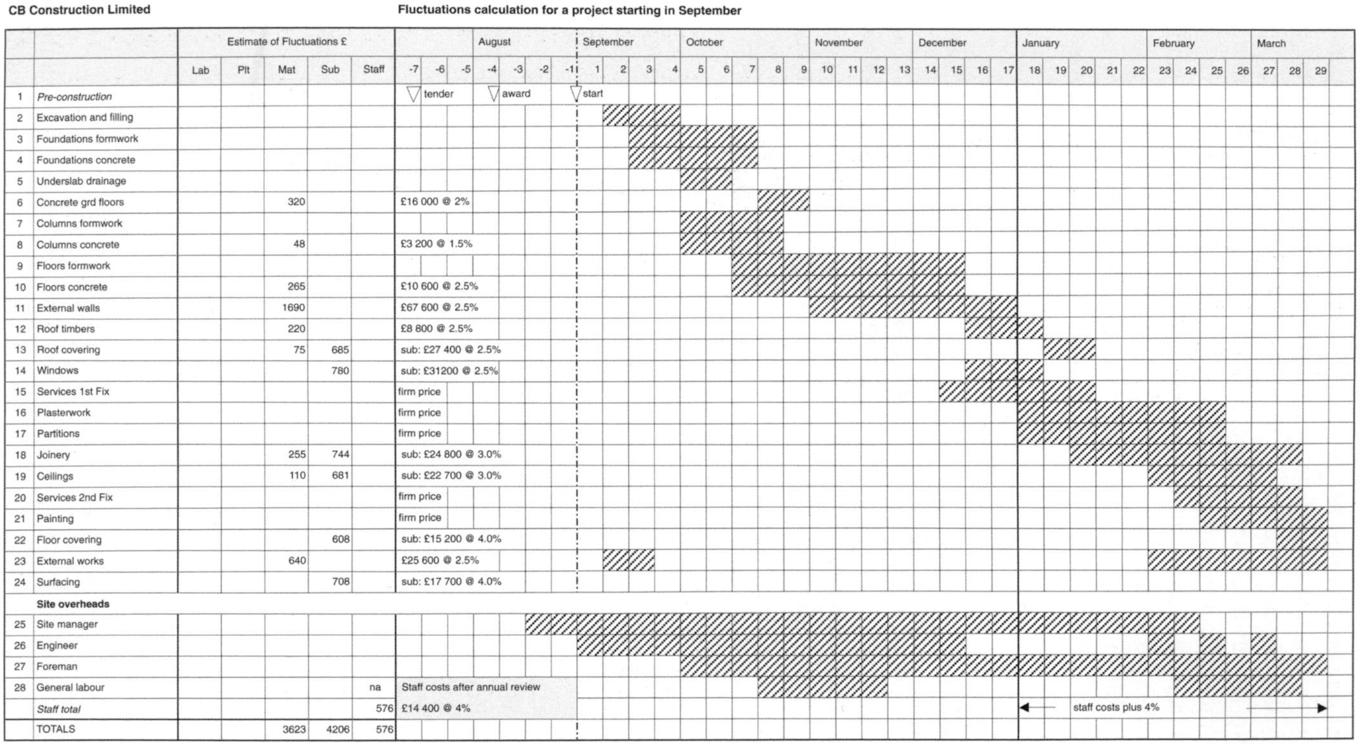

CB Construction Limited

Fluctuations calculation for a project starting in September

		Estimate of Fluctuations £					Schedule notes
		Lab	Plt	Mat	Sub	Staff	
1	*Pre-construction*						tender (-7); award (-4); start (-1)
2	Excavation and filling						
3	Foundations formwork						
4	Foundations concrete						
5	Underslab drainage						
6	Concrete grd floors			320			£16 000 @ 2%
7	Columns formwork						
8	Columns concrete			48			£3 200 @ 1.5%
9	Floors formwork						
10	Floors concrete			265			£10 600 @ 2.5%
11	External walls			1690			£67 600 @ 2.5%
12	Roof timbers			220			£8 800 @ 2.5%
13	Roof covering			75	685		sub: £27 400 @ 2.5%
14	Windows				780		sub: £31200 @ 2.5%
15	Services 1st Fix						firm price
16	Plasterwork						firm price
17	Partitions						firm price
18	Joinery			255	744		sub: £24 800 @ 3.0%
19	Ceilings			110	681		sub: £22 700 @ 3.0%
20	Services 2nd Fix						firm price
21	Painting						firm price
22	Floor covering				608		sub: £15 200 @ 4.0%
23	External works			640			£25 600 @ 2.5%
24	Surfacing				708		sub: £17 700 @ 4.0%
	Site overheads						
25	Site manager						
26	Engineer						
27	Foreman						
28	General labour					na	Staff costs after annual review
	Staff total					576	£14 400 @ 4%; staff costs plus 4% (January–March)
	TOTALS			3623	4206	576	

Fig. 13.2 *Calculations of fluctuations for a project starting in September*

CB Construction Limited **Fluctuations calculation for project starting in March**

		Estimate of Fluctuations £					
		Lab	Plt	Mat	Sub	Staff	
1	*Pre-construction*						
2	Excavation and filling						
3	Foundations formwork						
4	Foundations concrete						
5	Underslab drainage						
6	Concrete grd floors	174		320			Lab: £5 810 @ 3%
7	Columns formwork						
8	Columns concrete	36		48			Lab: £1200 @ 3%
9	Floors formwork						
10	Floors concrete	116		265			Lab: £3 850 @ 3%
11	External walls	1860		1690			Lab: £62 000 @ 3%
12	Roof timbers	74		220			Lab: £2 480 @ 3%
13	Roof covering	60		75	685		Lab: £2 000 @ 3%
14	Windows				780		
15	Services 1st Fix						
16	Plasterwork						
17	Partitions						
18	Joinery	132		255	744		Lab: £4 400 @ 3%
19	Ceilings	51		110	681		Lab: £1 700 @ 3%
20	Services 2nd Fix						
21	Painting						
22	Floor covering				608		
23	External works	708		640			Lab: £23 600 @ 3%
24	Surfacing				708		
	Site overheads						
25	Site manager					nil	
26	Engineer					nil	
27	Foreman					nil	
28	General labour	228					Lab: £7 600 @ 3%
	Staff total					nil	
	TOTALS	3439	0	3623	4206	0	

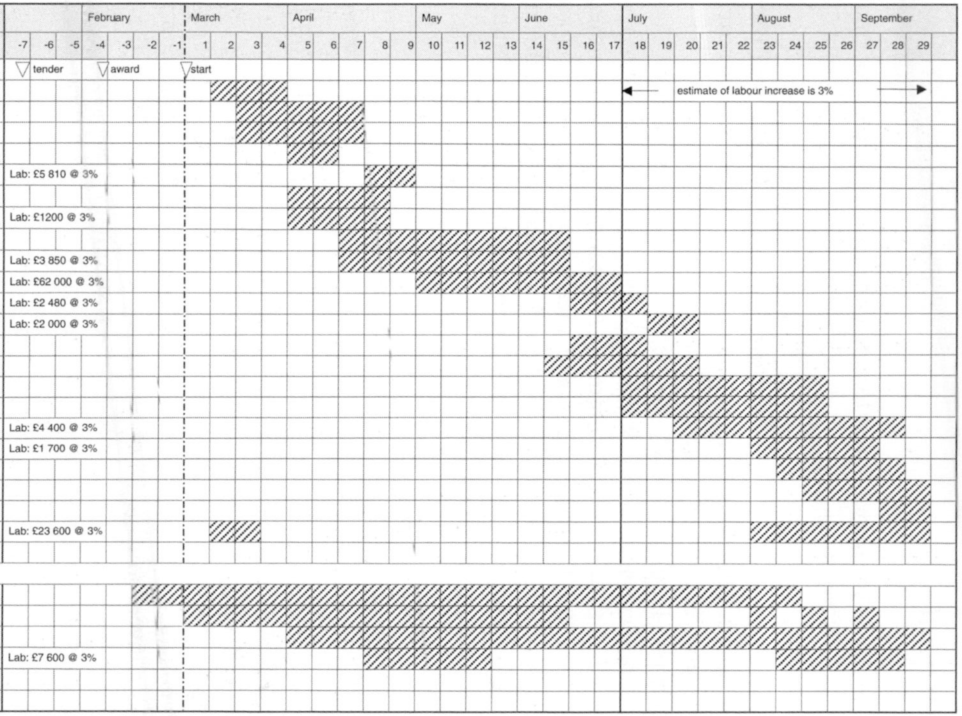

Weeks: February −7 to −1; March 1–4; April 5–9; May 10–13; June 14–17; July 18–22; August 23–26; September 27–29

tender (week −7) · award (week −4) · start (week 1)

← estimate of labour increase is 3% → (weeks 18–29)

Fig. 13.3 *Calculation of fluctuations for a project starting in March*

Materials and plant

There are two methods commonly used to assess the increased costs of materials and plant:

1. The estimator could assume a constant increase over the time that resources are purchased or hired, and determine the increases up to the mean purchase date. As a simple example, if the estimator uses a constant increase of 1% a month for a 12-month contract he might add 6% to the total value of materials.
2. A more accurate assessment can be made by looking at each material or item of plant abstracted in the estimator's summary. A separate decision can then be made for each resource for likely increases and their timing with respect to the programme. If concrete prices are due to increase, for example during the last third of a building project programme, there may be no need to add for inflation, because concrete is generally used for early activities.

Sub-contractors

All enquiries to sub-contractors must clearly state the rules for fluctuations. The estimator must check their quotations for compliance with the conditions. For some trades, such as surfacing, the estimator may have difficulty obtaining offers which fully comply with a request for a firm price tender. He will identify the problem and adjust the sub-contract value when he completes the domestic sub-contract register; an example is given on page 68 of the COEP. On the other hand, the firm price adjustments could be made on a sub-contract summary sheet (see Chapter 12), which lists all the sub-contract values used in the tender.

Firm-price tender and formula methods

For a firm-price tender, the estimator will add all the increases to his estimate, although in a strongly competitive market he may assume that certain price increases can be avoided or negotiated away.

For a contract which is based on a formula method, the estimator must add the non-adjustable element to his price.

During periods of low inflation (as experienced in the early years of the twenty-first century) allowances made by estimators for inflation have been re-assigned by site teams and added to margins. Unfortunately, when inflation rises faster than predicted at tender stage, losses can occur.

Inflation calculations for tenders with cost planning

In meeting a client's affordability, the estimator needs to be clear about when the budget was set. It might be necessary to apply an uplift to bring the budget up to current prices.

For example: the capital cost of a hospital has been published at £255m based on November 2005 prices and an index of 461. The indices for hospitals show that for January 2008 prices an index of 514 is required. So the adjusted price for a start in 1Q08 would be:

$$£255{,}000{,}000 \times 514/461 = £284{,}316{,}703$$

Year	Quarter	Indices quarterly	Indices % on year
2005	Q1	453	9.42
	Q2	461	8.47
	Q3	467	7.11
	Q4	461	3.83
2006	Q1	458	1.10
	Q2	467	1.30
	Q3	478	2.36
	Q4	483	4.77
2007	Q1	489	6.77
	Q2	494	5.78
	Q3	503	5.23
	Q4	508	5.18
2008	Q1	514	5.11
	Q2	519	5.06
	Q3	5.28	4.97
	Q4	533	4.92

In setting a timescale for a project, it has become common for indexing to be applied from 'start-to-start'. This means that the date used in the table of indices is for the start on site date. This assumes that the estimate includes sufficient inflation to achieve an out-turn price.

Standard forms of contract have terms for either a *firm* or *fluctuating* price. A firm price is one which will not be varied for changes in the cost of resources, although sometimes labour tax fluctuations can be reimbursed by the employer. In a fluctuating price (or 'variation of price') tender the price is agreed before the job starts but the contract sum can be adjusted for changes in the costs of resources. This is usually done using a nationally agreed formula applied to a breakdown of the original tender.

An estimator needs to understand the clauses dealing with fluctuations so that he can tell the sub-contractors of the risks and calculate his own forecast of increased costs.

Tender indexing is a well-established technique but suffers from small samples in certain sectors. For example, detailed tender breakdowns are not available for large PFI projects and framework/target prices produced at RIBA stage C. As a result, data for these projects can be unreliable.

In making adjustments for *forward* inflation, there are a number of methods used depending on the type cost information available. The following methods can be used:

Published indices	For example MIPS, PUBSEC, BCIS TPI	PUBSEC is a nationally published set of indices based on a statistical approach using data collected from a sample of public sector schemes. The index is smoothed in order to take out any irregular changes. MIPS based on PUBSEC and is used for hospitals. The BCIS undertake their own studies using data supplied by subscribers to the cost service.
Direct works	Using advice from suppliers	First build up rates using current rates. Then apply inflation percentages to all the items in the labour, plant and materials resource summaries.
Sub-contracted works	Obtain firm price tenders	Enquiries to sub-contractors include a programme for the works and a request for fixed prices.
Sub-contracted works	Advice from sub-contractors	The estimator might make inflation assumptions for each package using advice from sub-contractors.
Preliminaries	Adjust high cost items	Supervision cost increases are predictable. Sub-contracted items (see above) Insurances are usually based on current knowledge but future changes might be known.
Design fees	Design fees, LA charges and surveys	Design fees usually include increases for the proposed programme. A risk exists for slippage at tender stage and construction programme.
Overheads and profit	Applied after inflation has been added	Since inflation is applied before OH&P, it is not added to margin.

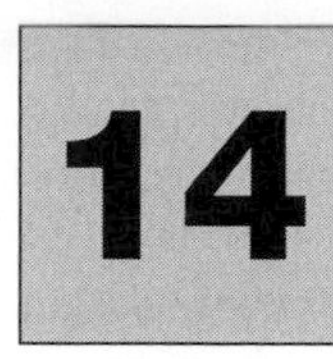

Provisional sums and dayworks

Introduction

Contractors have traditionally added the amounts for provisional sums and dayworks after the profit margin has been calculated. This is because, when provisional sums and dayworks are valued during the contract, the contractor receives reimbursement for overheads and profit. This changed in 1988 when SMM7 introduced the use of provisional sums for *defined work* and the JCT standard contract forms were amended accordingly.

Provisional sums for undefined work

Where the employer identifies that there is likely to be extra work for which there is no information at tender stage or it cannot be measured using the standard method of measurement, a provisional sum can be provided in the bill of quantities. The sum is spent at the direction of the architect (or engineer, ICE Conditions of Contract) and the work is valued in accordance with the valuation rules. There are two kinds of undefined provisional sum: a contingency sum, which is for work which cannot be identified at tender stage, usually for unforeseen circumstances; and sums for specific items, the extent of which is not known, such as more landscaping to a courtyard which has not been agreed with the client. The contractor adds these provisional sums to his tender after he has calculated his preliminaries and profit margin. SMM7 makes it clear that the contractor is entitled to any reasonable allowance for programming, planning and preliminaries. This is not just a financial compensation. JCT SBC (clause 2.29.2.1) gives the expenditure of a provisional sum for undefined work as a relevant event, which may lead to claim for an extension of time. Taken literally, this means that the estimator does not include undefined provisional sums when planning the work.

Provisional sums for defined work

SMM7 recognizes that there are certain items of work which cannot be measured using the standard method but could be taken into account by the contractor when he draws up his programme and calculates his preliminaries. It could be argued that a proportion of the provisional sum will be used to pay for head office overheads and profit. Simple examples would be providing a concrete access ramp for wheelchairs, or intumescent paint to roof trusses.

Contractors must be given more information about work in this category so that all the temporary works and overheads can be calculated. The question is: how much information must be given in the bill of quantities for a provisional sum for defined work? SMM7 states that the following must be provided:

1. The nature of the work.
2. How and where it is to be fixed.
3. Quantities showing the scope and extent of the work.
4. Limitations on method, sequence and timing.

Estimators have experienced problems with bills of quantities with provisional sums for defined work where the full extent of the temporary works is not clear. A typical case would be a provisional sum to replace defective windows in a multi-storey building. What assumptions should the contractor make for scaffolding? The defective windows could be found at high level, lower levels or throughout the building.

Dayworks

Construction contracts often involve changes from the original scheme. The term *variation* means alteration of the design, quality or quantity of the works, and can include changes in sequence or timing of the works. Where a variation occurs, the cost of the original work is deducted and new work is measured and priced by the quantity surveyor or engineer. The value of variations is determined according to the rules set out in the conditions of contract. The first method is to be by measurement using bill rates or a fair allowance added to the bill rates or by fair rates and prices. Where the work cannot be valued by measurement, it may be valued on a daywork basis, provided it is incidental to contract work. In practice, contractors are often asked to attach a value to a variation before the varied work is started and in some cases a term is incorporated in the contract so that agreement is reached before the work is carried out.

The daywork charges are calculated using the definitions prepared for building works by the RICS/Construction Confederation and civil engineering work by the Civil Engineering Contractors Association. The JCT Standard Building Contract states that where the valuation relates to additional or substituted work which cannot be properly valued by measurement, the valuation shall comprise: the prime cost of such work (calculated in accordance with the Definition of Prime Cost of Daywork carried out under a Building Contract issued by the Royal Institution of Chartered

Surveyors and the Construction Confederation which was current at Base Date) together with percentage additions to each section of the prime cost at the rates set out by the Contractor in the Contract Bills; or where the work is within the province of any specialist trade and the said Institution and the appropriate body representing the employers in that trade have agreed and issued a definition of prime cost of daywork, the prime cost of such work calculated in accordance with that definition which was current at the base date together with percentage additions on the prime cost at the rates set out by the contractor in the contract bills.

A footnote states that the RICS has agreement about the definition of prime cost with specialist trades associations – electrical and heating and ventilating associations. The contractor's daywork rates (or percentages) must take into account the rates required by the sub-contractors used in the tender.

The two industry definitions of prime cost of daywork state that the component parts which make up a prime cost are labour, materials and plant; with supplementary charges in the case of the civil engineering definition. The contractor adds for incidental costs, overheads and profit at tender stage, thus introducing competition into the daywork part of the tender:

- *Labour.* For building works, the hourly base rates for labour are calculated by dividing the annual prime cost of labour by the number of working hours per annum (see Fig. 14.1). The annual prime cost of labour comprises:
  (a) Guaranteed minimum weekly earnings.
  (b) Extra payments for skill.
  (c) Payments for public holidays.
  (d) Employer's National Insurance contributions.
  (e) Annual holiday credits.
  (f) Contributions to death benefit scheme.
  (g) Contribution, levy or tax payable by employer.
- *Materials.* The prime cost of materials is the invoice cost after deducting trade discounts, but includes cash discounts up to 5%. For civil engineering and government contracts the cash discount kept by the contractor cannot exceed 2.5%.
- *Plant.* The definitions include schedules for plant charges. They relate to plant already on site and the rates include the cost of fuel, maintenance and all consumables. Drivers and attendants are dealt with under the labour section.

In the case of the civil engineering definition the rates provide for head office charges and profit.

Overheads and profit

The anticipated value of daywork is included in a bill of quantities as provisional sums for labour, materials, plant – and supplementary charges in the case of civil

Calculation of prime cost of labour for daywork		days	hrs		
Working hours per week		39			
Working hours per year (x 52)			2028		
Annual holidays (enter days)		21	-164		
Public holidays (enter days)		8	-62		
Total working hours per annum			1802		
Calculation for 2003/04		Craft rate	Gen Op	Craft rate	General Operative
Weekly wage		379.08	285.09		
Annual costs for working hours		46.2	weeks	17 513.50	13 171.16
Extra for skill/hr	1802	0.30	0.40	540.54	720.72
Employer's Nat Ins (%) over ET		12.80	12.80	1 719.56	1 186.80
earnings threshold 46.2wks @ £100		*4 620*	*4 620*		
Holidays with pay (hrs x rate)	226	9.72	7.31	2 198.66	1 653.52
Maximum pension contribution	52	10.90	10.90	566.80	566.80
CITB levy (%) on PAYE payroll		0.50	0.50	101.26 #	77.73
		20 253	*15 545*		
Annual prime cost of labour				22 640.32	17 376.73
Hourly base rate divide by	1802	hours		**£12.57**	**£9.64**

In order to avoid the need for the labour rate to be built up during the project, some consultants ask contractors to submit an all-in rate for dayworks. This on-costs table allows the estimator to pick a rate which matches the on-costs requirement.

On-costs	Rates incl on-costs	
%	CRAFT	LAB
20	15.08	11.57
30	16.34	12.54
40	17.60	13.50
50	18.86	14.47
60	20.11	15.43
70	21.37	16.39
80	22.63	17.36
90	23.88	18.32
100	25.14	19.29
110	26.40	20.25
120	27.65	21.22
130	28.91	22.18

Fig. 14.1 *Estimator's spreadsheet for calculating the prime cost of building labour for daywork 2007/2008*

engineering work. The contractor is invited to add a percentage to each section for incidental costs, overheads and profit. The civil engineering definition states the percentage additions, but the contractor still has the opportunity to add or deduct from the percentages given.

Labour. The estimator must calculate the hourly base rate (see Fig. 14.1) and compare it with an 'all-in' rate which includes overheads and profit. The rate or percentage in a tender is an average for all types of labour regardless of trade or degree of supervision because the estimator has no idea what extra work will arise. One solution is to look at a similar job and compare the net cost of measured work with the tender sum. It might be found that the total mark-up including preliminaries was 30%. If the all-in rate for the current estimate is £16.23 then the gross all-in rate would be £16.23 × 1.30 = £21.10/hr. Figure 14.1 shows the comparable hourly base rate to be £12.57. This would suggest a percentage to be added to the hourly base rate of (21.10–12.57)/12.57 × 100 = 68%.

Figure 14.2 shows the incidental costs, overheads and profit items listed in Section 6 of the RICS definition and highlights the items which need to be added to the all-in rate. The overhead addition could also be found by comparing the daywork base rate with the all-in rate used in the estimate, as follows:

	Hourly base rate for labour (Fig. 14.1)	£12.57
	All-in hourly rate (Chapter 10, Fig. 10.1)	£16.23
		%
	% increase to recover the all-in rate	26
A	Head office charges	5
b	Site supervision	8
H	Travelling allowance	2
J	Contractors all risks insurance	2
o	Scaffolding	2
p	Site facilities and protection	3
r	Other liabilities	10
s	Profit	6
	Total to add to standard daywork rate	64%

The first calculation produced an answer of 68% and the second 63%. So why do some contractors want 110% and some specialists ask for over 150%? The reasons for such high percentages (given below) go some way towards answering this question:

1. The rates paid by contractors to labour-only sub-contractors are often higher than the all-in rate for direct employees, and, when the work is plentiful, the market rate for labour can be substantially higher.

Abstract from Section 6 of the RICS/Construction Confederation Definition of Dayworks – Incidental costs, overheads and profit

	Item	Included in all-in rate	Not included
a.	Head office charges	–	X
b.	Site staff including site supervision	–	X
c.	The additional cost of overtime (other than authorized)	–	X
d.	Time lost due to inclement weather	–	X
e.	Additional bonuses and incentive schemes	–	X
f.	Apprentices' study time	–	X
g.	Subsistence and periodic allowances	–	X
h.	Fares and travelling allowances	–	X
i.	Sick pay or insurances in respect thereof	X	–
j.	Third party and employer's liability insurance	–	X
k.	Liability for redundancy payments	–	X
l.	Employer's national insurance contributions	X	–
m.	Tool allowances	–	X
n.	Use, repair and sharpening of non-mech tools	–	X
o.	Use of erected scaffolding, staging, trestles and the like	–	X
p.	Use of tarpaulins, protective clothing, artificial lighting, safety and welfare facilities storage and the like that may be available on site	–	X
q.	Any variation to basic rates required by the Contractor in cases where the building contract provides for the use of a specified schedule of basic plant charges (to the extent that no other provision is made for such variation).	Applies to plant section	
r.	All other liabilities and obligations whatsoever not specifically referred to in this section nor chargeable under any other section	–	X
s.	Profit	–	X

Fig. 14.2 *Items to be added to the 'all-in' rates for labour*

2. Introducing variations into a normal work sequence can have a harmful effect on other work and the attitude of the workforce, particularly when changes make it difficult to earn the expected bonuses.
3. The rate quoted by the main contractor, for building work, must include the possibility of work being carried out by specialist sub-contractors; their operatives may be

earning higher rates of pay which are not recognized by the agreed definition, and specialist fitters often want a subsistence allowance while working away from home.

4. Contractors, again in the building industry, do not add the full percentages to materials and plant because they assume that labour is the major element of day-work; this means the labour percentage must carry the overheads which have not been added to materials and plant.

Materials. Most of the items for incidental costs, overheads and profit, listed in Fig. 14.2, do not apply to materials. Contractors therefore add a small percentage (usually between 10 and 15%) to cover head office overheads (a) and profit (s). The costs incurred in unloading and transporting materials around the site would be fully recoverable.

Plant. The rates for plant are given in the definition as provided by the contract. In building work, using the RICS/Construction Confederation definition, it is important to identify when the schedule of plant rates was produced in order to allow an additional percentage if the schedule is out of date. Section 6(q) of the RICS definition provides this opportunity. Schedules of plant charges usually cover a wide range of equipment, and apply to plant already on site. The costs which a contractor can claim are for the use of mechanical plant, transport (if plant is hired specifically for daywork) and non-mechanical equipment (except hand tools) for time employed on daywork. Labour operating plant is dealt with in the labour element.

The estimator must try to assess which are the most likely pieces of equipment to be used and compare the scheduled rates with those quoted by local plant hirers. The allowance for overheads and profit is commonly quoted between 10 and 15%.

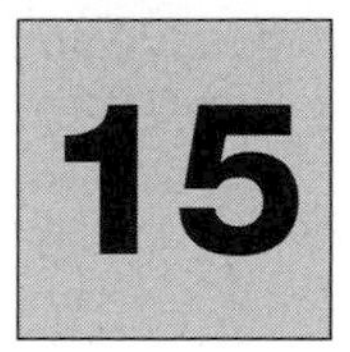

15 Preliminaries

Introduction

The preliminaries bill gives the contractor the opportunity to price preliminaries, which are defined in the Code of Estimating Practice (COEP) as: 'The site cost of administering a project and providing general plant, site staff, facilities and site-based services and other items not included in all-in rates.'

The standard methods of measurement for civil engineering and building give the general items which should be described in a bill of quantities, in two main parts: the specific requirements of the employer and the facilities which would be provided by the contractor to carry out the work. It could be argued that the latter is not really necessary in a bill of quantities because the contractor must provide general facilities whether they are measured or not. Presumably a simple breakdown of a contractor's general cost items is needed to make a fair valuation of the works during the construction phase. This approach to measurement can lead to duplication of descriptions, because an item may be required by a client and is something which a contractor would normally provide. It is common, in building for example, to find preliminary descriptions for security and protection of the works described twice.

Pricing preliminaries

For small repetitive works a contractor or sub-contractor may have a scale of overheads which he can apply to a new project. This may be calculated as a percentage of annual costs and adjusted where jobs deviate from the norm. Typically the site and office overheads for small houses and extensions may be 15%, and for sub-contractors who have facilities provided by main-contractors the figure would be nearer 10%.

Traditionally, in building, estimators have allowed for attendant labour, non-mechanical plant and certain items of mechanical plant in the rates inserted against measured work. It is becoming more common for these items to be considered as part of the general site overheads because very often these facilities are available to all trades and should be assessed using the tender programme.

A typical sequence of events for pricing preliminaries is:

1. Make notes of general requirements, such as temporary works and sub-contract attendances, when pricing the bill.
2. Prepare a site layout drawing showing the position of accommodation, access routes, storage areas, and services. Inspect site features and check feasibility of proposals during site visit.
3. Use the tender programme for planning staff, plant and temporary works requirements.
4. Read the client's specific requirements and all the tender documents.
5. Price the preliminaries sheets.

It is essential that the contractor has standard sheets which give all the main headings for pricing preliminaries. He cannot depend on the descriptions given in the tender documents because they are not necessarily complete. The COEP offers a comprehensive set of forms, but some estimators find there is too much detail for the average job and so a simplified checklist is given in Fig. 15.1. A detailed examination of the items in the checklist is presented in Figs. 15.2(a)–(j).

A resourced programme is an important aid to the accurate pricing of preliminaries because most of the general facilities are related to when the construction activities are carried out. The estimator or planning engineer can superimpose the main elements of the preliminaries on the tender programme. The example of a tender programme given in Chapter 9 shows staff and principal plant durations. Other items drawn from the programme are general plant, scaffolding, fluctuations, attendant labour, temporary works, traffic management and so on. There is an opportunity here for the contractor to be innovative and develop methods which might give a competitive advantage over other tenderers. A typical example is for wall cladding to be fixed by men working from mechanical platforms as opposed to standing scaffolding. This not only reduces the equipment costs but also cuts the overall contract duration with shorter erection and dismantling periods. Shorter programmes bring about further savings by reducing the staff, overheads and accommodation costs.

For large projects, contractors will obtain quotations for temporary facilities such as site accommodation, temporary electrics, scaffolding and cranage. Smaller and medium-sized projects can be priced using simple spreadsheet template – an example is shown in Fig. 15.3. The example of preliminaries given in Fig. 15.3 was produced by a regional contractor for a steel-framed three-storey student accommodation building in the UK. Net trade value is the direct cost of construction work excluding design fees, preliminaries, risk and margin.

Employer's requirements
SMM7 (A36) CESMM (A2)
Accommodation
Furniture
Telephone
Equipment
Transport
Attendance

Management and staff
SMM7 (A32,40) CESMM (A3.7)
Site manager
General foreman
Engineer
Planning engineer
Foreman
Assistant engineer
Quantity surveyor
Assistant quantity surveyor
Clerk/typist
Security/watchman

Facilities and services
SMM7 (A34,42) CESMM (A3.2)
Power/lighting/heating
Water
Telephones
Stationery and postage
Office equipment
Computers
Humidity/temperature control
Security and safety measures
Temporary electrics
Waste skips

Mechanical plant
SMM7 (A43) CESMM (A3.3)
Crane and driver
Hoist
Dumper
Forklift
Tractor and trailer
Mixer
Concrete finishing equipment
Compressor and tools
Pumps
Fuel and transport for plant

Temporary works
SMM7 (A36,44) CESMM (A2,3)
Access routes
Hardstandings
Traffic control
De-watering
Hoarding
Fencing
Notice board
Shoring and centring
Temporary structures
Protection

Site accommodation
SMM7 (A36,41) CESMM (3.1)
Offices
Stores
Canteen/welfare
Toilets
Drying and first aid
Workshops and laboratories
Foundations and drainage
Rates and charges
Erection and fitting out
Furniture
Removal
Transport

Attendant labour
SMM7 (A42) CESMM (3.7)
Unloading and distribution
Cleaning
Setting-out assistants
Drivers and pump attendance
General attendance
Scaffold adaptation

Contract conditions
SMM7 (A20) CESMM (A1)
Fluctuations
Insurances
Bonds
Warranties
Special conditions
Professional fees

Non-mechanical plant
SMM7 (A44) CESMM (A3.6)
External scaffolding
Internal scaffolding
Hoist towers
Mobile towers
Small tools and equipment
Surveying instruments

Miscellaneous
SMM7 (A33,35)
Setting out consumables
Testing and samples
Winter working
Quality assurance
Site limitations
Protective clothing

Fig. 15.1 *Preliminaries checklist*

Employer's requirements	
Accommodation	Offices, toilets, conference room, stores, laboratories and car parking space may be required depending on client's specific requirements
Furniture	If none stated, assume client providing own furniture
Telephone	Telephone and facsimile equipment can be specified including payment of standing charges (call charges are given as a provisional sum)
Equipment	Technical testing equipment and surveying instruments Protective clothing
Transport	Vehicles for employer's staff or consultants, fuel and maintenance Transport to suppliers to inspect production of components
Attendance	Drivers, chainmen, office cleaners, and laboratory assistants

NOTES:

SMM7 states that, where the employer requires accommodation on site, heating, lighting and maintenance are deemed to be included.

Notice boards are often given as a specific requirment but invariably will be provided by a contractor for information and advertising. (See Temporary Works)

Fig. 15.2a *Pricing preliminaries – employer's requirements*

Pricing the preliminaries bill

Both SMM7 and CESMM3 recommend that 'fixed' and 'time-related' charges are identified separately in a bill of quantities. SMM7 defines them as follows:

1. A fixed charge is for work the cost of which is to be considered as independent of duration.
2. A time-related charge is for work the cost of which is to be considered as dependent on duration.

Management and staff	
Site Manager	Required on most sites. Calibre of staff depends on size and complexity of project
General Foreman	Day to day management of labour and plant. Coordination of labour-only sub-contractors
Engineer	Analysis of building methods, setting-out and quality control. Services engineer to coordinate specialist services contractors
Planning Engineer	Master programme during mobilization. Up-dating exercises and short-term programmes
Foreman	Consider structure, finishings and snagging. Add non-productive time for trades foremen
Assistant Engineer	Setting out work, external works and internal fabric. Scheduling materials and attendance on sub-contractors
Quantity Surveyor	Some involvement on all jobs, particularly at beginning and end
Assistant quantity surveyor	For large contracts with complex valuations and control of sub-contractors' accounts and bonus payments
Clerk/Typist	General admin duties on site. Checkers. Telephone, post and reception duties
Security/Watchman	Usually employing security services. Important when fittings and furniture arrive

NOTES:

All-in average rates for each category of staff will be provided by senior management in a way which ensures that individual salaries are not identifiable. Employment costs for slaried staff are calculated on an annual basis with additional costs which arise from:

- pension scheme (employer's contribution)
- annual bonuses
- overtime
- computer and printer
- training levy
- car and expenses

The choice of site managers will depend on job size, complexity, duration, number of operatives and commitments to nominated and domestic sub-contractors.

Fig. 15.2b *Pricing preliminaries – management and staff*

Facilities and services	
Power/lighting/heating	Check availability of supplies, connection charges, temporary housings, fittings and consumption costs. Alternatively generators and gas bottles
Water	Provision of service to site, pipework from supply, distribution system, and charges. Bowsers required if no piped service available.
Telephones	Consider number of senior staff in deciding on number of lines, switchboard for larger systems and mobile phone for start of job
Stationery and postage	Average cost per week drawn from analysis of previous projects.
Office equipment	Photocopier, facsimile machine and typewriters are commonly required
Computers	Personal computers required for more complex projects. Security can be a problem; consider portable PCs or rubust equipment
Humidity and temperature control	Check specific requirements for de-humidifiers, heaters and attendance. Vulnerable materials include seasoned joinery and suspended ceilings
Security and safety measures	Security firm or own labour for watching site at night and at weekends Intruder alarms, traffic control, fire precautions and fire fighting
Temporary electrics	Transformers, distribution system, boards, loads and site lighting. Electrician may be resident on large building schemes
Waste skips	Regular collection of rubbish skips should be allowed Dustbins can be used to promote cleaner sites

NOTES:

Computers are used on site for material records, valuations and cost monitoring. Standard databases are often used to produce drawing registers and lists of instructions. Additional costs which include maintenance contracts, software and consumables can be as much as the value of the computers. The consumables and maintenance costs associated with photocopiers can be costly.

Humidity control is a complex requirement. If dehumidifiers are used the hire charges are high and additional costs include transport charges, electric power, attendance in removing water and daily monitoring of humidity levels. These measures can only be put in place when the building is enclosed and damp air is prevented from entering the building.

Fuels are deemed to be included in items for testing and commissioning mechanical and electrical work. (SMM7 Y51 and Y81). The contractor should liaise with specialists to ensure the fuel costs are included.

Fig. 15.2c *Pricing preliminaries – facilities and services*

Mechanical plant	
Crane and driver	Determine maximum lift in terms of weight, radius and height clearance Duration and location on site also needed to select a crane (see below)
Hoist	Usually required for multi-storey external access scaffolds. Consider type (goods or passengers), hire, transport, erect, adapt, and dismantle
Dumper	Difficult to dispense with dumpers for moving excavated material, stone, bricks, blocks and mortar. Include transport, hire and fuel
Forklift	Rough terrain forklift is a good all round tool used for unloading, distributing and hoisting palletised materials. May need attachments
Tractor and trailer	Popular for long drainage runs and kerb laying for roadworks
Mixer	Hire, transport and fuel for concrete and mortar mixers, silos and bunkers Minimum requirement is a mixer for brickwork and drainage
Concrete finishing equipment	Screeding rails and tamping bars Curing membranes and vibrators
Compressor and tools	Needed for demolitions and alterations, cleaning inside shutters, preparing stop ends in concrete, drilling and vibrating concrete
Pumps	Water pumps, hoses, fuel, transport and attendance (see attendant labour) Concrete pump hire for mobile or static equipment; check quote for extras
Fuel and transport for plant	Static fuel tank or fuel bowser may be required. Assess additional transport costs for plant from yard to site

NOTES:

Quotations should be obtained for long hire or large capacity plant, such as cranes, forklifts and concrete pumps. Consult plant suppliers for advice on running costs and fuel consumption.

Cranes need ancilliary equipment such as slings, concrete skips and lifting beams. Mobile cranes need sufficient space for outriggers, and hire charges are usually from time of leaving depot to return.

A tower crane will incur costs to transport, erect, adapt and dismantle, as well as a foundation or rail tracks, power source, fuel and operator. In some instances tower cranes may not be feasible owing to wind limitations or air space rights which cannot be infringed.

Fig. 15.2d *Pricing preliminaries – mechanical plant*

Temporary works	
Access routes	Plot layout on site plan taking advantage of hardcore under roads and buildings. Allow for maintenance and making-up levels on completion
Hardstandings	Additional areas are required for storage, site huts, and lay-down areas for materilas such as pipes and reinforcement. Reinstatement.
Traffic control	Check specification and statutory obligations. Use programme and site layout drawing to determine equipment needed and hire periods
De-watering	Establish type of system required. Quotation and advice from specialist needed for a well-point system
Hoarding	Serves to protect the public and forms a secure barrier around the site or contractor's compound. Costs to hire, buy, erect, adapt and dismantle
Fencing	Temporary or permanent fencing to maintain security at perimeter of site, protect trees, mark a boundary and to form site compound
Notice board	Notice board and local signage help drivers and visitors find the site, satisfy curiosity and provide cheap advertising. Check client's requirements
Shoring & Centring	Consider design, duration, and hire or buy calculations for falsework/shoring (temporary shoring will incur making good costs on completion)
Temporary structures	Temporary bridges, temporary roofs, façade supports, ramps, viewing platforms, accommodation gantries. May need specialist design input
Protection	High value components and all finishes need to be considered Decorated areas may require additional coats of paint

NOTES:

In poor soil conditions the loss of hardcore under access roads and hardstandings can be substantial. A ground improvement mat may be used with the approval of the contract administrator.

Excavation below ground water level or works affected by rivers or tidal water is normally identified in the bill of quantities. The contractor is threrefore responsible for finding an appropriate method for dealing with the problem. Where a full de-watering system is required by a client, the contractors are normally informed at tender stage.

The cost of protection is often undervalued; particularly in the case of building finishes which are difficult to protect during the commissioning and completion of a project. In a large building measures might include laying sheeting and boarding on floors, re-painting walls and woodwork, and a security system which detects responsibility for damage by allowing entry to finished areas with a written permit.

Fig. 15.2e *Pricing preliminaries – temporary works*

Site accommodation	
Offices	Mobile offices can be established quickly to high standard Sectional sheds require set-up costs, foundations and finishes
Stores	For secure protection of high value materials Hire or purchase accommodation, or in the building
Canteen/welfare	Use labour strength from programme and add for sub-contractors Add for equipment, cooking facilities, furniture and food subsidies
Toilets	Mobile toilet units – check drainage and services available Allow for sundries such as soap, towels and cleaning materials
Drying and first aid	Accommodation with lockers and heaters First aid room depending on number of people on site
Foundations and	Drainage concrete or sleepers to support cabins Wherever possible connect drainage to live sewer
Rates and charges	Local authority charges are payable on temporary accommodation May need to acquire land for site establishment
Erection and fitting out	Materials for erection and fitting out Labour may be here or as Attendant Labour
Furniture	Company owned desks, chairs and cabinets usually available Replacements can be provided from secondhand market
Removal	Consider the need to re-site the accommodation during the job Taking down can be priced as Attendant Labour
Transport	Often priced as transport to site only if follow on work expected Cranage off-loading and loading accommodation

NOTES:

The estimator should keep records of hire rates for mobile accommodation and compare with the purchase of sectional buildings. The capital cost of timber buildings can be divided by the life span (plus an allowance for repairs and renovation) to arrive at an equivalent hire rate. There is additional labour in erecting and dismantling timber buildings together with the provision of sundry materials such as sleepers for foundations, felt for roofing, glass, insulation board lining and so on.

Additional hire during the defects liability period may be required for a foreman's office and storage container. Minimum requirements for accommodation and health and welfare are given in the Construction Regulations.

Fig. 15.2f *Pricing preliminaries – site accommodation*

Attendant labour	
Unloading and distribution	Envisage an unloading and handling labour gang; adding drivers for dumper and forklift. Balance requirements with attendant labour in bill rates
Cleaning	Daily and weekly tidying-up and cleaning may be carried out by the general gang or casual labour
Setting-out assistants	Each setting-out engineer will need some assistance from a chainman.
Drivers and pump attendance	Each item of mechanical plant will need some operator time; for example: a water pump might require 1.5 man hrs/day
General attendance	Clearing away rubbish for sub-contractors, adapting scaffolding and materials distribution (may be priced elsewhere)
Scaffold adaptation	An assessment of time is required and rates should be obtained from the scaffolding specialist

NOTES:

A site which relies on scaffold hoists will need more labour than that which has a tower crane. Where multi-storey buildings have blockwork walls internally, a considerable amount of labour for distribution is needed.

It can be argued that general attendance for sub-contractors is already included in the tender by assessing all the site facilities in the preliminaries. Special attendance can also be dealt with in this section, although some estimators prefer to price specific attendances for nominated sub-contractors in the bill of quantities, which is presumably what is envisaged in the standard method of measurement.

Fig. 15.2g *Pricing preliminaries – attendant labour*

There are certain items that are difficult to allocate, such as the use of specialist plant. A crane for example may be on site for two weeks: should this be classed as a fixed or time-related charge? For many schemes, all general plant and facilities are divided by the duration to produce equal sums for monthly payments.

Contract conditions

Fluctuations	Calculation of fluctuation costs includes changes forecast for preliminaries See fluctuations chapter for factors to be considered
Insurances	A quotation (or guide rate) can be obtained for all-risks insurance. The all-in rate for labour usually includes the Employer's Liability insurance
Bonds	Cost of bond is based on contract value and duration inc defects liability period. Check that bond is acceptable and bank facilities are not exceeded
Warranties	More common with contractor-designed elements which are usually backed up with warranties from specialists and designers
Special conditions	Check changes to standard conditions, poor documentation and that form of contract, specifications and SMMs are up-to-date
Professional fees	Alternative bids which include a design element, tests on materials, QS services, and legal fees to check contractual arrangements. Land surveys

NOTES:

* Most fluctuations in prices will relate to production costs, not preliminaries. It is therefore preferable to calculate fluctuations independently and transfer the result to the tender summary.

Not all losses can be recovered under an insurance policy. There could be many losses on a site which each fall below the policy excess agreed with the insurers. An estimate of an average value of losses should be added to the tender.

Where applicable, a Parent Company Guarantee would provide a measure of protection, at little cost to the client.

Tendering costs are not normally added to an individual tender. They become part of the general company overhead.

Fig. 15.2h *Pricing preliminaries – contract conditions*

Non-mechanical plant	
External scaffolding	Quotation needed for erection and hire of scaffold for larger schemes Programme used to assess duration on each section of work
Internal scaffolding	Birdcage scaffold for large voids, decking for ceilings and ductwork Consider lift shafts, stair wells, high walls and inner skin of external walls
Hoist towers	Extra cost to provide enclosure for mechanical hoist Platforms to unload materials at each level
Mobile towers	For short duration and localised work Consider also platform hoists or scissor lifts (mechanical plant)
Small tools and equipment	Hand tools such as drills, power saws, picks and shovels, rubbish chutes, bending machines, bolts crops and traffic cones
Surveying instruments	Purchase, hire or internal charges, maintenance and consumables Check requirements of engineering staff listed in supervision section

NOTES:

For scaffolding the following should be considered:

a. Hire charges
b. Transport costs
c. Labour costs
d. Losses
e. Adaptation
f. Safety measures
g. Baseplate support
h. Debris netting
j. Polythene sheeting
k. Temporary roofing
l. Platforms to land materials

The loads imposed on scaffolding should be considered, in particular the use of scaffolding for short-term storage of block stone or bricks.

There is a relationship between the amount of labour in a job and the small tools and equipment costs. Some companies add a percentage to the all-in labour rate for small tools others include a sum in the preliminaries based on a percentage of total labour costs for the job. Clearly feedback from previous contracts is needed so that the estimator has a realistic guide for this item.

Fig. 15.2i *Pricing preliminaries – non-mechanical plant*

Miscellaneous

Setting out consumables	Pegs and profile boards, tapes and refills Larger projects need concrete and steelwork for site stations
Testing and samples	Concrete cube testing is calculated from the specified frequency of tests Samples of materials may be supplied free but composite panels have a cost
Winter working	Additional protection for operatives, heating, and site lighting Reduced productivity, location of work, heated concrete etc
Quality assurance	For large schemes, a significant proportion of the superviser's time may be dedicated to the control and monitoring systems. See Management and Staff.
Site limitations	Check employer's requirements such as access restrictions, control of noise, weight of vehicles, protection of services, etc. See Temporary Works.
Protective clothing	Staff and directly employed people will need protective clothing, depending on time of year. Safety hats/clothing for employees and labour-only s/c's

NOTES:

The responsibility for quality assurance remains with all the parties and everyone in the organization. The establishment of a quality system will produce quality statements, a set of company procedures, training, and control mechanisms. The cost is usually carried by the general off-site overhead. It is often argued that the cost of setting up and implementing a quality system is off-set by the benefits which result.

Fig. 15.2j *Pricing preliminaries – miscellaneous*

Similarly, staff costs are difficult to share between valuations because they are usually higher at the beginning of a contract and taper off gradually towards the end. Staff costs are often incurred before the start date, when mobilization activities – such as procurement of initial packages – take place.

When the estimate has been reviewed by management the estimator will allocate sums of money to the preliminaries bill. This is an opportunity to ensure that a satisfactory (and possibly positive) cash-flow position is secured. In particular, the

CB Construction Limited						**Student Accommodation**		**Preliminaries**	GFA (m^2)	1,697
						Steel frame solution		14.4%	Date	**11-Jul-08**
								net trade value £		**2250 000**
	Description	**note**	**quant**	**unit**	**factor**	**Fixed charges**		**Time charges**		**Sub-tot**
							133 523		**189 905**	**323 428**
SUPERVISION p=pre-start							0		0	
p	Site manager pre-constr		10	wk	0.5	1 300	6 500	0	0	
	Site manager constr		–	wk	1	0	0	0	0	
p	Design manager pre-constr		10	wk	0.5	1 220	6 100	0	0	
	Design manager constr		–	wk	1	0	0	0	0	
p	QS pre-constr		10	wk	0.5	1 300	6 500	0	0	
	QS constr		35	wk	0.25		0	1 300	11 375	
	Site Agent (roads)		–	wk	1		0	0	0	
	Site Manager (Building)		35	wk	1		0	1 300	45 500	
	Sen engineer (roads)		–	wk	1		0	0	0	
	Sen engineer (building)		–	wk	1		0	0	0	
	Engineer (building)		6	wk	1		0	950	5 700	
	Foreman		30	wk	1		0	850	25 500	
	Secretary		–	wk	1		0	0	0	
p	Legal advice		1	it	1	5 000	5 000	0	0	
	Visiting safety manager		15	vis	1	0	0	1 050	15 750	
									0	**127 925**
SITE FACILITIES							0		0	
	Notice boards		1	it	1	650	650		0	
	Fencing/hoarding		200	m	1	30	6 000		0	
	Gates 10 m wide		2	nr	1	2 000	4 000		0	
	Reinstatement		1	it	1	1 500	1 500		0	
	Set up - foundations		1	sum	1	1 000	1 000		0	
	Set up - transport/crane		4	nr	1	375	1 500		0	
	Set up - offices		80	hr	1	13	1 040		0	
	Set up - fitting out		20	hr	1	13	260		0	
	Dismantle/remove		40	hr	1	13	520		0	
	Install electricity		1	sum	1	3 000	3 000		0	
	Install water		1	sum	1	1 500	1 500		0	
	Install drainage		1	sum	1	1 200	1 200		0	
	Install telephone/IT line		1	sum	1	1 000	1 000		0	**23 170**
SITE RUNNING EXPENSES					1		0		0	
	Electricity charges		35	wk	1		0	75	2 625	
	Water charges		35	wk	1		0	45	1 575	
	Telephone charges		35	wk	1		0	165	5 775	
	LA rates		35	wk	1		0	15	525	
	Postage and stationery		35	wks	1		0	45	1 575	
	Photocg/printing consumbls		35	wks	1		0	45	1 575	
	Canteen/office cleaning		30	wks	1		0	200	6 000	
	Out-of-pocket expenses		35	wks	1		0	85	2 975	
	Settg-out consumables		1	sum	1	1 500	1 500	0	0	
	Protective clothing		15	sets	1	150	2 250		0	

Fig. 15.3 *Continued*

CB Construction Limited					**Student Accommodation**		**Preliminaries**	GFA (m^2)	**1,697**
					Steel frame solution		14.4%	Date	**11-Jul-08**
								net trade value £	**2250 000**
Description	**note**	**quant**	**unit**	**factor**	**Fixed charges**		**Time charges**		**Sub-tot**
Photographs plus camera		8	mth	1		200	10	80	
Materials testing		200	m^3	1	3	600		0	
Security services		–	wk	1		0	0	0	
Fire equipment		1	sum	1	1 000	1 000		0	
Site consumables		1	sum	1		0	2 500	2 500	**30 755**
GENERAL LABOUR				1		0		0	
Chainman		6	wks	1	390	2 340		0	
Gen gang - clear site				1		0		0	
Gen gang - mat distribution		20	wks	1		0	410	8 200	
Snagging		4	wks	1	600	2 400		0	
Protection		100	hr	1	13	1 300		0	
Final clean		0.001		1	2220 000	2 220		0	
				1		0		0	**16 460**
SITE SET UP				1		0		0	
Offices	2	35	wks	1		0	45	3 150	
Messrooms	1	35	wks	1		0	65	2 275	
Stores	2	35	wks	1		0	40	2 800	
Toilets	1	35	wks	1		0	55	1 925	
Surveying equip - level	1	35	wks	1		0	20	700	
Surveying equip - theo	1	35	wks	1		0	30	1 050	
Site vehicles				1		0		0	
Lab equipment				1		0		0	
Office equip - photocopier		35	wks	1		0	30	1 050	
Office equip - fax machine		35	wks	1		0	20	700	
Office equip - computer		2	nr	1		0	1 000	2 000	
				1		0		0	**15 650**
GENERAL PLANT						0		0	
Forklift+(0.5)driver+fuel		20	wks			0	480	9 600	
Skips - general		60	nr			0	150	9 000	
Skips - final clean		8	nr			0	150	1 200	
Skips - finishes		10	nr			0	150	1 500	
Goods and passenger hoist	setup	–	sum		1 260	0	0	0	
Goods and passenger hoist	hire	–	wks			0	165	0	
Goods and passenger hoist	rem	–	sum		1 250	0		0	
Generator		–	wks			0	80	0	
Mobile crane		3	vsts		450	1 350		0	
Mixers	ave	35	wks			0	25	875	
Mob/demob generally		1	sum		1 500	1 500		0	
Minor plant		1	it			0	5 000	5 000	
Road sweeper		5	wk			0	650	3 250	**33 275**
TEMPORARY WORKS						0		0	
Site hoarding		40	m		75	3 000		0	
Double gates		1	pr		750	750		0	
Scaffolding - ext ind		1,550	m^2		20	31 000		0	

Fig. 15.3 *Continued*

CB Construction Limited	**Student Accommodation**	**Preliminaries**	GFA (m^2)	**1,697**
	Steel frame solution	14.4%	Date	**11-Jul-08**
		net trade value £		**2250 000**

	Description	note	quant	unit	factor	Fixed charges		Time charges		Sub-tot
	Scaffolding - adaptions	incl	100	hr		15	1 500		0	
	Scaffolding - internal		1	it		2 000	2 000		0	
	Internal lighting		1	sum		2 200	2 200		0	
	Power distribution		1	sum		2 200	2 200		0	
	External lighting		1	sum		1 850	1 850		0	
	Tempscreens lower floors			m^2		50	0		0	
	Make up hardcore roads/site		1,100	m^2		0	0	6	6 600	
	Protection materials	GFA	1,697	m^2		3	5 091		0	
							0		0	
							0		0	**56 191**
COMMERCIAL/FINANCIAL					1		0		0	
	Insurance - all-risks		0.002		1	2220 000	4 440		0	
	Insurance - hired plant		0.019		1	100 000	1 920		0	
	Insurance - Emply liab		0.003		1	2220 000	6 660		0	
	Insurance - premium tax		0.050		1	15 240	762		0	
	Insurance excesses		2	nr	1	2 000	4 000		0	
	Professional indemnity		0.001		1	2220 000	2 220		0	
p	Legal advisors pre-constr		-		1		0		0	**20 002**
DESIGN AND BUILD FEES AND COSTS							0		0	
	Architectural fees						0		0	
	Structural engineer						0		0	
	MandE advice and coord						0		0	
	QS fees						0		0	
	CAD file on disk						0		0	
	Building regs						0		0	
							0		0	**0**

Fig. 15.3 *Continued Typical preliminaries spreadsheet*

contractor needs to identify setting-up costs that should be claimed in the first valuation as fixed charges. If all site and general overheads were reimbursed in proportion to time, the contractor would have more expenditure than income and this poor cash-flow position would persist during most of the contract duration.

The early fixed charges often include the following:

Employer's requirements

- Accommodation
- Furniture

- Install telephone
- Provide equipment
- Transport charges.

Supervision

- Hotel expenses
- Planning
- Procurement.

Services

- Installation charges
- Office equipment.

Mechanical plant

- Transport of plant
- Purchase of plant.

Temporary works

- Design and purchase of structures
- Access routes and hardstandings
- Enclosures
- Dewatering, piling and formwork.

Site accommodation

- Transport and cranage
- Purchase costs
- Foundations and furniture
- Erection and fitting out.

Contract conditions

- Insurance premium
- Bond
- Professional fees – in particular, design fees and surveys
- Initial land/building surveys.

Non-mechanical plant

- Scaffold erection
- Small tools and equipment.

Miscellaneous

- Setting-out consumables
- Samples
- Quality planning
- Protective clothing.

The costs of dismantling site facilities and cleaning are much smaller by comparison and are rarely priced as separate fixed charges.

Preliminaries for tenders based on cost planning

As with other aspects of tendering with cost plans, a target for preliminaries is established in the target cost plan by comparing the proposed project with similar schemes. The estimator must then show how the target can be split against the various headings so that the construction project manager can evaluate site overheads economically.

Benchmark data for preliminaries

A review of previous schemes will establish the range of preliminaries appropriate for each market sector. The data given in Fig. 15.4 has been obtained from a selection of tenders. In practice, the out-turn costs have been similar, but there have been examples of these being exceeded.

Sector	Prelims on net build cost
PFI hospital	17–19%
BsF schools	14–17%
MOD prime contracts	16–22%
Large D+B warehouse	8–11%
Excludes internal bid costs	
The range depends on project specific requirements such as tower cranes, scaffolding and a 'core' staff team	

Fig. 15.4 *Typical ranges for preliminaries*

Preliminaries workbook

A preliminaries workbook is needed for all projects in order to reflect programme durations, method-related costs such as temporary works, staff rates, bonds and insurances. In most cases this is the company standard pro forma. In some cases, where a number of options are being costed, preliminaries could be added to the electronic bill of quantities to assist with cost modelling.

					Note:
	Approximate net build cost			42,000,000	1
			£	% of net	
	Management		3 342 543	7.96%	
	Management accommodation		857 290	2.04%	
	Site establishment and security		1 208 197	2.88%	
	Health and safety		288 501	0.69%	
	Temp site services		340 000	0.81%	
	Plant tools and equipment		137 100	0.33%	
	Scaffolding and access equipment		765 900	1.82%	
	Clearance and cleaning		1 027 014	2.45%	
	Temporary works		12 000	0.03%	
	Sundries		3 600	0.01%	
	Insurances/finance		774 000	1.84%	2
	Preliminaries total		**8 756 145**	**20.86%**	**3**
	Design Fees and Charges		4 389 000	10.45%	4
	Fixed Price				
	Pre-commencement Costs		167 750	0.4%	5

Notes:

1 Net build cost is used as the 'base cost' for preliminaries percentages. Usually the sum of BCIS elements 1 to 6 or the equivalent trade packages.

2 For PFI projects check policy for project insurance by the Special Purpose Company (SPC).

3 Prelims % might look high but if expressed on gross £63m would be 13.9%.

4 Design fees and charges are from inception to completion. Fees and charges if expressed on gross of £63m would be 6.97%.

5 Construction staff for the preferred bidder and mobilisation period.

Fig. 15.5 *Typical breakdown of preliminaries*

Design fees and charges

Scope of services

Figure 15.6 shows a list of services which might be required for a large contract. The design manager will agree the scope of services under each heading, with the consultants.

Design fees	Target	Charges and surveys	Target
Architect (lead consultant)	5.00%	Planning application fees	0.25%
Structural engineer	1.80%	Building control	0.19%
Civil engineer		BREEAM advice	0.08%
Services engineer	1.80%	Planning supervisor/co-ordinator	0.18%
Services – transfer to M+E at FC		Special BESD/SEN input	
Landscaping architect	0.35%	Ground investigation	
Acoustic consultant	0.11%	Intrusive surveys	
Fire consultant	0.12%	Educationalist (function specialist)	
Geotechnical and contaminated land	0.25%	Dimensional survey	0.10%
Catering advice and equipment list	0.07%	Ecologist/sustainability	
FF+E schedule and advice	0.26%	Underground/infrastructure surveys	
Façade consultant	0.17%	Topographical	0.20%
Taking off services	0.07%	Drainage survey	0.09%
Expenses		Lighting survey	
Environmental impact assessment	0.09%	Condition survey	
DDA		Dilapidation survey	
Part L calculations		Archaeologist	
Daylight + thermal modelling		Acoustic survey	
Win bonus		Asbestos survey	
		Temporary works design	0.20%
		Gas monitoring earthworks advice	
		3D model	0.15%
		DQI facilitator and toolkit	0.06%
		Highways and transport	
Target % on net build cost (see definition below)			
Insert lump sums for fees and surveys when confirmed			

Fig. 15.6 *Design fees and surveys checklist (school project)*

Stage	% of total fee
Early bid stages PQQ and ITPD (at risk)	0%
Initial bid stage ITCD	15%
ITSFT	3%
Preferred bidder stage	25%
Production information (spans PB and construction)	35%
Construction stage	20%
Post construction stage	2%
	100%

Fig. 15.7 *Typical payment stages for professional fees*

A target for fees and surveys is established in the target cost plan by comparing the proposed project with similar schemes. The estimator must then show how the target can be split against the various headings so that the design manager can conclude the negotiations.

The basis for calculating professional fees and charges is a problematical topic. For traditionally procured projects, designers looked for fees based on a percentage of the out-turn price. This meant that fees would increase if the value of work increased and would reflect the level of margin added by the contractor. With the widespread use of the design-and-build form of procurement, fees are more commonly related to the package values. A sensible arrangement would be to create a benchmark for fees based on package values plus a percentage for some design work needed for contingencies and inflation, to reflect the eventual package values.

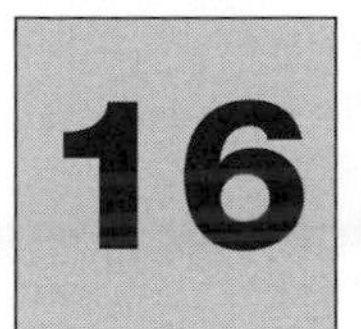

Cashflow forecasts

Introduction

At tender stage, a contractor sets up his financial and time objectives by calculating construction costs and producing a project programme. By linking the two sets of data, an estimator can firstly help a client produce his forecast of payments and secondly compare this with his likely payments (to suppliers and sub-contractors) to produce his own cashflow forecast. In this way a contractor is in a unique position to give accurate information to the building team.

There is seldom enough time at tender stage to produce detailed cashflow forecasts. The contractor knows the objective is to win the contract, so there must be good reasons for putting in this effort. Obviously if a client has asked for a tender-stage programme and cashflow forecast it must be done. The contractor may also need to assess the cashflow benefit of taking on a job, because this is part of his assessment of risk. A spreadsheet model may be able to answer questions such as: are there any sudden cash commitments? How much early money will be needed to make the contract self-financing?

Construction costs for PFI projects are produced by an estimator usually working for one of the parties making up a consortium. Construction is only one constituent of the PFI bid, the others being facilities management, equipment, project management and finance. The estimator must produce his capital costs in the form of a monthly draw-down chart. The first payment is usually very large because considerable design and bidding costs need to be refunded early.

Cashflow calculations

There are two methods commonly used to predict the value of project work over time. Firstly, cost plans can be used at various pre-contract stages to produce approximate forecasts; and secondly, the estimator's more detailed calculations form the basis of an accurate method.

If a client needs a schedule of payments, the simplest model would be a straight-line relationship of value against time from which the client's commitments can be shown (see Fig. 16.1). The assumption being that all payments are of an equal amount based on the full value divided by the number of payments, with a small adjustment for retention money.

A slightly more sophisticated technique assumes that a construction project accumulates costs in a way which can be represented by an S-curve graph (Fig. 16.2). This model is based on the presumption that only one quarter of the costs are incurred during the first third of a project duration, half the costs arise from the second third (in a linear fashion) and the remaining quarter of costs occur in the final third of the project duration. The S-curve can be created manually by drawing a straight line between the first and second 'third' points and sketching the parabolic end portions. The contractor's cumulative cost curve can be superimposed on the chart by deducting the profit margin from the cumulative value curve. The S-curve method is of course a theoretical technique which is difficult to change to take account of the nature of individual projects and contractors' pricing methods, but is successfully used at early stages when detailed pricing information is not available.

The GC/Works/1 Edition 3 form of contract has introduced the S-curve principle as a basis for stage payments. The printed form gives charts for projects with contract values over £5.5 million, and others are available for smaller jobs. Fig. 16.3 shows the S-curve produced from the data given for a 100-week project. Clearly, the project manager is able to predict the client's payments and a great deal of time is saved each month in producing valuations. The same is true for the contractor, but there will be a severe shortfall in payments if the establishment costs are high. An allowance is also needed for design work carried out at the start of the project.

An estimator's calculations and programme are the best starting point for a contractor's cashflow forecast. The rates calculated by the estimator can be linked with the relevant activities on a tender programme. The total costs associated with each activity are divided by the duration, to arrive at a weekly cost. The information used to produce the value curve is simply taken from the sums inserted in the client's bill with adjustments for retention. The contractor's income can be predicted by taking the value at each valuation date and allowing a delay for payments.

To calculate the weekly cost commitment (the contractor's outgoings), each element of cost should be viewed separately. This is because spending on labour, materials, plant and specialist contractors develops in different ways. Direct labour, for example, is a weekly commitment, and credit arrangements for materials can delay payments for up to nine weeks. Expenditure and income can be plotted on a graph against time. The combined effect is a cumulative cashflow diagram which will show the extent to which the job needs financing by the contractor.

Example of a contractor's cashflow forecast

An estimator who was successful in winning a contract for Fast Transport Ltd. carried out the following analysis. He received the enquiry, in the form of drawings and a

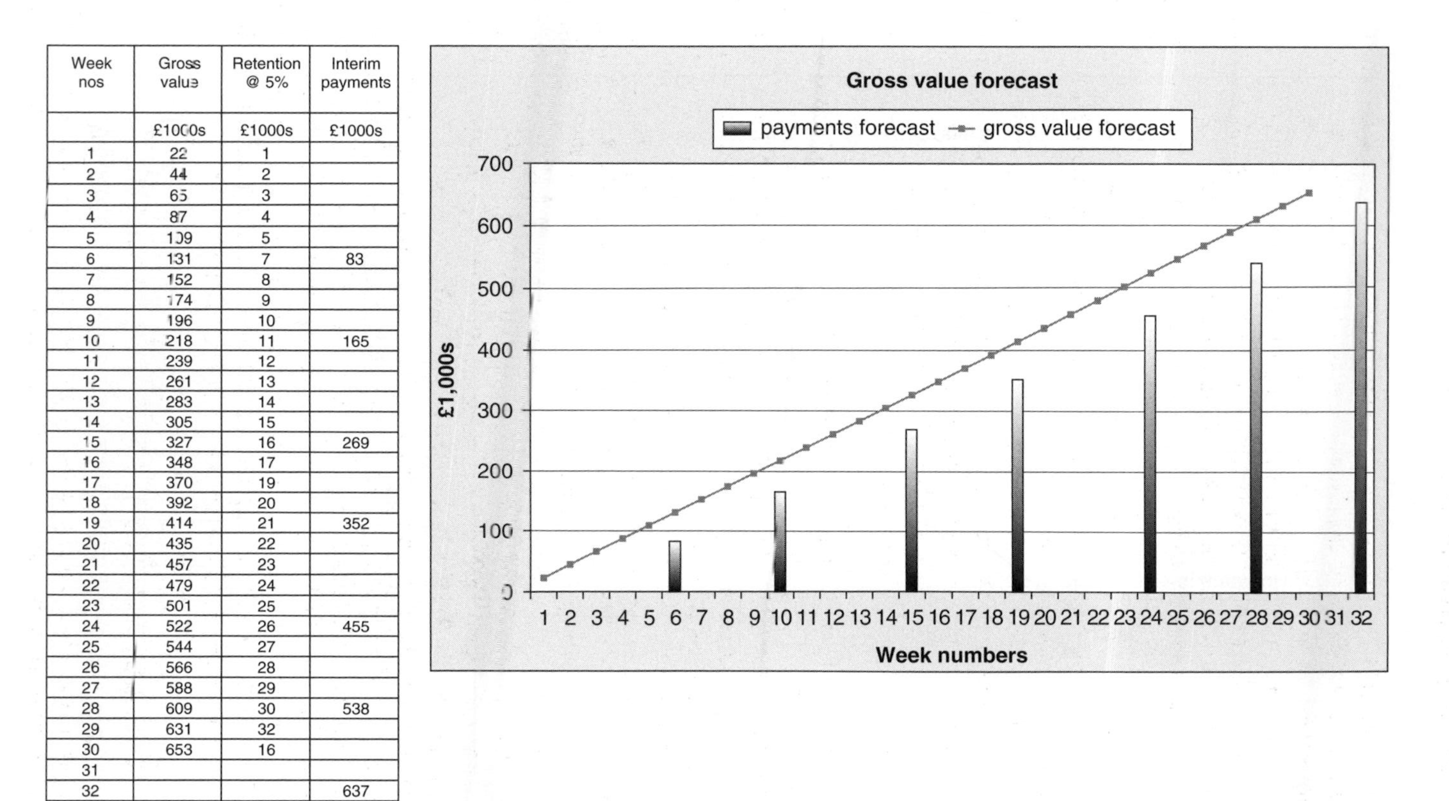

Week nos	Gross value	Retention @ 5%	Interim payments
	£1000s	£1000s	£1000s
1	22	1	
2	44	2	
3	65	3	
4	87	4	
5	109	5	
6	131	7	83
7	152	8	
8	174	9	
9	196	10	
10	218	11	165
11	239	12	
12	261	13	
13	283	14	
14	305	15	
15	327	16	269
16	348	17	
17	370	19	
18	392	20	
19	414	21	352
20	435	22	
21	457	23	
22	479	24	
23	501	25	
24	522	26	455
25	544	27	
26	566	28	
27	588	29	
28	609	30	538
29	631	32	
30	653	16	
31			
32			637

Fig. 16.1 *Simple linear plot of cumulative value*

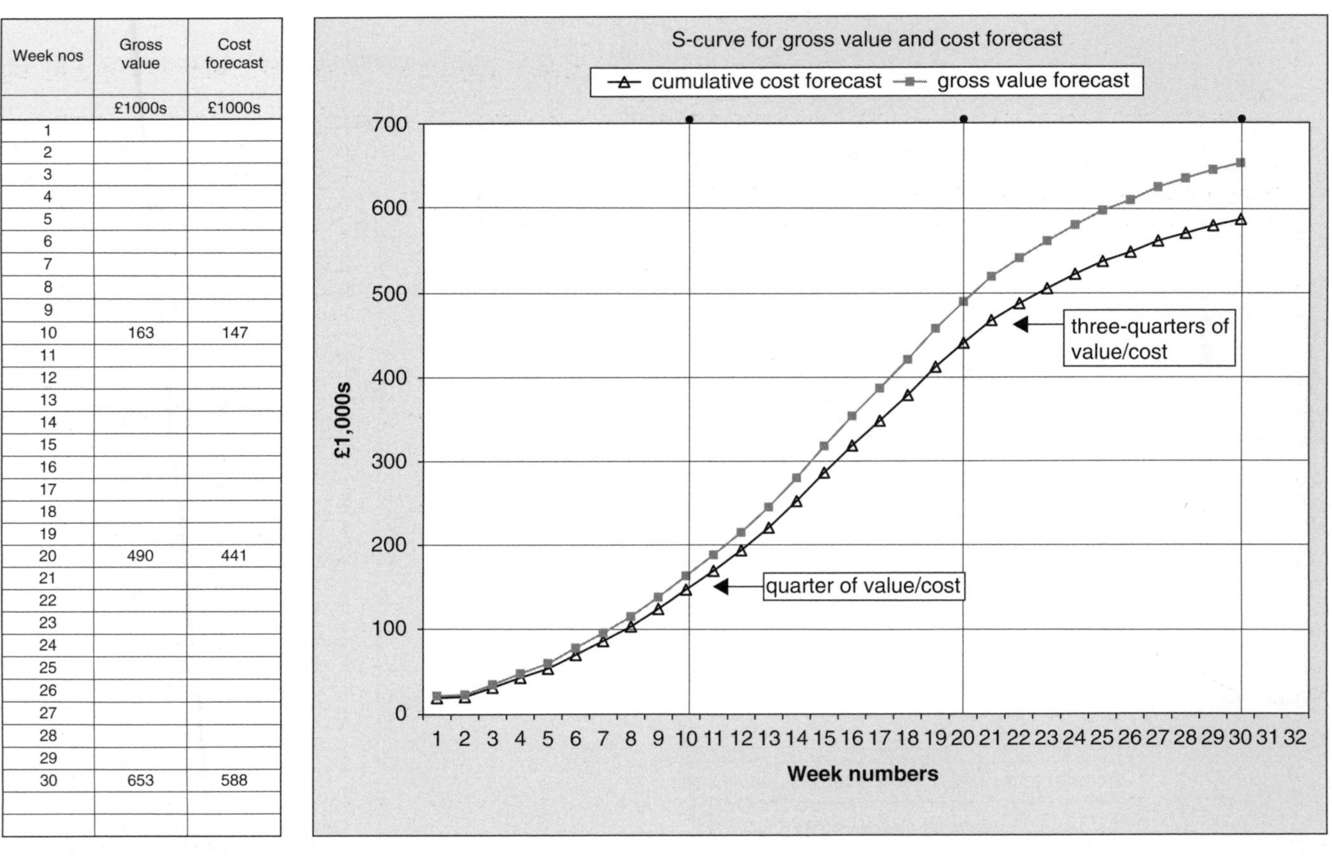

Week nos	Gross value	Cost forecast
	£1000s	£1000s
1		
2		
3		
4		
5		
6		
7		
8		
9		
10	163	147
11		
12		
13		
14		
15		
16		
17		
18		
19		
20	490	441
21		
22		
23		
24		
25		
26		
27		
28		
29		
30	653	588

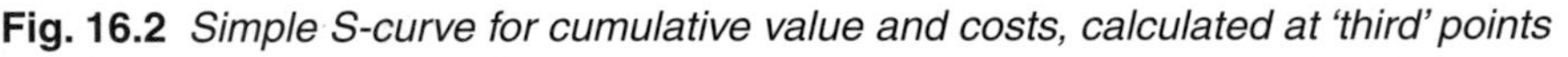

Fig. 16.2 *Simple S-curve for cumulative value and costs, calculated at 'third' points*

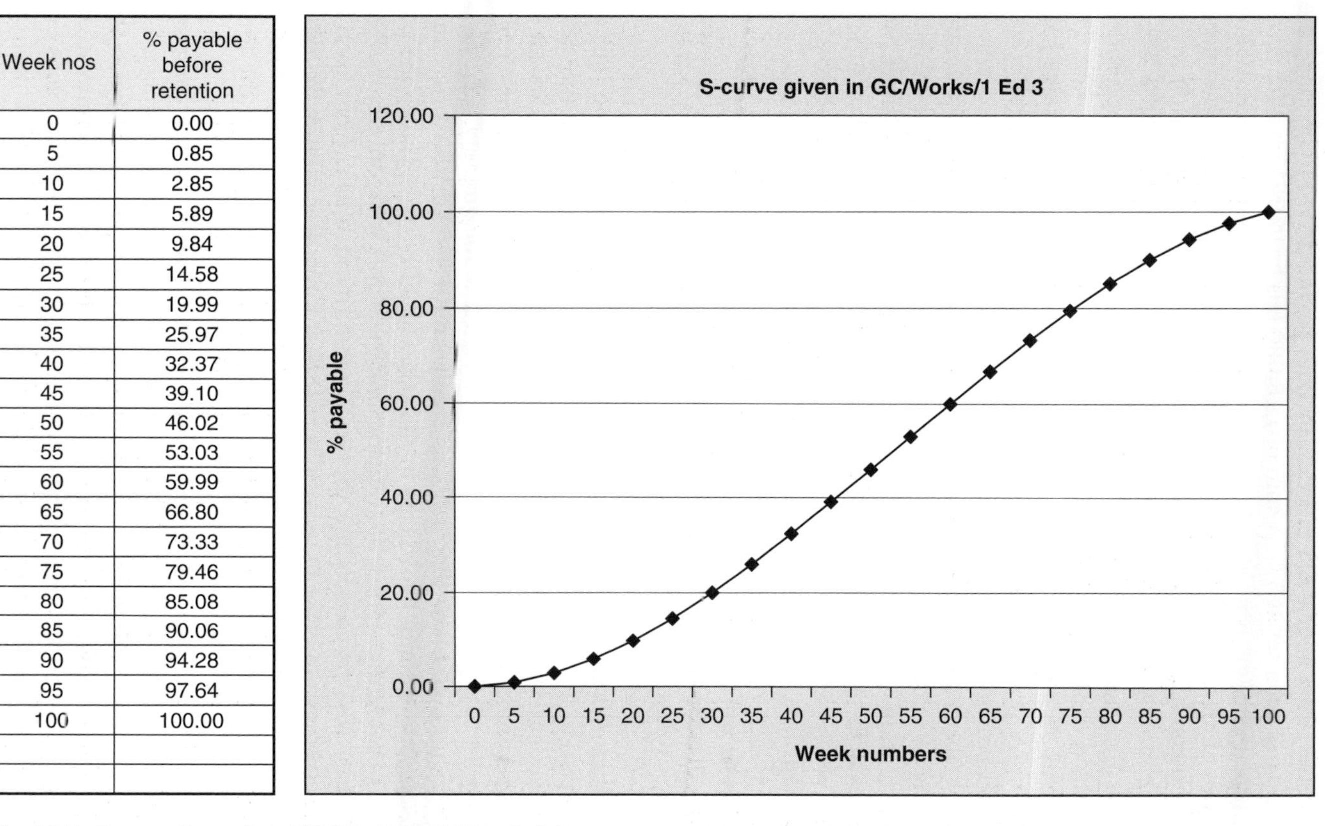

Week nos	% payable before retention
0	0.00
5	0.85
10	2.85
15	5.89
20	9.84
25	14.58
30	19.99
35	25.97
40	32.37
45	39.10
50	46.02
55	53.03
60	59.99
65	66.80
70	73.33
75	79.46
80	85.08
85	90.06
90	94.28
95	97.64
100	100.00

Fig. 16.3 *S-curve based on GC/Works/1 Edition 3 data*

bill of quantities, from a local QS practice. Interim payments are to be made monthly and retention has been set at 5%. A sum of £18,000 has been included for profit, and the rates in the bill of quantities exclude any on-site or off-site overheads. Following a careful review of the estimate, the construction director asked for all sub-contract discounts to be taken from the estimate but made an adjustment to the profit margin.

During the mobilization period, the estimator was asked to produce a forecast of payments for the client and a cashflow forecast for the commercial department. The estimator priced the bill of quantities analytically using estimating software. The first task was to ensure that all the adjustments made at the final review stage had been made to the rates in the bill of allowances. This fully adjusted bill would become the budget for the construction team to monitor the financial progress of the job.

The next stage was to assign sums of money to each activity on the tender programme. This was done using a spreadsheet program because each cell on the screen can hold text, graphics, numbers or formulae. Each row on the spreadsheet can show weekly values with shading used to locate the activity bars. Fig. 16.4 shows the tender programme with contract values taken from the bill of quantities, which was submitted to the client. The main assumptions used for this graph were that:

1. Each activity will be completed on time.
2. Sums are divided equally for the duration of an activity.
3. Provisional sums and daywork are not included.

The total for preliminaries has been split equally over the whole duration. This is not a good interpretation for the contractor because he will incur more expenditure setting up his site facilities at the start of a job. The final presentation to the client did not include the costed programme or a graph. The commercial department felt that the client would want a simple list of payments, and included it in a letter to the quantity surveyor.

The contractor's cashflow diagram was created on a spreadsheet program which had the facility to use multiple sheets in the same file. Fig. 16.5 shows the contents of the costed programmes for labour, plant, materials and sub-contractors. The bottom sheet was used to consolidate the cost commitment drawn from each of the other sheets with the forecast income. The cashflow forecast is simply the difference between income and expenditure. The results were plotted, by the program, on a single diagram (Fig. 16.6) and submitted to the commercial manager in the form of a graph (Fig. 16.7) linked to the consolidated costed programme.

Since all the information was held in a single spreadsheet file, the computer was able to answer 'what if' questions, which allowed the commercial manager to minimize the borrowing requirement. Clearly this information can be the basis of a cost monitoring system (which is beyond the scope of this book). The cumulative value curve can help the construction team in monitoring progress of their work and that of the sub-contractors. This technique is sometimes adopted by clients' project managers as a check on the progress of contractors.

CB CONSTRUCTION LIMITED | Proposed Offices for Fast Transport Limited | Tender Programme No: T354/P1

Activity	contract Value £	2008 January				February				March					April				May					June				July				August	
		1	2	3	4	5	6	7	8	9	10	11	12	13	14	15	16	17	18	19	20	21	22	23	24	25	26	27	28	29	30	31	32
1 Mobilization and set up	6 450	3 225	3 225																														
2 Excavation and filling	21 480		7 160	7 160	7 160																												
3 Foundations formwork	19 260			3 852	3 852	3 852	3 852	3 852																									
4 Foundations concrete	27 800				6 950	6 950	6 950	6 950																									
5 Underslab drainage	5 210					2 605	2 605																										
6 Concrete ground floors	16 470							8 235		8 235																							
7 Columns formwork	6 007					1 502	1 502	1 502	1 502																								
8 Columns concrete	2 700						1 350		1 350																								
9 Floors and beams formwork	22 220							2 778	2 778	2 778	2 778	2 778	2 778	2 778	2 778																		
10 Floors and beams concrete	21 140									7 047			7 047		7 047																		
11 External walls	78 580										9 823	9 823	9 823	9 823	9 823	9 823	9 823	9 823															
12 Roof timbers	9 350																3 117	3 117	3 117														
13 Roof covering	19 030																		9 515	9 515													
14 Windows	25 500																6 375	6 375	6 375	6 375													
15 Services 1st fix	43 000															7 167	7 167	7 167	7 167	7 167	7 167												
16 Plasterwork and partitions	43 500																		5 438	5 438	5 438	5 438	5 438	5 438	5 438	5 438							
17 Joinery	41 670																				5 209	5 209	5 209	5 209	5 209	5 209	5 209	5 209					
18 Ceilings	23 000																								4 600	4 600	4 600	4 600	4 600				
19 Services 2nd fix	28 000																									5 600	5 600	5 600	5 600	5 600			
20 Painting	12 300																						2 050	2 050			2 050	2 050	2 050	2 050			
21 Floor coverings	12 300																												4 100	4 100	4 100		
22 External work and drainage	60 920		6 092	6 092																			6 092	6 092	6 092	6 092	6 092	6 092	6 092	6 092			
23 Preliminaries	107 450	3 582	3 582	3 582	3 582	3 582	3 582	3 582	3 582	3 582	3 582	3 582	3 582	3 582	3 582	3 582	3 582	3 582	3 582	3 582	3 582	3 582	3 582	3 582	3 582	3 582	3 582	3 582	3 582	3 582	3 582		
Gross weekly value forecast	653 k	6.81	20.1	20.7	21.5	18.5	19.8	26.9	9.21	21.6	16.2	16.2	23.2	16.2	23.2	20.6	30.1	30.1	35.2	32.1	21.4	14.2	22.4	22.4	24.9	30.5	27.1	27.1	26	21.4	7.68		
Gross interim valuation	653 337 k					69 095				74 440					93 414				103 924					125 261				104 942				82 261	
Net interim payments	**620 670**					**65 640**				**70 718**					**88 743**				**98 728**					**118 998**				**99 695**				**78 148**	
After completion	**32 667**																															**32 667**	
week number		1	2	3	4	5	6	7	8	9	10	11	12	13	14	15	16	17	18	19	20	21	22	23	24	25	26	27	28	29	30	31	>>

Fig. 16.4 *Client's cashflow forecast produced by the estimator*

CB CONSTRUCTION LIMITED **Labour cost forecast**

Activity	lab	plt	mat	sub	contract Value	duration	2008 January 1	2	3	4
1 Mobilization and set up	2 800				2 800	2	1 400	1 400		
2 Excavation and filling	7 800				7 800	3		2 600	2 600	2 600
3 Foundations formwork	12 480				12 480	5			2 496	2 496
4 Foundations concrete	4 600				4 600	4				1 150
5 Underslab drainage	1 920				1 920	2				
	0				0					
6 Concrete ground floors	3 420				3 420	2				
7 Columns formwork	3 907				3 907	4				
8 Columns concrete	750				750	2				
9 Floors and beams formwork	14 570				14 570	8				
10 Floors and beams concrete	3 750				3 750	3				

CB CONSTRUCTION LIMITED **Plant cost forecast**

Activity	lab	plt	mat	sub	contract Value	duration	2008 January 1	2	3	4
1 Mobilization and set up		2 400			2 400	2	1 200	1 200		
2 Excavation and filling		7 980			7 980	3		2 660	2 660	2 660
3 Foundations formwork		0			0	5			0	0
4 Foundations concrete		0			0	4				0
5 Underslab drainage		1 750			1 750	2				
		0			0					
6 Concrete ground floors		1 750			1 750	2				
7 Columns formwork		0			0	4				
8 Columns concrete		250			250	2				
9 Floors and beams formwork		0			0	8				
10 Floors and beams concrete		1 750			1 750	3				

CB CONSTRUCTION LIMITED **Material cost forecast**

Activity	lab	plt	mat	sub	contract Value	duration	2008 January 1	2	3	4
1 Mobilization and set up			1 250		1 250	2	625	625		
2 Excavation and filling			5 700		5 700	3		1 900	1 900	1 900
3 Foundations formwork			6 780		6 780	5			1 356	1 356
4 Foundations concrete			23 200		23 200	4				5 800
5 Underslab drainage			1 540		1 540	2				
			0		0					
6 Concrete ground floors			11 300		11 300	2				
7 Columns formwork			2 100		2 100	4				
8 Columns concrete			1 700		1 700	2				
9 Floors and beams formwork			7 650		7 650	8				
10 Floors and beams concrete			15 640		15 640	3				

CB CONSTRUCTION LIMITED **Sub-contract cost forecast**

Activity	lab	plt	mat	sub	contract Value	duration	April 14	15	16	17	
12 Roof timbers				0	0	3			0	0	
13 Roof covering				16 400	16 400	2					
14 Windows				25 500	25 500	4			6 375	6 375	
15 Services 1st fix				43 000	43 000	6		7 167	7 167	7 167	
				0	0						
16 Plasterwork and partitions				43 500	43 500	8					
17 Joinery				34 000	34 000	8					
18 Ceilings				19 800	19 800	5					
19 Services 2nd fix				28 000	28 000	5					
20 Painting				12 300	12 300	6					

Fig. 16.5 *Multiple sheets for cashflow analysis*

CB CONSTRUCTION LIMITED **Priced Programme** **Proposed Offices for Fast Transport Limited**

Activity	lab	plt	mat	sub	contract Value
1 Mobilization and set up	2 800	2 400	1 250		6 450
2 Excavation and filling	7 800	7 980	5 700		21 480
3 Foundations formwork	12 480	0	6 780	0	19 260
4 Foundations concrete	4 600	0	23 200	0	27 800
5 Underslab drainage	1 920	1 750	1 540	0	5 210
6 Concrete ground floors	3 420	1 750	11 300	0	16 470
7 Columns formwork	3 907	0	2 100	0	6 007
8 Columns concrete	750	250	1 700	0	2 700
9 Floors and beams formwork	14 570	0	7 650	0	22 220
10 Floors and beams concrete	3 750	1 750	15 640	0	21 140
11 External walls	36 700	0	41 880		78 580
12 Roof timbers	4 140	0	5 210		9 350
13 Roof covering	1 350	0	1 280	16 400	19 030
14 Windows				25 500	25 500
15 Services 1st fix				43 000	43 000
16 Plasterwork and partitions				43 500	43 500
17 Joinery	2 360		5 310	34 000	41 670
18 Ceilings	1 800		1 400	19 800	23 000
19 Services 2nd fix				28 000	28 000
20 Painting				12 300	12 300
21 Floor coverings				12 300	12 300
22 External work and drainage	16 700	12 550	19 870	11 800	60 920
Total for measured work	119 047	28 430	151 810	246 600	545 887
23 Preliminaries	57 430	35 490	11 230	3 300	107 450
Resource cost analysis					
Cumulative labour value		delayed	1 wk		176 477
Cumulative plant value		delayed	4 wks		63 920
Cumulative materials value		delayed	6 wks		163 040
Cumulative sub-contract value		week after payment received			249 900
24 **Weekly cost commitment**					
Gross weekly value forecast					653.3 k
Gross interim valuation					**£653 337**
25 **Net interim payments**					**£620 670**
26 **After completion**					**£32 667**
week number					

Activity	2008 January 1	2	3	4	February 5	6	7	8	March 9	10	11	12	13
1 Mobilization and set up	3 225	3 225											
2 Excavation and filling		7 160	7 160	7 160									
3 Foundations formwork			3 852	3 852	3 852	3 852	3 852						
4 Foundations concrete				6 950	6 950	6 950	6 950						
5 Underslab drainage					2 605	2 605							
6 Concrete ground floors							8 235		8 235				
7 Columns formwork					1 502	1 502	1 502	1 502					
8 Columns concrete						1 350		1 350					
9 Floors and beams formwork							2 778	2 778	2 778	2 778	2 778	2 778	2 778
10 Floors and beams concrete									7 047			7 047	
11 External walls										9 823	9 823	9 823	9 823
12 Roof timbers													
13 Roof covering													
14 Windows													
15 Services 1st fix													
16 Plasterwork and partitions													
17 Joinery													
18 Ceilings													
19 Services 2nd fix													
20 Painting													
21 Floor coverings													
22 External work and drainage		6 092	6 092										
Total for measured work	3.2	16.5	17.1	18.0	14.9	16.3	23.3	5.6	18.1	12.6	12.6	19.6	12.6
23 Preliminaries	3.6	3.6	3.6	3.6	3.6	3.6	3.6	3.6	3.6	3.6	3.6	3.6	3.6
Resource cost analysis													
Cumulative labour value	-	3	11	20	28	35	43	53	58	65	73	82	91
Cumulative plant value	-	-	-	-	2	9	14	18	20	22	24	25	28
Cumulative materials value	-	-	-	-	-	-	1	6	12	21	30	39	54
Cumulative sub-contract value	(retention ignored)				-	-	3	3	3	3	3	3	3
24 **Weekly cost commitment**	-	3.31	10.9	19.6	30.1	43.9	50.7	79.5	92.2	111	130	150	176
Gross weekly value forecast	6.81	20.1	20.7	21.5	18.5	19.8	26.9	9.21	21.6	16.2	16.2	23.2	16.2
Gross interim valuation					£69 095				£74 440				
25 **Net interim payments**					**£65 640**				**£70 718**				
26 **After completion**													
week number	1	2	3	4	5	6	7	8	9	10	11	12	13

Activity	April 14	15	16	17	May 18	19	20	21	22	June 23	24	25	26
1 Mobilization and set up													
2 Excavation and filling													
3 Foundations formwork													
4 Foundations concrete													
5 Underslab drainage													
6 Concrete ground floors													
7 Columns formwork													
8 Columns concrete													
9 Floors and beams formwork	2 778												
10 Floors and beams concrete	7 047												
11 External walls	9 823	9 823	9 823	9 823									
12 Roof timbers			3 117	3 117	3 117								
13 Roof covering					9 515	9 515							
14 Windows			6 375	6 375	6 375	6 375							
15 Services 1st fix		7 167	7 167	7 167	7 167	7 167	7 167						
16 Plasterwork and partitions					5 438	5 438	5 438	5 438	5 438	5 438	5 438	5 438	
17 Joinery							5 209	5 209	5 209	5 209	5 209	5 209	5 209
18 Ceilings											4 600	4 600	4 600
19 Services 2nd fix												5 600	5 600
20 Painting									2 050	2 050			2 050
21 Floor coverings													
22 External work and drainage									6 092	6 092	6 092	6 092	6 092
Total for measured work	19.6	17.0	26.5	26.5	31.6	28.5	17.8	10.6	18.8	18.8	21.3	26.9	23.6
23 Preliminaries	3.6	3.6	3.6	3.6	3.6	3.6	3.6	3.6	3.6	3.6	3.6	3.6	3.6
Resource cost analysis													
Cumulative labour value	100	109	116	123	131	135	138	140	142	146	150	154	159
Cumulative plant value	29	30	32	33	35	36	37	39	40	41	42	43	46
Cumulative materials value	57	69	76	82	94	100	112	118	125	133	135	136	137
Cumulative sub-contract value	3	3	4	4	4	4	38	38	38	38	38	133	133
24 **Weekly cost commitment**	189	212	227	243	264	276	326	335	346	358.1	365.9	467	474
Gross weekly value forecast	23.2	20.6	30.1	30.1	35.2	32.1	21.4	14.2	22.4	22.37	24.92	30.5	27.1
Gross interim valuation	£93 414				£103 924					£125 261			
25 **Net interim payments**	**£88 743**				**£98 728**					**£118 998**			
26 **After completion**													
week number	14	15	16	17	18	19	20	21	22	23	24	25	26

Fig. 16.6 *Contractor's cashflow forecast using priced programme*

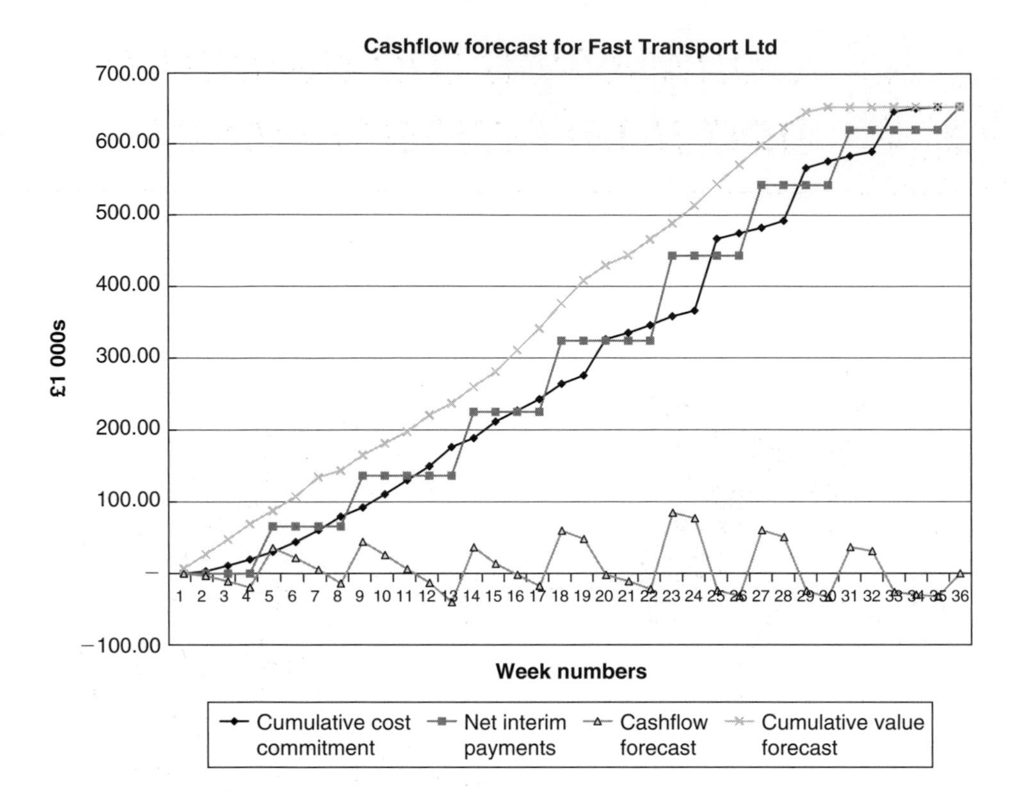

Fig. 16.7 *Cashflow forecast graph*

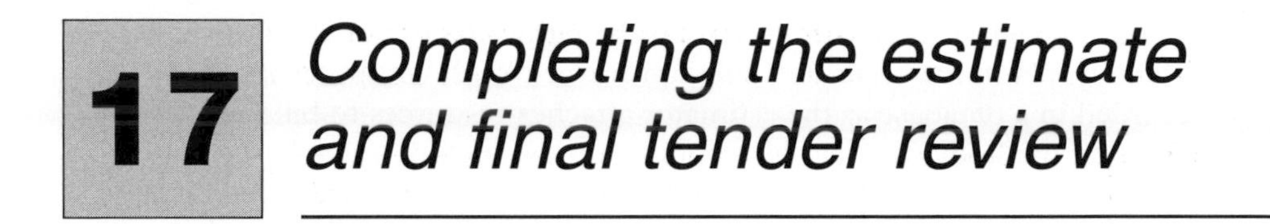

17 Completing the estimate and final tender review

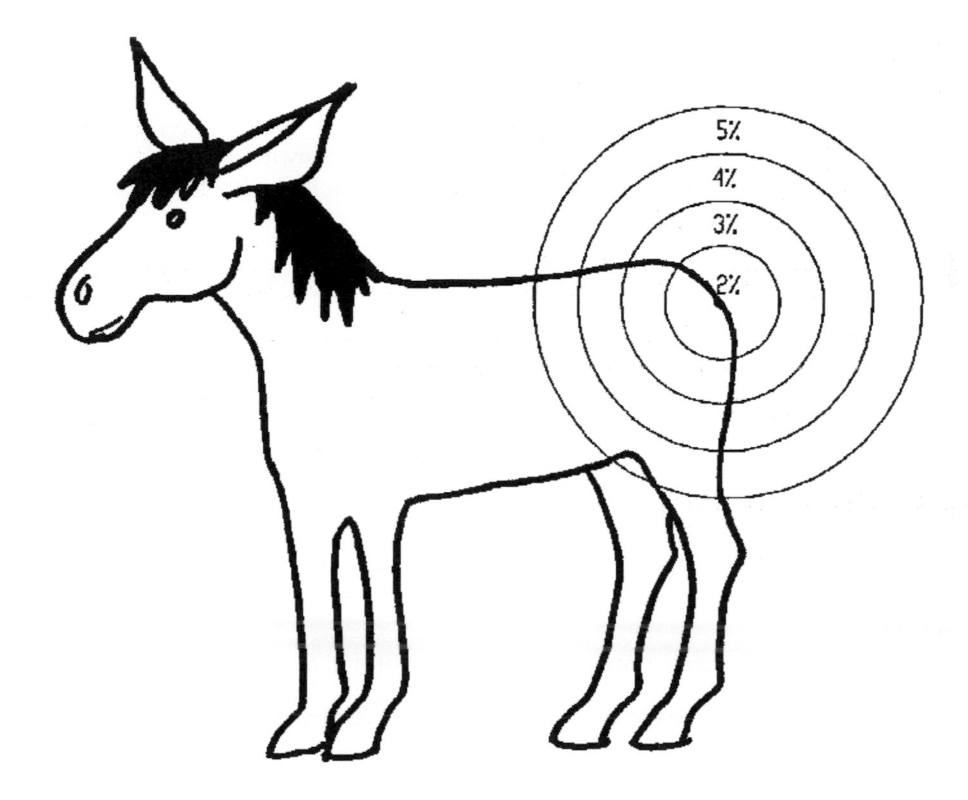

Converting an estimate into a tender

Completing the estimate

The estimator must assemble clear pricing information so that all the build-ups, assumptions and underlying decisions can be seen and understood by the final-review team, and the construction staff if the tender is successful. Bills of quantity should be extended and totalled, with separate subtotals being produced for the four basic elements of labour, plant, materials and sub-contractors. With computer methods, rates will be recorded in a database as the estimator attaches resources to bill items.

All estimating methods (manual or computer) must include reliable techniques to ensure the analysis of the estimate presented to management is correct. Mathematical checks are relatively simple and can be carried out by estimating clerks or by putting check totals in a spreadsheet for example. It is perhaps more important to develop a method which can identify a mistake in resourcing (particularly the main items), collecting to summaries, and input of quantities if computers are used. Other pricing checks include:

1. Rates must apply to the unit of measurement (for example, an item measured in m^2 is not priced as m^3).
2. All items on each page have been priced or included elsewhere.
3. Major items needing costly resources should be reconciled with resource summary sheets.

The resource summaries given in the CIOB Code of Estimating Practice illustrate the information needed by management to assess the costs, which have the most affect on the tender. For example, there is no point discussing formwork valued at £1,850 when there are concrete blocks worth £68,000.

With the use of computers for building up rates, it has become more difficult to identify mistakes at the final review stage. An experienced manager would employ some coarse checks such as making a comparison between labour and materials in each trade. For example, if labour costs for formwork amount to £27,700 and this is approximately 60% of the total formwork costs, then the figures may appear to be correct. There are 55,000 facing bricks which might cost £480 per thousand to lay, i.e. £26,400. The labour summary shows that the estimator has allowed £15,200, so a closer examination is called for.

Resource summary sheets provide vital information for three reasons:

1. Reconciliation can be made between resource schedules and bill totals for labour, plant, materials and sub-contractors.
2. They give management a breakdown of resources so they know where to focus their efforts at the final review stage.
3. Adjustments to resource costs can be calculated on the sheets before being taken to the tender summary. (This could be done quickly using a computer estimating package but some people prefer to have the full picture on paper before changing the estimate held on the computer.)

A tender programme must be prepared so that the estimator can examine the resources estimated in time on the programme with those that he has expressed in cash terms in the estimate. A common problem is the need to maintain excavators on site during groundwork operations. The planner may believe that an excavator is needed on site for ten weeks whereas the estimator has allowed enough money for only six weeks. This reconciliation needs to be carried out jointly by the estimator and planner, and clearly presented to management for final review.

Estimator's report

Once the bill of quantities has been priced and checked, cost summaries and reports can be filled in. Each company has its own standard layout for these forms and a comprehensive example is given in the CIOB Code of Estimating Practice (COEP). It would be wrong to make any changes after this stage to the bill, whether produced manually or by computer, because the analysis sheets and summaries would also have to be changed. As long as adjustments are carefully recorded, they can be built back into the bill of quantities later; in particular when the inked-in copy of the bill is called for. This is one reason why priced bills of quantities for building projects are not submitted with the tender.

The estimator's report must show a full breakdown of costs for management. Decisions can then be reached and the summary of the tender can be compiled. Summary for a tender with a detailed build-up (Fig. 17.2) and summary for a cost-planned tender (Fig. 17.3) will be satisfactory for all sizes of contracts. Each item has a reference number for easy identification within the text. If an estimate report has been produced by a computer system, there should still be an independent check of the major elements using the summary sheets.

Computer-aided estimating systems will generate resource summaries and bill totals. Now that most procurement arrangements are based on a contractor's own quantities, a contract sum analysis is usually generated by the estimating software. BCIS (Building Cost Information Service) elemental cost plan headings have become the industry standard for listing building costs and form the structure of a contract sum analysis for design and build, lump sum, PFI, and many other contracts. An elemental analysis provides management with elemental costs as lump sums and costs per m^2, and this means that data can be compared with other comparable schemes.

There are of course many other supporting documents which are included in the estimator's report, and are crucial for management to understand the full technical and commercial requirements of the project. Very often contractors use the final review agenda to list the contents of the report, and again there is a good example of an agenda in the COEP.

Figure 17.1 is a domestic sub-contract summary reproduced from Chapter 12 in order to illustrate how costs from an estimate are transferred to tender summary forms.

CB CONSTRUCTION LIMITED **DOMESTIC SUB-CONTRACTOR'S SUMMARY**

Project	Fast Transport		
Ref.No.	T354	Date	27.6.08

Ref.	Trade	Company	Quotation	Discount offered (%)	Net amount	Used in estimate	Firm price allowance	Alternative/late quotations: Company	Net amount	Saving
S1	Roof covering	Beaufort Roofing	17 672	2.5	17 230	17 230	nil			
		Brett Technologies	19 560	nil	19 560					
S2	Windows	Valley Fabrications	30 641	2.5	29 875	29 875	1 225	Archiglass	27 550	2 325
S3	Plumbing	Consort	4 550	nil	4 550	4 550	nil			
		Fisher Ltd	5 890	2.5	5 743					
S4	Plastering and partitions	Swift Services	57 990	nil	57 990	57 990	nil	Oscar Finishes	46 760	11 230
S5	Joinery	Projoin Site Services	41 900	nil	41 900	41 900	1 935	L.P.Monk	38 450	3 450
S6	Suspended ceilings	Hill Systems	19 882	2.5	19 385	19 385	nil			
		Busby Site Services	21 320	2.5	20 787					
S7	Painting	Tudor Decorations	12 659	2.5	12 343	12 343	nil			
S8	Floor coverings	Freedom Finishes	12 615	2.5	12 300	12 300	nil			
		Baker Group	14 870	nil	14 870					
S9	Electrical installation	Comech Engineering	35 887	2.5	34 990	34 990	nil	Beta Technologies	22 860	12 130
		Shelton	36 500	2.5	35 588					
S10	Mechanical installation	Comech Engineering	25 667	2.5	25 025	25 025	nil			
		Shelton	27 410	nil	27 410					
S11	Surfacing	W. Smith Contracting	11 800	nil	12 450	12 450	nil			
S12	Scaffolding	CCG Scaffolding	see prelims							
		Totals			£	268 037	3 160		£	29 135

Fig. 17.1 *Summary of sub-contractors (reproduced from Chapter 12)*

CB Construction Limited		Project:	Fast Transport Ltd
TENDER SUMMARY		Date:	Jul 2008
		Tender no:	T354

			Estimate	Review changes	Tender	Analysis
1	Own measured work	Lab	115 627		115 627	20% of 7
2		Plt	31 330	-5 300	26 030	4% of 7
3		Mat nett	145 502	-2 411	143 091	25% of 7
4	Domestic sub-contractors	nett	268 037	-29 135	238 902	41% of 7
5	Named sub-contractors	nett	33 610		33 610	6% of 7
6	Provisional sums	Defined	26 500		26 500	5% of 7
7		Work sub-total	620 606	-36 846	583 760	
8	Preliminaries	Fixed	12 960		12 960	
9		Time-related	46 180	1 800	47 980	
10		Insurances	5 950	5 000	10 950	
11		Bond	3 300		3 300	
12		Prelims sub-total	68 390	6 800	75 190	13% of 7
13	Prof fees and surveys	10% *	62 061	-3 685	58 376	10% of 7
16	Price fluctuations	Lab *	3 500	0	3 500	
17		Plt *	2 950	-2 950	0	
18		Mat *	1 675	0	1 675	
19		Subs *	3 160	-3 160	0	
20		Prelims fixed *	1 620	0	1 620	
21		Prelims time *	3 650	0	3 650	
22		Fees	0	0	0	
23		Inflation sub-total	16 555	-6 110	10 445	1% of 7,12,13
24	Risk/opportunity	Risk register	12 000	-2 000	10 000	2% of 7
25	Overheads on turnover	2.5% of 31 less 'below-the-line' sums				
26	Profit on turnover	3.5% of 31 less 'below-the-line' sums				
27		On-costs sub-total	49 762	-2 671	47 092	6%
28	Dayworks	**	14 600		14 600	
29	Provisional sums	Undefined **	30 500		30 500	
30		'Below line' items	45 100	0	45 100	
31		TENDER £	874 474	-44 511	829 963	
* may be priced in prelims		** include OH+P	net to gross	32%		

Fig. 17.2 *Tender summary form*

CB Construction Limited		Project:	Primary school
SUMMARY - COST PLANNED TENDER		Date:	Nov 2008
		Tender no:	T354

	Gross floor area (m²)	2 450				Review		
	Date of benchmark data	1Q08			Estimate	changes	Tender	Analysis
				£/m²	£	£	£	
1	Trade packages	Frame+upper floors		195	477 750		477 750	13% of 7
2		Civils		205	502 250	-5 300	496 950	14% of 7
3		M+E		350	857 500	-2 411	855 089	24% of 7
4		Cladding		245	600 250	-29 135	571 115	16% of 7
5		Dry wall		95	232 750		232 750	6% of 7
6		FF+E		145	355 250		355 250	10%
7		Other		265	649 250		649 250	18% of 7
8		Work sub-total		1,500	3675 000	-36 846	3638 154	
9	Preliminaries	Staff	7%	105	257 250		257 250	
10		Other	6%	90	220 500		220 500	
11		Insurances	1%	15	36 750	-12 000	24 750	
12		Bond	1%	15	36 750		36 750	
13		Prelims sub-total	15%	225	551 250	-12 000	539 250	15% of 7
14	Prof fees and surveys	10%	*	150	367 500	-3 685	363 815	10% of 7
15	Price fluctuations	Frame	*		14 333	0	14 333	
16		Civils	*		15 068	-3 160	11 908	
17		M+E	*		25 725	-25 725	0	
18		Cladding	*		18 008	-1 550	16 458	
19		Dry wall	*		11 638	0	11 638	
20		Other	*		38 955	0	38 955	
21		Prelims staff	*		7 718	0	7 718	
22		Prelims other	*		6 615	0	6 615	
23		Fees + surveys			0	0	0	
24		Inflation sub-total		56	138 058	-30 435	107 623	2% of 7,12,13
25	Risk/opportunity	Risk register		39	95 000	-2 000	93 000	3% of 7
26	Overheads on turnover	2.5% of	32	less 'below-the-line' sums				
27	Profit on turnover	3.5% of	32	less 'below-the-line' sums				
28		On-costs sub-total		126	308 094	-5 423	302 671	6% of 32
29	'Below-the-line' sums	Loose furniture	**				0	
30								
31		'Below line' sums			0	0	0	
32		TENDER £		2 096	5134 902	-90 389	5044 513	
	* may be priced in prelims	** include OH+P		nett to gross	0	£/m²	2 059	

Fig. 17.3 *Tender summary form for a cost-planned tender*

Comments on tender summary form

Before the final review meeting the estimator will complete the left-hand column, 'estimate' of the tender summary, which gives management an outline of the whole job. In this example the estimator has perhaps gone further than necessary and suggested a total margin of £49,762 for overheads and profit. In the event, the margin was reduced to £47,092 because there were reductions in the costs. The 'review changes' column shows the financial effect of the decisions made by management during the meeting. The most significant changes are to the contractor's own work and domestic sub-contractors, where potential savings have been identified. On the other hand, senior managers are also prepared to add to the estimate and in this case have increased staff costs and insurances. These last two points may have been the subject of a company policy change, which the estimator would not have known. Clearly an aggressive stance has been taken with sub-contractors, because the inflation allowance made by the estimator has been removed.

Items 1–5 are taken from a computer report and preliminaries (items 8–11) come from a separate summary, often from a preliminaries workbook. This summary has separate totals for fixed and time-related costs. This is to enable management to see the effect of changing the programme duration on cost. So, for example, if the duration is extended by two weeks, only the time-related costs need to be applied.

Item 6. Provisional sums must now be split between those which are defined (contractor to allow for programming, planning and preliminaries) and those (item 29) which are undefined, where the contractor will be able to recover overhead costs.

Item 16–22. Price fluctuations are allowances for inflation. Again, management have taken an aggressive view of inflation for the plant costs by assuming the plant rates can be held for the duration of the project.

The insurances and bonds items 10 and 11 may have been priced in the site overheads schedule, but can be more accurately calculated when the full estimate is known. Each of these items is governed by the contract sum.

Item 10. Insurances are priced in the preliminaries workbook but need to be checked when the tender total is known. The employer's liability insurance is rarely included in the average labour rate today. This is because labour can be drawn from several sources and a global calculation is carried out in calculating preliminaries, or at the final review stage when all the parts of the estimate are known. Contract conditions should be examined carefully and changes to the standard conditions noted. Since insurance provisions are notoriously difficult to interpret, the estimator will ask for help from his insurance adviser; and in some cases obtain a quotation before completing the estimate. When a job is on site, there are often many small losses suffered by the contractor, which are not recovered by an insurance policy. The estimator should ask for records of what the average shortfall may be for the work envisaged. In the example, management decided to add £5,000, which was the excess given in the all-risks policy.

Item 11. Performance bonds and parent company guarantees are often required in today's construction market. The cost of a bond will depend on the suggested wording, duration and value, and creditworthiness of the contractor. The quotation for a bond is usually expressed as a percentage of the contract value per annum, and extends into the defects liability period. This means that a bond for a twelve-month contract with a twelve-month defects liability period will require twice the annual cost.

Item 24. Risks are usually calculated separately in a risk and opportunity register.

Items 25 and 26. Overheads and profit should be sufficient to produce a return on turnover. As an example, if the contractor needs overheads and profit of 5%, the total net cost is divided by 0.95 to produce the tender total.

Items 28 and 29. Dayworks and provisional sums (undefined) are added afterwards because they carry their own overheads.

Tender summary for a cost-planned tender

At a review meeting for a cost-planned tender, management expect to see cost plan parameters such as rates per m^2 and percentages for each component of the bid. In the example given in Fig. 17.3, the main trade packages have been market-tested. A spreadsheet workbook has been used to build up job specific preliminaries. Professional fees and surveys have been targeted at 10% of trade costs. Specialist contractors have given advice about inflation. From experience of similar projects, risk has been set at 3% of trade total.

Overheads and profit

There are three main stages in reviewing a tender. It is management's responsibility to:

1. Understand the nature and obligations of the work.
2. Review the costs given in the estimate, and if necessary adjust the costs for market conditions and errors.
3. Add to the estimate sums for general overheads and profit.

Overheads and profit should be evaluated separately because they are calculated in different ways for different purposes.

The term 'overheads' relates to off-site costs which need to be recovered to maintain the head office and local office facilities. Items to be covered include:

- Salaries and costs to employ directors and staff;
- Rental fees, rates and maintenance of offices, stores and yards;
- Insurances;
- Fuel and power charges;
- Cars and other vehicle costs for office staff;
- Printing, stationery, postage and telephone;
- Advertising and entertainment;

- Canteen and consumables;
- Office equipment including computers;
- Finance costs and professional fees.

These charges are compared with turnover to arrive at an overhead percentage. Most organizations will know the figures for previous years, but both overheads and turnover should be predicted for the future when the project is under way.

Unfortunately, when work is scarce and turnover drops, contractors look for ways to reduce tender mark-ups. The temptation is to reduce the amount for overheads at a time when they are rising in proportion to turnover. The alternative is to win less work and suffer large losses. Another solution would be for some contracts to make a greater contribution to head office costs than others.

The profit figure is a combination of discounts and additional profit required by management. Long gone are the days when supplier discounts could be thought of as a small reserve fund. In a competitive market all discounts are taken out before a small profit margin is added, to help win the work.

The profit calculation is the responsibility of senior managers (and ultimately the directors). In fact it is not strictly a calculation but a view or hunch about what margin would give the maximum profit for the company with the likelihood of winning the contract.

There are, nevertheless, some important issues which must be considered before a tender can be completed. These include:

1. The desire to win the contract; perhaps to increase turnover, or the job might be the first in a number of similar schemes.
2. Whether the project will involve contractor's finance; cashflow calculations will show the net finance needed or benefit available.
3. The effect of winning the contract on the present workload; is there sufficient turnover to meet the company's objectives and are the company's resources being used efficiently?
4. Knowledge of the client and his consultants; the attitude and competence of other parties can have an impact on the smooth running of a project.
5. The local market conditions; consider the strength of competition for the type of construction in the area (this is often the single most important criterion for choosing a winning profit margin).
6. An evaluation of previous bidding performance; knowledge of profit margins is gained by an examination of results from previous tenders.
7. There is a theory that contractors may be influenced by the client's budget; this target is found from 'intelligence' information, it may be given in the preliminary invitation to tender, or deliberately released by the client to keep the price down; in practice this seldom changes the profit margin.

Where the tender appears to exceed the sum that would win the contract, there are two refinements that can be used:

1. Re-examine the suppliers' and sub-contractors' quotations for any evidence that lower prices may be available after the main contract is awarded.
2. Consider different profit margins for direct work and that for which sub-contractors will be responsible.
3. Consider a different profit margin for loose items delivered at the end of a project, such as loose furniture and costly equipment.

Further approaches can be made to 'preferred' suppliers and trade specialists to negotiate and secure the best market price. This arrangement would include an undertaking that should the contractor be successful then the supplier or specialist would have no further negotiations and would be awarded the contract.

Finally, once the overheads and profit are settled, the amounts can be put on the summary form. All that remains is to add daywork and undefined provisional sums to arrive at the overall total.

When the final review meeting is over, it is important for the estimator to return to his desk and check all the figures on the summary sheet once again. Even better, enter the estimator's adjustments and final review changes into the electronic estimate and ensure the bill total is the same as the agreed tender figure. This fully adjusted bill will form the basis of a set of allowances for the construction team and can be adapted for presentation to the client.

18 Tender submission and results

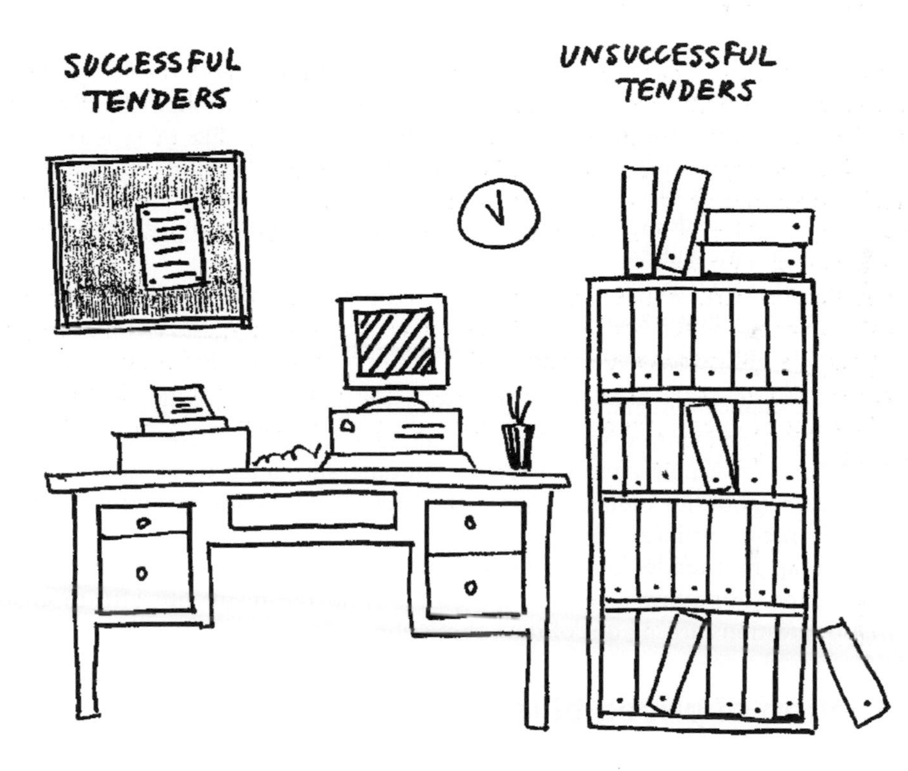

An assessment of tender performance

Introduction

How do you submit a tender? The answer used to be simple: fill in the tender form and make sure that it is delivered on time. Today, however, the winning of a contract can be more than simply giving the lowest price. The content and style of a submission for a design-and-build contract is important, and detailed programmes and method statements are commonly required by construction managers. There is, of course, a balance to be achieved between what the client wants and appearing to be too clever; different clients have different expectations.

Tender proposals for large public projects usually comprise responses to a set of questions. This has the dual benefits of focusing submissions along clearly defined elements which then permit a fair evaluation of each bidder.

The contractor needs to consider the criteria the client will use for selection. These can be:

1. Price – will the lowest price alone be the basis for selection?
2. Time – will a programme show the client that the contractor has thought about how the job can be finished on time, or ahead of time?
3. Allocation of money – will the way in which money is distributed in the priced bills help or irritate the client?
4. Method statement – would the client wish to know the methods to be adopted before accepting the offer?
5. Safety and quality – does the client expect a statement of safety or quality showing how the contractor will manage this particular contract?
6. Construction team – is the contractor proposing to supervise the job with experienced staff who will work as an effective team with the consultants?
7. Presentation – how important is an accurate, well-presented offer?

With private finance tenders, where the contract is for design, construction, finance and facilities management, a wider range of factors are part of the scoring system. It is becoming increasingly common for the funding allowance to be known and so there is an expectation for bidders to offer the best product and services for the sum (or annual payment) available.

The follow up to a tender can affect the outcome – the estimator will contact the client soon after submitting a tender, not only to find the result but also to ensure there are no questions arising out of the submission.

Completion of priced bills

The tender summary form produces a tender sum which must be transferred to the bills of quantities for submission. If a priced bill of quantities has to be submitted with the tender, then the way in which money is spread in the bill should be decided at the final review meeting. The contractor often changes the actual breakdown of prices in a priced bill of quantities, to:

1. Produce a reasonable cashflow arising from interim valuations;
2. Apportion monies in a way which the client will find acceptable;
3. Increase the money set against undermeasured items and decrease the price of overmeasured items.

The example tender analysis forms given earlier show the first bill total was £664,705 before the preliminaries were added and any adjustments were made. The tender sum arrived at after final review was £698,437. If the bill was inked in using the first pricing level, the amount remaining for preliminaries would be:

$$£698,437 - £664,705 = £33,732$$

This is not the true preliminaries sum but is the amount needed to bring the bill total up to the tender sum. If the parties agreed to proceed on this breakdown (and they probably will), a part of the preliminaries, overheads and profit would remain in the measured work portion of the bill. In particular, the items to be carried out by domestic sub-contractors would carry a considerable mark-up.

Taking the example a stage further, if the estimator had the use of a computer and had made the tender adjustments *before* inking in the bill for the client, the breakdown would be in line with the tender summary below:

	£	£
Measured work and provisional sums (including £620 fluctuations on labour)		**529,398**
Preliminaries		
Labour	12,960	
Plant	37,890	
Materials	5,780	
Sub-contracts	3,300	
Staff	37,619	
Fluctuations	1,675	
Water	650	
Insurances	5,000	
Risk/opportunity	10,000	
Bond	1,050	
Overheads and profit	18,000	**133,924**
Provisional sums and dayworks		45,100
Tender total		**708,422**

The client should not be surprised to see this large sum (£133,924), which is 25% for preliminaries because it is based on the true allowance. If the contractor anticipates a problem with this breakdown, he can move some money, either:

1. Into 'safe' items in the measured work portion of the bill, looking for work which will be carried out early in the contract (safe items are those which appear to be measured correctly or are judged to be under-measured at tender stage); or
2. By using a computer system to add a percentage to all the rates in a bill of quantities.

The preliminaries total should be broken down in the bill of quantities with sums for fixed and time-related items. A surprising number of contractors ignore this breakdown and prefer to insert a lump sum in the collection; they assume that if their tender is the lowest, more details can be submitted to meet the needs of the quantity surveyor (or engineer) for valuation purposes. The contractor also knows that if there are some small queries raised by the client then his tender is (probably) being considered for acceptance.

When contractors are tendering in a competitive market, in which work is scarce, they know that their bids must be close to the predicted cost of carrying out the work, with little mark-up. Sometimes tenders can be slightly below cost. The contractor, in taking a calculated risk on how the contract will turn out, may price some items in a way which appears to be inconsistent. As an example, assume that a bill of quantities has two equal amounts in items for breaking out rock: one in reducing the site levels and the other in excavations for drains. Contractor A priced both items at £28.00/m^3 and contractor B priced the rock in open excavations as nil and in drains at £36.00. The overall effect on the tender sum was the same but contractor B had discovered a serious under-measurement in the drainage bill; he was therefore hoping the drainage bill would be re-measured and valued at the higher rate. This might appear to make sense but, as many contractors have learned to their cost, plans can go wrong. If the quantity of rock in open excavation increased substantially, the contractor would suffer a serious financial loss.

Tender submission for cost-planned tenders

The estimate is used to populate a tender cost plan which can be used to generate:

- Cash flow – a schedule of payments often called a cash flow or 'value-drawdown' schedule.
- Life cycle inputs – an elemental cost plan, in the BCIS format, is sent to the life cycle surveyor for him to value the capital replacement programme.
- Submission cost plan – the internal cost plan is modified to give the best commercial position in an external submission.

Outputs from estimate

In most framework and partnering contracts, there are open-book principles. In these cases there is only one cost plan which is shared with the client during all stages of the tender period. Furthermore, when payments are made to the contractor to produce a more detailed tender, there is usually an obligation to share all information as it arises.

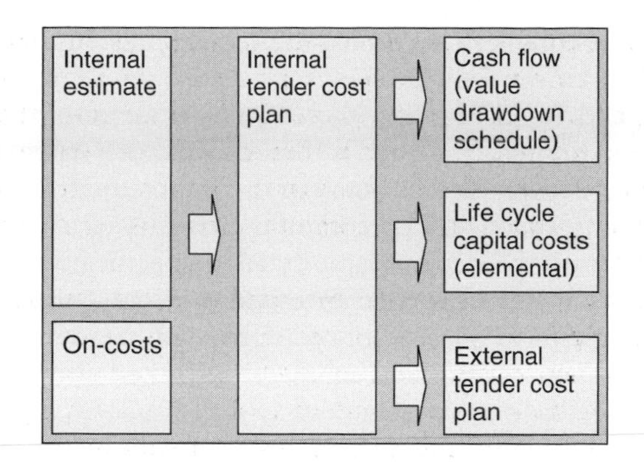

Detailed submissions

The estimator needs to be aware of the submission requirements from the start of a bid. Tenders based on cost plans are often produced with broad cost plan data and so detailed priced build-ups and bills of quantities may not be available. Examples of requests which are onerous are:

- Priced schedules of fittings, furniture and equipment;
- Bills of quantities for external works;
- Costs of functional areas, such as hospital departments or rates for clinical space as opposed to circulation;
- Fully priced preliminaries which can be especially tedious if the client has a different pro forma.

The solution is to submit information that *has been produced.* It is important to note that the estimator should familiarize himself with the submission requirements at the start. This makes it much easier to complete the forms.

In order to assist the client's team in understanding the bid, notes can be attached to cost plans which explain the way in which the forms have been filled in. For example, an estimator might say: 'Element 5B of the elemental cost plan includes costs for catering equipment in the coffee shop and examination lamps in the treatment rooms', and 'The heading "Professional Fees" includes fees for the design team, the cost of surveys and charges for planning permission and building regulations.'

Tender presentation

Most tenders are submitted on a pre-defined form which has the effect of standardizing the offers and discouraging exclusions, alternative bids and other qualifications.

For traditional procurement methods, the tender presentation normally includes the form of tender and a covering letter.

With design and build projects, the contractor submits his tender in the form of 'the contractor's proposals'. This is the contractor's response to the employer's requirements and is explained in detail in the Supplement Number Two to the CIOB Code of Estimating Practice. The common elements of a design and build offer are: drawings to illustrate the proposals, a detailed specification, and a tender sum broken down into its major elements. In order to avoid confusion later, the employer should stipulate the form and extent of information needed. The contract sum analysis should be adequate for both valuing work executed and changes after the contract is awarded.

Instructions to tenderers should include a date, time and location for submitting a tender. The contractor is responsible for presenting the documents by the time given and in some cases may be permitted to send a tender by email or facsimile transmission, followed up by first-class post. Contractors rarely submit their tenders early for two reasons:

1. They might receive a lower quotation from a sub-contractor or supplier which could improve the bid;
2. A tender price cannot be communicated to a competitor in time for him to better the price.

Clearly, if a form of tender is required, the contractor must enter the price in the space provided, and ensure that the document is signed and dated by a person authorized to act for the tenderer. The estimator must carefully check the instructions to tenderers for any other documents to be submitted with a tender. Many public organizations would not expect to see additional documents, but sometimes ask for an outline programme and method statement. Contractors commonly attach a letter to the form of tender and priced bills of quantities, but are careful not to add statements or conditions to their offer which might be seen as qualifying the tender.

There are times, of course, when qualifications are unavoidable. The following examples show how a contractor may have no choice but to bring matters to the client's attention either before submission or in the tender:

1. A contractor may decide that the wording of a performance bond is unacceptable.
2. Having examined all the resources needed for a job, the contractor may find the contract duration is too short.
3. Late amendments can impose extra responsibilities which the contractor is unable to resolve.
4. The contractor may wish to add to the list of named or approved sub-contractors following the failure of one of the firms on the list.

There is a theory (which is not sensible in a competitive market) that all such problems can be resolved by 'throwing money at them'. A short duration, for example, can be overcome by adding liquidated damages for the expected overrun. The JCT Practice Note No. 6, 'Main Contractor Tendering' replaces the Code of Procedure for Single Stage Selective Tendering. It suggests that the tenderer should tell the client if there are any matters needing clarification as soon as possible and preferably not less than 10 days before the tenders are due. If the tender documents need to change, the tender date may be extended. The Practice Note takes a strong line on qualified tenders by stating that qualifications should be withdrawn otherwise the tender may be rejected. With this in mind, qualifications are sometimes written in general terms so the contractor can delay his decision on which issues he wants to qualify. Typical statements used are:

1. 'During the tender period, we identified some savings which can be brought about by small technical changes ...';
2. 'We would need to clarify some of contractual matters before entering into an agreement, but do not expect this to affect our price ...'

Another approach, more common in civil engineering contracts, is to submit an alternative tender. In this way a contractor is able to comply with the tender conditions by submitting a 'clean bid' and at the same time reveal an alternative offer which usually reduces the construction costs with only minor specification/contractual changes. An alternative tender may also be the vehicle to propose a shorter duration, submit a programme and impress the client with technological expertise. Another form of alternative tender is to offer an amended contract where the contractual risks can be shared by open and frank problem-solving in an atmosphere of trust. This produces a contractor-led partnership or alliance between the parties as envisaged by Sir Michael Latham in his 1994 report, *Constructing the Team*.

There is a growing practice of submitting company brochures, technical literature and other publicity material with the tender. This 'window dressing' is often unnecessary. The client is more interested in the price and approach to *his* job; in any case the company profile has been examined at the pre-selection stage.

The letter accompanying the tender can be used to confirm the amendments to the tender documents received during the tender period. The basis of the offer is after all the tender documents and all amendments received by and not sent to the contractor before the tender date. The main rule for this letter is to keep it short, no more than one page. The rules for writing letters or emails are often set out by the client's team. Correspondence should be channelled through a named contact so that the client is not bombarded with questions from different team members. Any clarifications for large projects are numbered and made available for all bidders unless there is a request to keep the question confidential for commercial reasons.

Vetting of tenders

If bills of quantities are not required with the tender form, the contractor who has submitted the lowest bid is asked to submit his priced bill of quantities for examination (and adjustment where errors are found). The unsuccessful contractors should be told immediately that their tenders were unsuccessful. Once the contract has been let, all the tenderers should be notified of the results so that they can measure their performance against others in the industry. The results must remain confidential before a contract is awarded because the lowest contractor could negotiate higher prices if he knows the tenders made by his competitors. There have also been instances of higher tenders being reduced below the lowest price received on the tender date.

The JCT practice Note No. 6 2002 has introduced the concept of best value, as an alternative to lowest priced bid. For a best-value tender, the criteria to be used to assess best value should be stated in the pre-qualification or tender documents. Once a choice has been made using this assessment criterion (sometimes referred to as 'score card') the pricing documents of the preferred tenderer can be opened and checked.

An examination of a contractor's bill of quantities may reveal different kinds of errors. Some errors should be corrected using the JCT Practice Note, Alternative 1 or 2. Alternative 1 gives the contractor the choice of standing by his tender or withdrawing it. Alternative 2 gives the contractor the opportunity to confirm his offer or amending it to correct genuine errors. The term 'genuine errors' is not defined in the Practice Note, but normally means:

1. 'Errors of computation' caused by mistakes in multiplying quantities by rates or totalling pages;
2. Patent errors in pricing such as pricing hardwood joinery at the same rates given for softwood earlier in the bill, or pricing steel reinforcement at a rate per kilogram where the unit is tonnes;
3. Inking-in errors are usually simple to correct because the summaries will be correct even if an individual rate has been entered wrongly.

Some patent errors can be difficult to spot because the contractor may have his own commercial reasons for distributing the money in a certain way. A common difficulty is the pricing of similar items at different prices. A quantity surveyor may wonder why concrete in a ground floor slab is priced at £71.55 per m^3 in the workshops and at £102.15 in the office area of a factory development. There are many logical reasons for this apparent mistake. It could be to do with continuity of work; perhaps the latter case involved a high extra cost for part load charges from the concrete supplier. If the office area was programmed for the beginning of the contract, the quantity surveyor might think the contractor had 'front-loaded' the bill to produce an early income. The contractor cannot be required to change the rates but the quantity surveyor may not be able to recommend the tender to the client. If the contractor has priced the bill in such a way that considerable sums of money are overpaid at the beginning of a project, the client might be at risk if the contractor fails to meet his obligations, or becomes insolvent.

Post-tender negotiations and award

Each tenderer will want to know the result as soon as possible in order to plan the construction phase of successful bids, or file away the documents and redeploy resources involved in unsuccessful ones. The direct approach is usually the most productive. A telephone call to the person carrying out the vetting process is often enough to know whether the tender documents can be archived and the computer files backed up on CD ROM.

The next stage for the lowest tenderer could be meeting the client, or his advisers. This is often necessary before an award can be made. The matters discussed are mainly financial and contractual although methods can be important. Any errors or discrepancies in the bill of quantities can be resolved at the meeting and contractual details can be discussed and agreed. This allows both parties to understand their obligations before the formal agreement comes into effect.

The contractor should be represented by the estimator and senior construction staff. The estimator should ask for an agenda and a list of those attending. The agenda will allow him to brief his team in advance and take relevant documents to the meeting. The list of client's representatives and advisers is important. The contractor will try to respond to questions with staff who have the necessary specialization. Above all, an estimator must avoid the situation where he alone enters a room where all the consultants and client's representatives are assembled, confidently expecting to get the best deal for the client.

There are some pre-award meetings where the estimator may not be the best person to lead the contractor's team. The approach must be robust with a firm commitment to carrying out the work to a high standard and on time. Unfortunately estimators often get bogged down in detail and have been known to highlight small errors in the documents or be pessimistic about aspects of the programme. If this happens, a senior manager can present a wider view and suggest positive remedies which have been successful on other projects. Above all the client must have confidence and believe the contractor can carry out the work with a willingness to solve problems and work closely with the client's team at all times.

Tendering performance and analysis of results

There are several ways in which the performance of an estimating department can be measured. The simplest method would be to count the number of successful bids

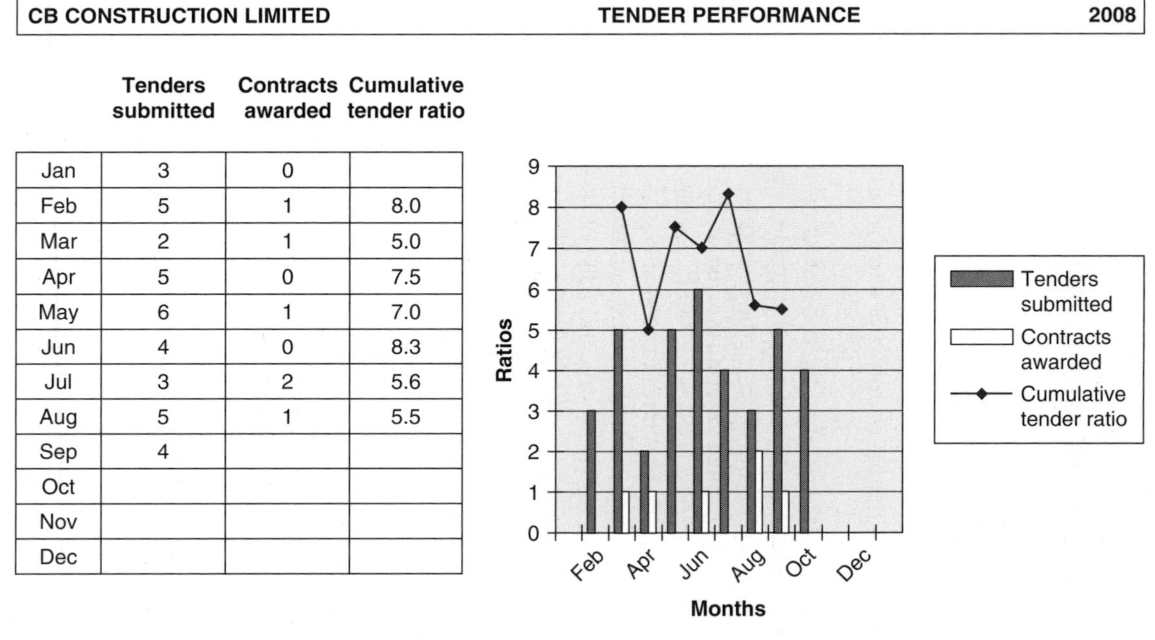

CB CONSTRUCTION LIMITED	TENDER PERFORMANCE		2008
	Tenders submitted	Contracts awarded	Cumulative tender ratio
Jan	3	0	
Feb	5	1	8.0
Mar	2	1	5.0
Apr	5	0	7.5
May	6	1	7.0
Jun	4	0	8.3
Jul	3	2	5.6
Aug	5	1	5.5
Sep	4		
Oct			
Nov			
Dec			

Fig. 18.1 *Cumulative tender ratio*

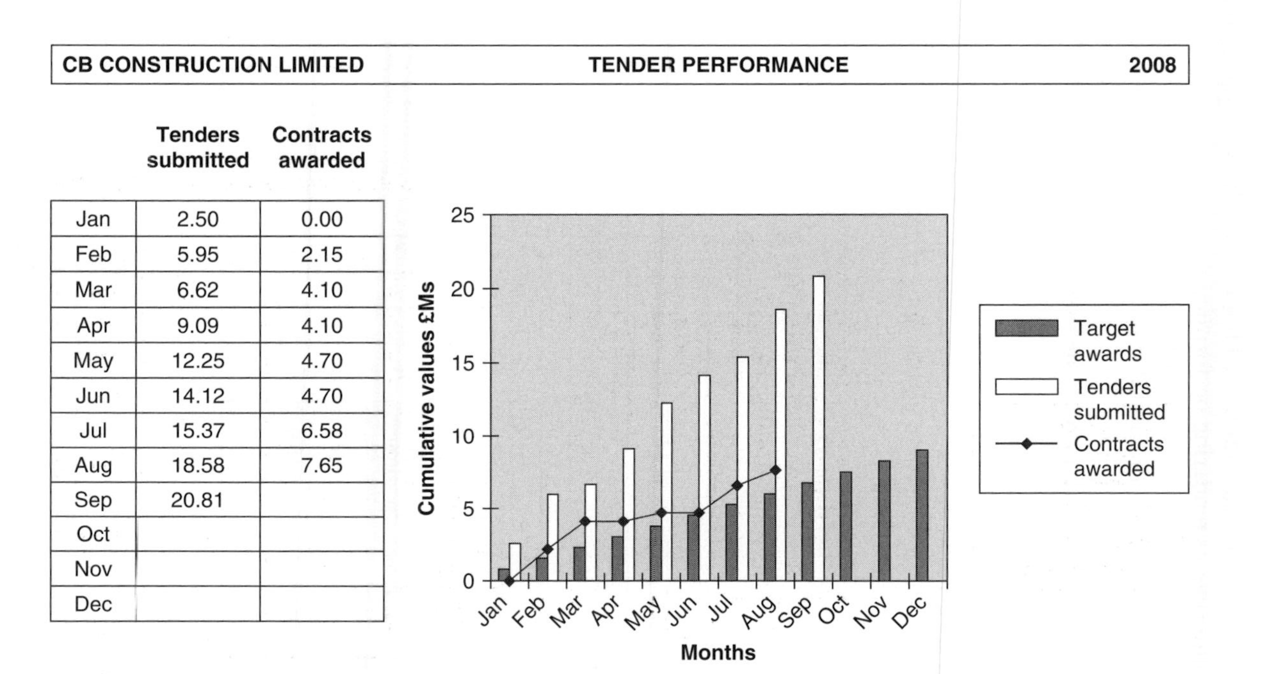

CB CONSTRUCTION LIMITED	TENDER PERFORMANCE	2008

	Tenders submitted	Contracts awarded
Jan	2.50	0.00
Feb	5.95	2.15
Mar	6.62	4.10
Apr	9.09	4.10
May	12.25	4.70
Jun	14.12	4.70
Jul	15.37	6.58
Aug	18.58	7.65
Sep	20.81	
Oct		
Nov		
Dec		

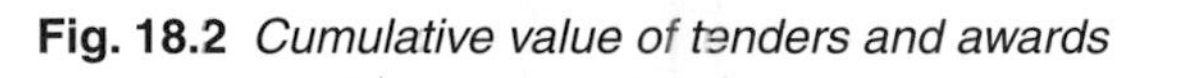

Fig. 18.2 *Cumulative value of tenders and awards*

compared with the number of tenders submitted. Fig. 18.1 shows the cumulative ratio of tenders to contracts won in eight months. This is a crude technique, which does not help the firm to improve its tendering performance.

If a contractor's business strategy is to increase turnover, then a simple graph showing the value of contracts awarded would be useful. Fig. 18.2 shows the value of contracts won, with the total for tenders submitted. This graph is made more effective by showing the performance related to a target set by the board of management.

The CIOB Code of Estimating Practice offers a suitable form for recording the results and subsequent analysis of tenders. Fig. 18.3 illustrates three methods which can be used to produce a ratio analysis for a tender; each contractor will pick the method which helps him to evaluate his tender performance. The underlying principle is that, in general, cost category ratios of similar jobs remain approximately the

CB CONSTRUCTION LIMITED	**Tender analysis**	Project name	Fast Transport Limited
		Tender date :	Jul 2008
		Tender no :	T354

	Tender summary Description		£	Column A % of 13	Column B % of 6	Column C % of 9
1	Own measured work Labour		115 627	16.6	22.8	18.3
2	Plant		26 030	3.7	5.1	4.1
3	Materials	nett	143 091	20.5	28.2	22.7
4	Domestic sub-contractors	nett	220 935	31.6	43.5	35.1
5	Price fluctuations		2 295	0.3	0.5	0.4
6	Direct costs		507 978		100	
7	Preliminaries		104 249	14.9		16.5
8	Overheads, profit and discounts		18 000	2.6		2.9
9	Tender less fixed sums		630 227			100
10	PC Sums - Nom sub-contractors		19 500	2.8		
11	PC Sums - Nom suppliers		3 610	0.5		
12	Provisional sums and contingencies		45 100	6.5		
13	TENDER		698 437	100		

Fig. 18.3 *Alternative methods for tender ratio analysis (relates to data given in Tender Summary Form)*

same. If there are significant changes in these ratios, the estimator should be able to explain the reasons for the deviations at the final review meeting.

Column A of Fig. 18.3 has been calculated with each element being expressed as a percentage of the total tender sum. This is the least refined measure because the elements are being compared with a figure which includes overheads and profit. Column B shows the elements expressed as percentages of direct costs excluding PC and provisional sums. This has the advantage of removing the sums fixed by the client but fails to bring the project (site) overheads into the calculation. Perhaps the best solution would be to include preliminaries and take out all sums set by the client (see column C).

The next stage is to examine bid results in more detail. Figure 18.4 lists the tenders submitted and shows the effect of subtracting the sums set by the client. The

CB CONSTRUCTION LIMITED	Project name:	Fast Transport Ltd
	Tender date :	Jul 2008
	Tender no :	T354

Tender list	Tender	Tender less set sums		% over lowest bid	% over mean bid
1 Baldwin Bros	764 780	696 570		14.4	6.2
2 Javelin Construction	707 990	639 780		5.0	-2.4
3 Barry and Hardcastle	677 250	609 040	L	0.0	-7.1
4 CB Construction	698 437	630 227		3.5	-3.9
5 Wessex Contracting	718 415	650 205		6.8	-0.8
6 Newfield Building	756 410	688 200		13.0	5.0
7 Simpson	743 382	675 172		10.9	3.0
Mean tenders	723 809	655 599		7.6	0.0

Sums set by client	
Statutory undertakings	19 500
PC Sums - nom suppliers	3 610
Prov sums	45 100
Total	68 210

Margin lost by	£21 187

Fig. 18.4 *Tender results*

percentage over lowest bid gives a measure of the margin by which the job was lost, and the percentage over mean bid provides the contractor with a guide to the deviation from the average prices set by other contractors. As a rough rule of thumb, many contractors would feel confident in their estimating performance if their percentage over mean bid fell within the range ±10%. Figure 18.5 gives a summary of tender results for a three-month period.

It would be unwise to change a bidding strategy using a small amount of data. The analysis of performance must start with:

1. A period of consistent net pricing;
2. A steady tendering policy in terms of adjustments to direct costs and the percentage mark-up;
3. A determination to obtain tender results from consultants and clients.

CB CONSTRUCTION LIMITED **Summary of tender results** 2008

	Project title	Tender date	No. of tenderers	Rank	% over lowest bid	% over mean bid
T351	Renton Cakes	Jun 08	3	1	0.0	-2.3
T352	Warland Access Road	Jun 08	6	5	11.0	4.9
T353	Retail units, Swindon	Jun 08	6	1	0.0	-3.1
T354	Fast Transport Limited	Jul 08	7	2	3.5	-3.9
T355	Star Tyres	Jul 08	5	2	5.0	-3.8
T356	Cinema Roof	Jul 08	6	4	12.3	2.2
T357	Fire Station	Aug 08	6	3	6.7	1.2
T358	Railway workshops	Aug 08	7	1	0.0	-7.7
T359	British Coal underpinning	Aug 08	13	5	15.0	1.0

Fig. 18.5 *Summary of tender results*

Bidding strategy

A bidding strategy can be defined as a broad framework of methods and timing to achieve stated objectives. It is interesting to note that in military terms, the word 'strategy' means the skilful management of an army in such a way as to deceive an enemy and win a campaign. In business the stated objectives can sometimes be achieved by deceiving the opposition but principally the specific objective is to be successful in winning contracts at prices which would allow the organization to carry out work profitably. A tendering strategy can be developed as a statement of aims, as follows:

1. To identify a suitable market in terms of type of work, size of contracts and geographical location.
2. To develop a reputation for safety, quality and speed of construction within economic limits.
3. To secure stated targets for turnover.
4. To evaluate the company's performance, and compare with that of its competitor.
5. To compare the financial performance of a project with the costs predicted at tender stage.

A contractor can improve his tendering efficiency with better marketing, and only accepting invitations to tender which meet clear guidelines. Most companies have criteria which include type of work, size and location of the project; risk to be transferred to the contractor; and status of the client.

Finally, it is worth considering a different approach; that is, to ignore the tender levels set by your competitors. Produce pricing levels which are right for your company and avoid playing the market. If prices fall below what you consider to be an economic level, look for other markets where margins can be preserved.

19 Action with the successful tender

Introduction

Gone are the days when the award of contract signalled the end of the estimator's contribution to a project. There are two important tasks to be performed: firstly, helping the construction team with procurement and technical advice during the mobilization period; and secondly, producing cost information which will form the budget for the job. The transfer of information has been improved dramatically with the introduction of computers. There are many estimating packages available, which can be used to produce tender allowances and later assist during the construction phase to control sub-contractors' payments and produce valuations for the client. The main advantage of using a computer is that an estimator can adjust his estimate to take account of the decisions made at the final review meeting and make post-tender adjustments which are sometimes agreed with the client following the vetting stage. For close financial control of a project, the site manager should be trusted with the detailed budget; he can then control the resources efficiently and contribute to a simple feedback system.

Information transfer

The estimator must take great care to produce accurate information in a form that is simple to understand. A checklist can be used to ensure the handover information is complete, as follows.

The estimate

- Correspondence with client and the client's team
- Survey reports
- Quotations and advice from the supply chain
- Analysis of quotations
- Tender programme and method statements
- Preliminaries workbook with review adjustments

- Cost plan/bills of quantities with internal allowances
- Tracking of changes made at the final review meeting
- Estimator's notes
- Further information received after tender submission (if any)
- Any other findings or assumptions which might include temporary works drawings, photographs, site layouts, minutes of meetings with specialists and the client, technical literature and site visit reports.

The offer

- Invitation to tender (ITT) documents including surveys and drawings
- Form of tender
- Submitted cost plan and bills of quantities
- Contractor's proposals/responses
- Correspondence and clarifications forming part of the offer.

The item that needs the most effort is the contractor's bill of allowances because this gives the site manager the fully adjusted rates for the work. The changes made during the tender period must be applied to all the rates affected. It is not satisfactory, for example, to say to a site manager that the budget for concrete is as given in the supplier's quotation less 5%, which was an additional discount taken during the final review meeting, plus a 20% addition to the waste allowance for concrete poured against the sides of excavations. No site manager has the time to trace such changes made by the estimator.

There has been a cynical view in the industry that an estimator's figures should not be made available to construction staff on site. There are three main reasons for this attitude:

1. The construction team should get the resources for the job at the lowest possible rates and not just beat the tender target figures.
2. There may be a security problem if the rates are open to view on site; a competitor may steal an advantage and sub-contractors could be upset if they had sight of the real allowances.
3. There may be problems with interpreting data, and staff can become confused when confronted with bills produced for different purposes.

A more subtle reason might be that some organizations like to adjust the financial targets as more is known about the job. As an example, a contracts manager might decide that a concrete pump will not be necessary for placing concrete and hopes the site manager has not seen this provision in the tender. This clearly shows a lack of confidence in the site manager and could adversely affect the financial control of the project.

The important point is that the site manager must appreciate the cost implications of the decisions made on site, particularly with the use of direct labour, the costs of materials and equipment and the value of sub-contracted work. The information provided by the estimator can be used:

1. to set bonus levels;
2. to produce forward costing data;
3. to quantify resources for planning exercises;
4. to examine the financial performance of a contract through forward costing methods;
5. to compare the final building costs with the tender budget.

The person on site will perform better if trusted with important information; he is then more likely to accept greater responsibility for the financial success of a project and contribute more effectively to the evaluation of future projects.

The extract from a bill of allowances given in Fig. 19.1 was produced by a computer package that was used by the estimator during the tender stage and the quantity surveyor during construction. It gives the true allowances to the site manager (columns a to d) and the rates submitted to the client in column e. The rates given to a client will differ from the tender allowances for many reasons. These include:

1. Changes are made to rates after the submission bill is inked in. For example, a site might be found for tipping surplus material, which reduces the rate for disposal. In extreme cases, the bill can be inked in before the final review meeting. This would lead to many differences.
2. A proportion of overheads and profit may be included in the rates submitted to the client.
3. Money may have been moved to ensure early payments or to take advantage of mistakes in the bill; the true rates must be given to the site manager for cost control purposes.

There are, in effect, two bills of quantities – the client's bill for valuation purposes, sometimes called the 'selling' rates, and the contractor's bill for costs control, referred to as the budget, site allowances, buying rates or internal bill.

Before work can start on site, the construction manager will bring together all those associated with the contract. This internal pre-contract meeting is an opportunity for the estimator to introduce the scheme to the construction team and personally explain the contents of the handover package, and expand on the methods of construction, resources and organization used as the basis of the tender. The estimator will be given certain duties at this meeting, mainly to do with the transfer of information described earlier, checking the contract documents, and he may be asked to take part in negotiations with sub-contractors appointed at the start of the project.

CB CONSTRUCTION LIMITED	**BILL OF ALLOWANCES**	Project name:	Fast Transport Ltd		
		Contract no:	94022	Date:	Aug 2008

Bill Ref:		Short description	Quantity	**Site Allowances** Lab (a)	Plt (b)	Mat (c)	Sub (d)	Rate	Total	Margin/ adjustmt	**Client's bill** Rate (e)	Total
3/9	A	Filling to excavations	18 m^3	2.66	2.58	17.79		23.03	414.54	-2.00	21.03	378.54
	B	Filling to soft spots	100 m^3	1.88	2.19	17.79		21.86	2 186.00	-2.74	19.12	1 912.00
	C	Level and compact exc	224 m^2	0.70	0.39			1.09	244.16	0.03	1.12	250.88
	D	Dust blinding and compact	2 890 m^2	0.38	0.38	0.75		1.51	4 363.90	-0.62	0.89	2 572.10
	E	Grout under baseplate	21 m^2	37.50		5.00		42.50	892.50	-1.08	41.42	869.82
	F	Plain conc 15N trench	76 m^3	19.06		65.25		84.31	6 407.56	1.72	86.03	6 538.28
	G	Plain conc 15N isol founds	4 m^3	20.31		67.00		87.31	349.24	0.31	87.62	350.48
		Page total	3/9	£3 808	£1 451	£9 599			£14 858	-£1 986		£12 872
3/10	A	Reinf conc 35N founds	52 m^3	17.89		75.94		93.83	4 879.16	-12.50	81.33	4 229.11
	B	Reinf conc 35N isol founds	5 m^3	19.77		78.16		97.93	489.65	-14.69	83.24	416.20
	C	Reinf conc 35N col casing	12 m^3	41.09		81.50		122.59	1 471.08	-13.38	109.21	1 310.50
	D	Fabric D49 wrapping	40 m^2	2.19		1.17		3.36	134.40	1.18	4.54	181.62
	E	Fabric A393 in slab	2 765 m^2	1.56		5.16		6.72	18 580.80	0.53	7.25	20 042.71
		Page total	3/10	£5 923		£19 632			£25 555	£ 625		£26 180

Fig. 19.1 *Extract from contractor's bill of allowances*

The CIOB Code of Estimating Practice recommends procedures for checking contract documents. The estimator must ensure the contract documents received for signature are identical to those which were used as the basis of the tender, and any amendments issued by the client during the tender stage have been correctly incorporated. Correspondence confirming post-tender negotiations may also be included if it helps clarify the basis of the agreement.

Feedback

Feedback is the weakest part of the estimating function. In practice, an estimator receives information about the actual costs of construction in a haphazard way and usually hears, through a third party, about underestimated costs – rarely will he be told about high rates. So why are companies slow to set up procedures which would ensure feedback information is available to estimators? The following problems are often quoted:

1. Feedback information is historical.
2. Each project is different from the next.
3. Financial performance is determined by the effectiveness of the site management team.
4. An estimator uses constants, which are not always job-specific.
5. A feedback system would be expensive to implement.
6. Market prices can change dramatically with little notice.
7. Confidential information is not available to site staff.

Buyers and site staff are sometimes reluctant to divulge the low prices they have achieved through aggressive procurement. Their fear is that if the estimator priced further work at these levels, it would be difficult to improve on the budget if the tender was successful.

It may be that there is a middle course whereby a company can report on certain aspects of a job in progress, as follows:

1. The actual costs associated with preliminaries could be written in a spare column on the schedule produced by the estimator.
2. Individual investigations can be carried out to find the actual waste of high-value materials.
3. The average cost of employing certain categories of labour could be compared with the all-in rate used at tender stage.
4. The value of sub-contracts (and major material orders) which have been let can be entered on a comparison sheet (see Fig. 19.2). This would give management, and estimators, evidence of the buying margins which are available in the current market.

CB CONSTRUCTION LIMITED				Project name :	Fast Transport Limited		
Sub-contracts placed				Date :	Mar 2009		
				Contract no:	94022		

	Trade	Tender allowance name	allowance	Contract placed name	order value	projected value	buying gain
1	Roof covering	Grange Roofing	20 500	Grange Roofing	18 450	18 450	2 050
2	Windows	Aliframe	31 875	Westpoint Windows	29 319	30 188	1 688
3	Plasterwork	McLaughlin	29 313	McLaughlin	32 088	32 088	-2 775
4	Partitions	Port Drylining	25 063	Port Drylining	25 063	25 063	
5	Joinery	Robin Joinery	42 500	Robin Joinery	36 938	36 938	5 563
6	Ceilings	Wignall Hampton	24 750	Shrimpton Ceilings	23 250	23 250	1 500
7	Painting	T & G Jackman	15 375	not yet placed			
8	Floor coverings	ABA Furnishings	15 375	not yet placed			
9	Surfacing	Gatwick Plant	5 250	not yet placed			
10	Landscaping	no quote	9 500	not yet placed			
11	Mechanical	Moss and Lamont	48 875	Hutley Engineers	44 859	46 431	2 444
12	Electrical	Tate Electrics	39 875	Tate Electrics	39 875	39 875	
		total	**£308 250**			total	**£10 469**

Fig. 19.2 *Comparison of sub-contracts placed with sub-contract allowances*

5. For small repetitive jobs, where detailed feedback is needed, an extra column could be inserted in the bill of allowances for the eventual costs to be added to each item. This is an obvious application for a computer using a tailor-made package or a spreadsheet program.

The relative importance of these investigations will depend on the estimator's need for information and the size of each contract. The benefits are that future estimates become more reliable and more accurately reflect the cost of construction work.

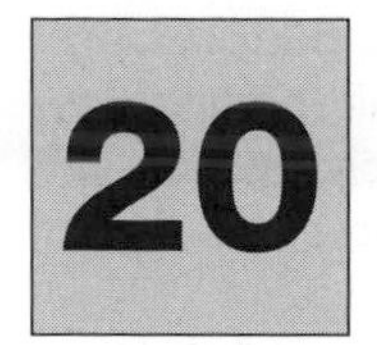

20 Computer-aided estimating

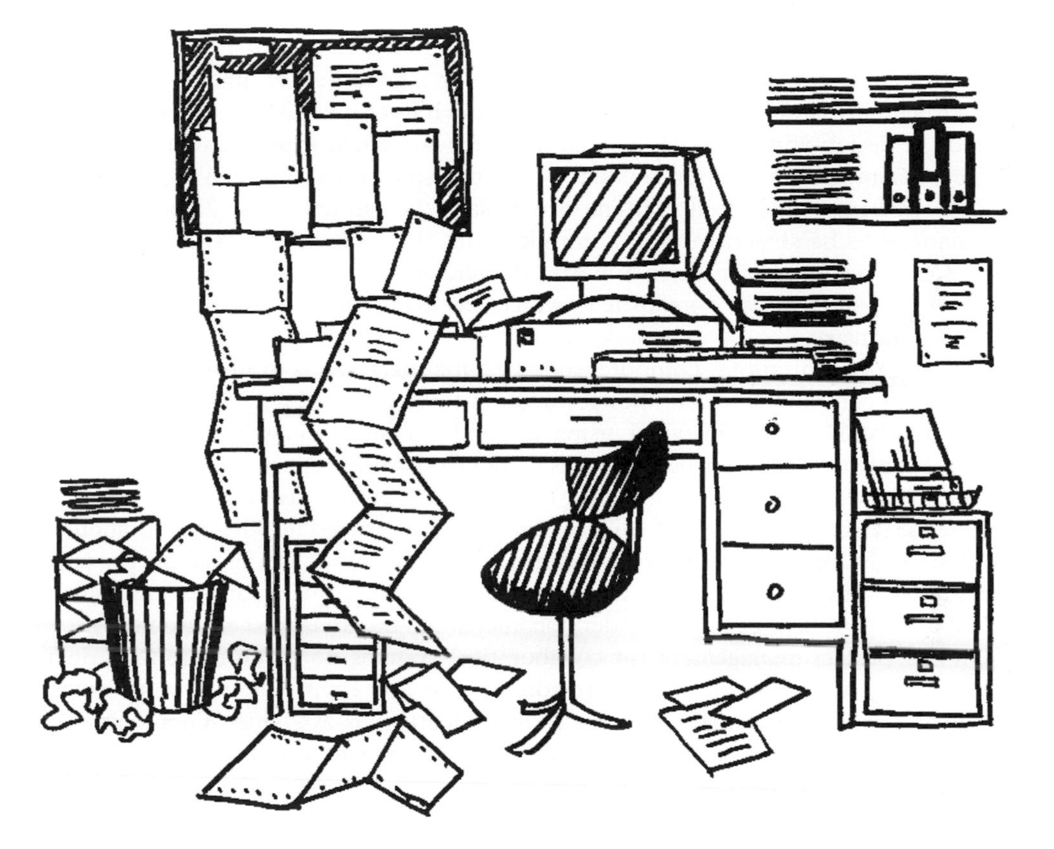

The paperless office

Introduction

The use of computers by estimators has grown steadily since the early 1980s when stand-alone computers were introduced to the desks of estimators. It has been difficult to measure the improvements that computers have brought to estimating performance. This is probably because the benefits come from additional facilities that manual systems cannot provide. For example, quantities or rates can quickly be changed at any time with a computer system. Summaries can be produced by using sort codes representing elements or sub-contract packages.

For many projects, the estimating process begins with the receipt of tender documents. The estimator will now receive drawings, written documents and where appropriate bills of quantities or cost plan templates in an electronic format. File formats often cause confusion. For example, drawings are commonly sent in PDF format without scale bars, which makes measurement very difficult. Even worse, some cost plan templates are issued in PDF files which can take up a great deal of time converting to a spreadsheet format. When contractors commission bills of quantities for plan and specification or design-and-build tenders, the estimator lays down appropriate protocols.

Early examples of computer-aided estimating software have been replaced with flexible systems, which do not attempt to replace the estimator's skills but allow the calculations to be structured and controlled with the added benefits of rapid calculations and computer-generated reports. If there is a standard piece of application software, it would be the spreadsheet, which has been adopted throughout the construction industry.

For larger organizations, computer-aided estimating systems are best implemented on 'central' computers, which allow estimators to work simultaneously on a project. Local networks can be installed using a Windows interface, which runs general-purpose software and specialist packages linked to shared printers.

The most exciting opportunities will come from a greater use of the Internet and on-line services. The estimator will no longer be restricted to the information on his desktop PC. Day-to-day correspondence is sent by email, lists of suppliers can be accessed from interactive business directories, up-to-date technical libraries are available on a 'pay-as-you-view' basis and tender documents will be exchanged electronically.

On-line project management (or collaborative) systems do much more than document management. In addition to providing an on-line environment – or extranet – for storing project information, a wide range of features is available. When combined with telephone systems, such as video conferencing and now web conferencing, the opportunities to communicate effectively are extensive, with the additional benefit of reducing travel time between meetings.

Aims of computer-aided estimating

The computing debate has raised questions about the role of the estimator and whether estimators should change their methods to conform to computing techniques or should computers be used to mimic the way estimators have worked in the past. What appears to have happened is that estimators have developed their computing skills, not just in using estimating systems but adopting spreadsheet and database packages where appropriate; and the software specialists are beginning to respond to the needs of estimators with more flexible systems. There is still a market for the large database of standard items, probably in the bill production phase whether created by the private QS or contractor.

So why has it been so difficult to implement computer-aided systems in construction? There are certainly less packages available and many estimators have migrated to spreadsheets.

Estimators need to make decisions throughout the pricing stage, mainly because each project is different. The standard labour and plant outputs that the estimator has in mind are adjusted to suit the circumstances. The circumstances might include many variables such as distance from compound, ground conditions, depth below ground, plant available on site, item sizes, quality of workmanship, quantity (scope of work), degree of repetition, access and so on. The estimator clearly needs skills which are judgement-based, as well as the ability to work consistently at mechanical processes. The software must allow the estimator to exercise his judgement particularly when he is building up his rates for each job.

The general aims of computer-aided estimating are:

- To provide the estimator with a kit of tools which will enable him to save time and exercise his personal judgement within a given framework, with reasonable scope for flexibility and user ingenuity.
- To help the estimator in his role as the person who calculates the total net cost of the project, and those who have to make decisions based on the estimator's reports and allowances.
- To provide the opportunities for contractors to gain a commercial advantage over their competitors.
- To handle information electronically in order to produce less paperwork, provide faster access to data and costs summaries.
- To give access to up-to-date information from internal and external networks.
- To implement company procedures through standardization.

Communications and collaborative systems

When bidders produce a tender for large projects, teams are created from internal and external businesses, sometimes involving over 50 people. Communications take place, in order of use, as shown in Fig. 20.1.

With so much information being exchanged electronically, there is a need for common file formats which are readily available to the whole team. A protocol is required and simple to implement.

	Type	Advantages	Disadvantages
1	E-mails	Rapid despatch from computer; attachments enable transfer of all types of files; secure within business	Tendency to send and receive too much information; information not filed for general use
2	Meetings	Consensus gained from discussion; explanation of information given	Very time-consuming; impact of travel on time, cost and environment
3	Collaborative tools	Information available to many users in maintained searchable data base; available on Internet; secure; viewers available for CAD files	New skills needed to upload files and messages; people download files to achieve faster access; reluctance to replace use of e-mails
4	Telephone	Speed; feelings; test ideas; conference calls	No record; people not always available
5	Post	Large files can be sent on CD-ROM; essential for hard copy documents	Slow
6	Facsimile	Faster than post	Slower than emails; needs to be given to recipient; no electronic file transfer

Fig. 20.1 *Communication systems*

Drawings	PDFs can be opened by anyone. For measurement exercises a scale bar is essential. AutoCAD™ DWG and DXF. The DXF Drawing Exchange Format ASCII file is designed for the interchange of drawings between AutoCAD systems, but has become widely used as a transfer mechanism between graphics programs
Spreadsheets	Microsoft Excel
Texts	Microsoft Word
Presentations	PowerPoint

In a modern estimating office, information technology systems provide many opportunities for doing more in less time. In addition, these systems save money and are an important part of a sustainability plan. Fig. 20.2 lists the collaboration tools available in most offices.

Collaboration tools	Description
Video conferencing	Telephone and IP-based systems are available for PC, meeting room, board room and large auditorium. There is a huge saving by reducing travel (and travel time) costs. Early systems used a camera and television combination but today the preferred solution is an integrated system with links to a projector. Part of a sustainability solution
Teleconferencing	By dialling a single number, people can join a conference call. It is robust, simple to use and available 24 hours a day with no need to book ahead. Part of a sustainability solution
Web-based conferencing and file sharing	Web conference that allows you to present PDF, PowerPoint, spreadsheet or document files to other participants, and even show what is on your screen to other participants. Can be used for training, estimate reviews or discussions with clients.
On-line discussion	A collaboration solution which provides a secure, Web-based workspace for people spread around the country. You can share documents, calendars, information, and conduct live meetings from the office or the road.
Discussion groups	This is an Internet forum for holding discussions and posting user generated content. Internet forums are also commonly referred to as web forums, message boards, discussion boards, (electronic) discussion groups, discussion forums, bulletin boards, or simply forums. A forum administrator typically has the ability to edit, delete, move or otherwise modify any thread on the forum
Project management and document control	Construction industry tools provided using hosted services: 4Projects and ASITE for example
Calendar management	Format for exchanging scheduling and activity-recording information electronically. Also online PA services.

Fig. 20.2 *Collaborative tools*

On-line project management systems do much more than document management. In addition to providing an on-line environment – or extranet – for storing project information, a wide range of features is available. Features commonly include:

- Notification of new documents available, and decisions awaited, by extranet ‘in-trays’, or text message;

- Management and tracking of document approval processes, providing a clear audit trail of who has read what, and when, which can help resolve disputes should they occur;
- On-line document mark-up, versioning control and document history;
- The ability for all parties to view any document within the extranet, regardless of whether or not they have the software in which it was originally created;
- Query and action tracking modules, especially useful for stages such as snagging;
- Project calendars and personal diaries, so that users can keep track of the progress of the whole project, and their own key tasks;
- Contact directories;
- Custom-variable levels of access and permissions for each user;
- Full document search facilities;
- 'Offline' modes, allowing users to synchronize overnight and then work offline during the day.

Some of the online collaboration tools on the market include:

Asite: www.asite.com
BuildOnline On-Demand: www.buildonline.com/en/index.php
Causeway Technologies: www.causeway.com
Dochosting: www.dochosting.co.uk
Microsoft SharePoint: www.microsoft.com/sharepoint/
4Projects: www.4projects.com
CJ Collaboration: www.cjcollaboration.4projects.com/

The Network of Construction Collaboration Technology Providers (NCCTP) at www.ncctp.net provides links to further construction-specific products within the UK.

As an example of a collaborative solution, 4Projects offers far more than standard document management systems, by providing access to many users in different organizations using disparate IT systems. Information is available from any computer with Internet access 24 hours a day and every day of the year. The 4Projects system has been optimized through its design to work effectively in a web-based environment, and ensures that users on connections from 56K modem dial-up speed and above can use the system productively.

Other features include:

- Multiple/batch upload and download capability.
- Log-on screen with links to single or multiple projects.
- Simple interface and intuitive navigation to hide what are actually more complex database searches.
- Consistent interface across all modules enabling 'out of the box' implementation.
- Visual feedback of read and unread items.

- Simple process for inviting new users into the system.
- Powerful searching tools including free-text searching in uploaded files (including DWG, DGN, PDF, DOC, XLS files, etc.)
- Comprehensive audit trails on individual users and stored information.
- Access to every revision of every drawing/document uploaded to the project.
- Whilst some extranet products require multiple software downloads to work effectively, 4Projects requires only a standard web browser.
- A measuring tool to calculate lengths and areas on the screen image.

E-tendering

E-tendering provides electronic procurement for the whole tendering process, from advertising the opportunity to the award of contract. This includes the exchange of documents and communications in electronic format. The benefits of e-tendering to all those in construction are:

- Reduced tendering periods
- Fast and accurate pre-qualification and evaluation
- Faster response to questions and points of clarification during the tender period
- Reduction in the labour-intensive tasks of receipt, recording and distribution of tender submissions
- Reduction of the paper trail on tendering exercises, reducing costs to both clients and suppliers
- Improved audit trail, increasing integrity and transparency of the tendering process
- Improved quality of tender specification and supplier response
- Provision of quality management information.

E-tendering replaces manual paper-based tender processes to save time and money. Those procuring services can manage the tenders coming in, with all tenders stored in one place. They can cut and paste data from the electronic tender documents for easy comparison in a spreadsheet. Evaluation tools can provide automation of this comparison process. Suppliers' costs in responding to invitations to tender (ITT) are also reduced as the tender process cycle is significantly shortened.

E-tendering is a relatively simple technical solution based around secure email and electronic document management. It involves uploading tender documents on to a secure website with secure login, authentication and viewing rules.

Tools available in the current market offer varying levels of sophistication. A simple e-tendering solution may be a space on a web server where electronic documents are posted with basic viewing rules. This type of solution is unlikely to provide automated evaluation tools; instead users are able to download tenders to spreadsheet and compare manually, but in an electronic format. Such solutions can offer valuable improvements to paper-based tendering.

More sophisticated e-tendering systems may include more complex collaboration functionality, allowing numbers of users in different locations to view and edit electronic documents. They may also include email trigger process control which alerts users for example of a colleague having made changes to a collaborative ITT, or a supplier having posted a tender.

The most sophisticated systems may use evaluation functionality to streamline the tender process from start to finish, so that initial ITT documents are very specific and require responses from vendors to be in a particular format. These tools then enable evaluation on strict criteria which can be completely automated.

Buyers need to consider if the market sector they are trying to source from is ready for e-tendering because a loss of bidders due to perceived complexity may be a problem.

RICS Guidance Note October 2005: 'E-tendering' is a publication giving aspects of best practice for RICS members and others engaged in the construction procurement process. This document deals with the preparation and standardization of documents together with the processes that can be undertaken electronically.

4Projects tendering solution uses their usual interface and database system, ensuring that all information is easily transferable under their unique archive solution. Tender managers group related drawings, documents, photos, tasks and discussions into distinct 'work packages' using the binder module within the 4Projects Extranet. The binders are 'exported' from 4Projects extranet as individual tenders into a separate tendering site. Different access permissions prevent bidders from seeing who else is bidding on the tender.

Within the tendering site the tender manager sets dates for bidders to accept the tender and by which to submit their tender. Each displays a countdown to alert the bidders of the deadline. Each of the bidders is then invited to join the tendering site and decides whether to tender for the work.

Binders within the 4Projects extranet are placed under version control and automatic updates notify extranet users when a binder has been modified/revised, ensuring that they are aware of the very latest document and drawing revisions. Dependent upon whether the revisions will impact on the tender, the tender manager chooses to re-export the binder into the tendering site.

A simple interface for those tendering for a project or works package avoids being overwhelmed by technology. Each of the bidders joins the tendering site and on reviewing the tender documents decides whether to tender for the work.

Having accepted the invitation to tender, the organization tendering can use 4Projects facilities, at no cost to themselves, to collaborate with others whilst preparing their tender documentation. A user inbox, coupled with notifications, ensures that those tendering are well aware of the latest tender documentation.

Electronic exchange of information

For most construction projects, documents are exchanged electronically. Nevertheless, contractors are still asked to submit 'paper' copies of their proposals and forms of tender are still required in hard copy.

For computer-aided estimating, most effort has been concentrated on the main pricing document – the bill of quantities. This is because bills are generally produced on a computer and time can be saved, in the short tender period, by loading bill pages directly into contractors' systems.

In order to improve compatibility between PQS and estimating systems at tender stage, a collaborative project was set up in the mid-1990s to establish tender exchange standards. The Construction Industry Trading Electronically (CITE) initiative publishes a very simple set of rules for those writing (and electronically reading) bills of quantities. The format is based on a plain text file which can be loaded into general-purpose software and specialist packages. CITE now provides standards for a range of applications including invoices, orders and despatch notes.

In the absence of 'electronic' bills of quantity, contractors can feed pages into a scanner, and using text recognition software, load a complete document in far less time than can be achieved manually. Fortunately, most bills are produced electronically, so this facility is little used. If an estimator does not need descriptions on his screen at tender stage, a printed bill can be input manually with the minimum amount of information needed to build up and control the estimate. The essential data are bill and page references, item references, quantities and units. It is common to enter a 'trade' or 'sort' code at the same time. The trade codes developed for co-ordinated project information (such as E10 for *in situ* concrete) are very convenient because a bill of quantities measured under SMM7 uses the same system. A sort code gives the estimator the facility to print similar items, analyse them and price them together. This procedure is often referred to as 'trade' pricing, and allows others to help an estimator by pricing different trades for a project.

Some clients produce protected spreadsheets, which must be returned in support of a bid. The contractor must input his selling rates but is unable to change the text or formulae. In a similar way, cost plans submitted with bids for schools under the Building Schools for the Future (BsF) programme, NHS Procure 21 and MOD Prime contracting must be based on standard spreadsheet templates. This ensures a consistent approach by all bidders, tenders are easier to compare and the templates can be used to develop cost plans, guaranteed maximum prices and cost control on site.

For design-and-build and plan-and-specification contracts a contractor will ensure that a bill of quantities is produced in a form which is wholly compatible with his estimating package. In many cases, the computer-aided estimating system will be used to generate the bill of quantities – a paper copy will not exist.

On-line auctions

Organizations reduce costs through online auctions. The auctions work by inviting suppliers to bid for contracts online. Bidders then try to undercut each other's offers, while maintaining the technical requirements of the products or services.

The construction industry has been criticized in the past for being slow to adopt electronic solutions to business needs, but in the case of auctions construction companies have been purchasing goods and services since the late 1990s. Examples include the supply of company cars, mobile phones, and stationery. For large projects, this bidding method has been used for plant, building materials, joinery and standard components such as doors and windows.

Another leader in on-line bidding is the MoD, which spends £9bn a year through its logistics organization, and could become one of the largest procurers to use on-line auctions. The Royal Mail uses electronic auctions to cut costs and improve value. The extract shown in Fig. 20.3 is for hand-held devices for real-time tracking of house repairs.

Reverse auctions

What are the features of reverse auctions?

- Reverse auctions are on-line competitions, with the bid prices (or relative positions in the bid) visible during the auction.
- Simple products or services where the marketplace is highly competitive are most suitable for reverse auctions, yet any item with clearly defined requirements and more than one source of supply should be considered.
- It is essential that advertisements for competitions to be run on a reverse auction basis state this clearly, along with the criteria for selection.
- The auction, when it takes place, should be conducted on the basis of price only.

Reverse auctions, also known as 'on-line bidding', are a means of buying items or services against a published specification where pre-selected supply chain partners are invited to bid in an on-line auction. All bids made during the auction are published anonymously on-line, in the expectation that competitive pressure, when bidders see the prices bid, will force prices lower as the auction proceeds. The exception is the ranked auction, in which the bid amounts are not known to other bidders. The auction is time-limited, but arrangements may be put in place to ensure that if a 'leading' bid is made very close to the timed completion of the auction further time is provided to allow other bids to ensure that the lowest price is obtained. A contract is then awarded to the lowest bidder based on the terms and conditions published at the outset, during the contractor pre-selection phase of the reverse auction.

2007/S 76-879564
CONTRACT NOTICE
Supplies

SECTION I: CONTRACTING AUTHORITY

NAME, ADDRESSES AND CONTACT POINT(S):
Direct Repairs PLC. Tel. 01234 987654
Internet address:
General address of the contracting authority:
http://www.dirrepairsplc.com/home/tender11
Address of the buyer profile:
Specifications and additional documents (including documents for competitive dialogue and a dynamic purchasing system) can be obtained at: As in above-mentioned contact point.
Tenders or requests to participate must be sent to: As in above-mentioned contact point(s).

SECTION II: OBJECT OF THE CONTRACT

DESCRIPTION
Hand-Held Devices.
Type of contract and location of works, place of delivery or of performance:
Purchase
Main place of delivery: UK Wide
The notice involves:
A public contract.
Short description of the contract or purchase(s):
Hand-held scanning devices and peripherals, required by Direct Repairs to provide real-time job tracking status and door-step signature capture, and other functionality in support of the Tracked+ project.
Variants will be accepted:
Yes.
QUANTITY OR SCOPE OF THE CONTRACT
Total quantity or scope:
The requirement is estimated at up to: 8,000 Hand Held devices including battery 8,000 4 slot ethernet cradles including 240 V charger 8,000 vehicle charging cradles and vehicle install kits 8,000 carrying solutions 8,000 device screen guards 8,000 spare stylii 8,000 spare batteries 5 day break/fix warranty for all hardware.
Options:
No.
DURATION OF THE CONTRACT OR TIME-LIMIT FOR COMPLETION:
Duration in months: 60 (from the award of the contract).

SECTION III: PROCEDURE

TYPE OF PROCEDURE
Type of procedure:
Restricted.
Limitations on the number of operators who will be invited to tender or to participate:
Envisaged number of operators 5
Objective criteria for choosing the limited number of candidates: As detailed above.
Reduction of the number of operators during the negotiation or dialogue:
AWARD CRITERIA
Award criteria:
The most economically advantageous tender in terms of the criteria stated in the specifications, in the invitation to tender or to negotiate or in the descriptive document.
An electronic auction will be used:
Yes.
This requirement may be offered to tender by electronic means using the internet, and may also be through the medium of a reverse auction.
ADMINISTRATIVE INFORMATION
Time-limit for receipt of tenders or requests to participate:
31.10.2007 - 17:00.
Language(s) in which tenders or requests to participate may be drawn up:
English.

Fig. 20.3 *Example of an invitation to an electronic tender auction*

Reverse auctions rely on competition driving prices down and it therefore follows that the less complex or specialized the goods or service being procured, the greater the chance of a successful auction. Simple commodity items or services which can be clearly defined and have a wide range of potential suppliers will be best suited to the auction process. However, in considering the use of reverse auctions, it is important to ensure that the principles underlying the existing procurement process, namely those of confidentiality, fairness and equity, are maintained.

It is essential that an advertisement for goods or services, where a reverse auction is being considered, clearly state:

- That the ultimate selection may be made on the basis of a reverse auction;
- The evaluation criteria, including any weighting between fixed elements and the variable element of price;
- Information on the process itself including details of any third party service provider;
- Conditions of bidding including the minimum decrements permitted;
- Equipment/technical issues.

Prior to conducting an auction it is necessary to state clearly the specification of the goods or services to be purchased and to pre-select supply chain partners. Pre-selection should cover issues such as technical ability, financial viability, previous industrial supplier history, quality etc. The purchaser must ensure that they are confident that any industrial supplier taking part in the auction will be able to meet their business commitments should they win the auction. Since it would be unreasonable to conduct further checks or negotiations once the auction commences this pre-selection process is crucial and should be undertaken with considerable rigour and well before the auction is due to take place. The terms and conditions that will apply to the prospective contract must be stated at the outset and accepted by all prospective bidders. For overseas industrial partners, particular attention will be needed to deal with the issues of currency and timing. If the bid is not to be in pounds sterling, the exchange rate will need to be agreed in advance of the auction using an exchange rate calculated in accordance with a pre-agreed mechanism.

EU directives tend to discourage repeat tendering, although it is unclear how this method should be classified. When the EU updates its policies for e-procurement, it is likely that they will make changes to directives to recognize this process.

The customer and supplier should be aware of the benefits of an electronic trading environment but recognize the commitment and responsibilities that arise from a powerful form of procurement. The customer needs to act fairly with accurate information and provide assistance with the invitation to tender; and the supplier must understand the process, and commitment if successful. It is therefore important that the customer selects an experienced IT service provider to assist in the conduct of the auction.

Software

Spreadsheets, word processors and databases are commonly used in an estimating office. Spreadsheets provide the framework for price lists, estimating calculations and cost planning. A range of software is needed for tender presentations, which are much more sophisticated today. Clients expect contractors to demonstrate their capacity to work not only to a fixed price but also to a resourced programme and quality plan. Increasingly, clients call for these details at tender stage. Fortunately this information can be produced quickly with word-processing, desktop publishing and graphics packages. But the real advantage is the facility to edit text and graphics in order to produce polished presentations.

In the past, the best buying advice given to anyone entering the computer marketplace was to choose the software first and then find the hardware to run it. Personal computers are now able to run most applications and are pre-loaded with standard office applications. The list in Fig. 20.4 shows how software is used for common estimating tasks.

Computer-aided estimating packages

Estimators and their managers have not been slow to recognize the potential of computers to increase the efficiency of the estimating process, but during the 1980s and early 1990s were disappointed with the systems on offer and the problems of implementation. It could be said that contractors and software providers were looking at the problems from opposite perspectives. Contractors wanted software which would mimic their methods when in fact the systems were being developed to make best use of the hardware and programming techniques available. The result was a false start in computing because computers did not match users' expectations.

Some argue that software providers tackled the estimating challenge in the wrong way by creating huge databases of work items in order to mirror all the possible items that could be envisaged in a bill of quantities. This 'price library' approach is fine for taking off and pricing but is in conflict with the way estimators work, particularly when faced with pricing printed bills of quantity.

It is important to distinguish between contracts based on bills of quantities prepared by the client's quantity surveyor and those where only drawings and specifications are available and the estimator needs to assemble his own pricing document.

Printed bills of quantities often have thousands of work items which need to be matched with the coded descriptions in a computer library. This tedious task must be done by an experienced estimator who is also able to recognize and deal with rogue items. The library method can delay the pricing of printed bills of quantities. The counter argument is that where a contractor carries out work in a certain sector of the construction market many of the work items repeat. The library can be used to build

Application	Spread-sheet	Word processor	Database	Specialist package	Web-based tools
Cost planning	×		×	×	
Post box to communicate with team					×
Tender register	×	×	×	×	×
List sub-contractors and suppliers	×	×	×	×	×
Enquiry letters		×	×	×	×
Resource price lists	×		×	×	
Calculate all-in rates	×			×	
Produce standard bills for repetitious work	×			×	
Bar schedules	×			×	
Store and post documents					×
Measure drawings digitally				×	×
Rate build-ups	×			×	
Extend and total bills of quantity	×			×	
Lists of company staff costs and plant	×			×	
Calculate costs of fluctuations	×			×	
Adjust for late quotations	×			×	
Calculate/plot cash flow analysis	×				
Reports for management	×			×	
Adjust individual resources				×	
Adjust/distribute mark-up on rates	×			×	
Gross bill for client	×			×	
Bill of allowances for construction	×			×	

Spreadsheet (limited scope in comparison with a specialist package for estimating, but more flexible)
Word-processor package has limited calculation functions
Database programme or package with database facilities
Specialist package for computer-aided estimating
Web-based tools such as collaborative systems and e-tendering

Fig. 20.4 *Software used for common estimating tasks*

up a series of standard bills of quantities which would need minor changes on each estimate. Perhaps the best use of the library is the production of bills where the tender is based on a specification and drawings. The database of work items prompts the estimator with descriptions and guide prices. The items can be taken off in any order because the software will put the items into the sequence recommended by the standard method of measurement, and print trade bills for sub-contract enquiries.

Most of the criticisms have been answered, as follows:

1. The software reverted to a 'shell' arrangement whereby estimators could create a database for the items in the current job only. They no longer attempt to build a comprehensive database for future projects (some standard bills will be kept if similar projects are expected).
2. The complex price build-ups have been replaced by simple resource calculations which have a common method of entry for all items of work, for example:

Amount	×	Resource type	×	Cost	=	Rate
0.14 (hours)		carpenter		17.00	=	2.38
1.00 (m)		100 × 50 wallplate		1.45	=	1.45

   The individual resources can be priced either during the entry of the work item or all the resources can be dealt with separately.
3. A system of menus is used to guide the estimator around the system and context-sensitive help screens bring relevant advice to the user whenever he needs it. 'Context-sensitive' refers to the way in which software will display helpful information about the command that is about to be carried out or may advise a user who is uncertain about how to continue.
4. More powerful computers are available with the latest microprocessors and high-capacity storage devices with fast access times.
5. Menus, help screens and utilities can be called up on the screen using windows of information. Switching between windows allows the estimator to undertake various subsidiary tasks while he is working on the estimate. Data can be copied from one 'Windows' application to another.

The number of packages available to estimators has reduced since the start of the new millennium: there were too many providers in the market. Choosing software is still a very difficult process. There are few independent test reports. Claims made in the construction press for estimating packages give similar specifications but users know that their functionality, speed and reliability vary considerably and are very difficult to confirm. A short demonstration by a salesperson is not a reliable way to evaluate a system because deficiencies will be glossed over. It is important to ask about the facilities being offered by various suppliers, and write down which facilities will be of most benefit (see Fig. 20.5). It is unlikely that the software will meet all the needs of a company but some are more flexible in use than others. For example, there are some systems which store their data in database format, which can be output to other databases or sent to a word processor to produce high-quality presentations.

Main characteristics	Subsidiary characteristics
Hardware requirements	• Compatibility • Memory requirements • Multi-user systems • Optical character recognition • Digitizer facilities
Price	• Initial and annual charges • Upgrades • Telephone support
Method	• Entry of bill items (speed and convenience) • Data can be imported from spreadsheets • Library of standard text and build-ups • User-definable libraries
Help	• Free on-line help • Context-sensitive help screens • User manual
Reports	• Outputs fixed by software • Output designed by estimator • Elemental summaries • Speed of recalculation
Specific facilities	• Sub-contract and materials comparisons • Labour-only rates substitution for direct labour rates • Data can be transferred to a spreadsheet • On-screen calculator • Sort items by trade or user codes • Windows environment • Detection of unpriced items • Nested work assemblies • Recalculation/reporting speeds • Checking procedures

Fig. 20.5 *Features checklist for estimating packages*

Estimating systems may not be able to offer all the features listed in Fig. 20.5 and many estimators will not need all the features. An optical character recognition system will not be needed, for example, by an estimator who inputs bill items by reference to page and item numbers; a Direct Labour Organization may not need facilities for adding labour-only sub-contractors; a small company might not need a multi-user system; a trade specialist may not need a powerful sub-contract comparison system.

Clearly, estimating systems store a great deal of information about a project which can be linked to project planning, buying, valuation and accounting packages; in many ways this is what makes the use of computers worthwhile. The information produced at tender stage may need to be changed once a tender is successful. It would be unsafe to order materials, for example, from tender stage bills of quantities because the drawings may have changed. A project quantity surveyor would wish to insert item descriptions where the estimator did not.

Although some packages use a text-based environment, the windows interface has become the accepted standard and for some organizations is the principal criterion in selecting a system. Estimators have now accepted the change to mouse-driven software, because they benefit from the advantages, such as:

1. The facility to pick work items from a list, which has been created for standard building types.
2. Multiple windows which can be opened within a programme to show how changing the figures in one part of the programme affects another part. An estimator can see in one window the items he can select for his tender; in the other the selected items are growing into a bill of quantities.
3. A number of applications can be 'open' at any time. This means that information produced in one program can be copied into another. For example, an estimator may price a bill of quantities using estimating software and list the items needed for preliminaries in a spreadsheet designed for the purpose.
4. Estimators in future will develop skills in working within a single user interface where the instructions, menus and help facilities have a similar 'look and feel'.

There is a danger that people can be too concerned with generating data for its own sake. It is commonly said that: 'information is an organization's most valuable asset', but do we always need so much information? Estimators are resigned to the fact that most of their reports are ignored once a job is won. There is nevertheless a clear advantage to be gained by making pricing data available to construction staff. The computer will produce the properly priced (net) bill, the commercially priced (gross) bill, and any number of package (or trade) bills for negotiations with subcontractors. A schedule of material resources can be used by the purchasing department to set up orders with suppliers, always noting that the actual quantities will be determined on-site when the construction drawings have been issued. It is worth remembering that computers reduce the clerical effort, reproduce data in a sorted form, but above all cannot do anything that you cannot do manually (in time).

General-purpose software

A desktop computer can handle a vast range of programs and carry out thousands of different tasks. Most software can be purchased off-the-shelf either from a specialist producer or retailer. There are two main types of software: applications programs and systems programs. The kinds of programs available in each category are as follows.

Applications programs

1. General-purpose packages (e.g. word processors, spreadsheets, databases and document readers for pdf files);
2. Specialist packages (e.g. estimating, accounts and expert systems).

Systems programs

1. Operating systems (e.g. MS-DOS, Windows and UNIX);
2. Utility programs (e.g. anti-virus software).

One of the most difficult decisions for computer users is whether to buy specialist software, which has been tailored for a particular application, or use general-purpose programs such as spreadsheets and databases. For example, if an estimator wants to produce a small bill of quantities he can use a word processor, database or spreadsheet. He could alternatively decide to use a specialist bill production package.

Word-processors, spreadsheets and databases are the most popular products in the general-purpose software market, closely followed by computer-aided design and project planning.

Word processors

Most well-known word-processor packages are suitable for day-to-day use and offer many of the features required for bidding teams. Choosing word-processing software is not difficult; often the safe answer is to choose the market leader because the files will be compatible with many other systems and staff prefer to develop skills which will be widely accepted.

The estimator can make use of the basic word-processor features for letters, method statements, quality statements, safety plans, specification writing and bill production. In building, the National Building Specification is available as simple text files from NBS Services. Small bills of quantities are easier to produce using spreadsheet software but where a large unpriced bill is to be written, a word processor is often used. Normally, bill production packages use a database method which overcomes size limitations, and can be output to word-processor and spreadsheet files.

Spreadsheets

The key ingredient which has led to the widespread acceptance of the personal computer as something more than a clever typewriter is the spreadsheet. It is simple to use, and does not try to change the way people undertake their calculations. Most of the repetitious work of an estimator could be computerized without the help of a programmer. The immediate benefit is the fast recalculation of cost plans, priced bills of quantities and data tables. Estimators can test the effect of changing parameters, often referred to as 'what if' calculations.

Some clients issue bills of quantities, or activity schedules, in spreadsheet format so that priced tender documents can be returned with the tender on CD-ROM or by email.

Someone new to spreadsheets should start with those applications that lend themselves to the tabular presentation of data. The most elementary would be:

1. Look-up charts for reinforcement, brickwork, drainage and fixing ironmongery.
2. Small bills of quantities for composite items such as manholes, kerbs and simple house extensions.
3. Domestic sub-contractors' quotation analyses.
4. Early cost plans using costs from previous schemes.

After a little practice, the following could be attempted:

1. All-in hourly rate calculation.
2. Plant rate build-ups.
3. Bills of quantities for standard house types.
4. Look-up charts for more complex rates, such as formwork and disposal of surplus excavated material.

More advanced applications include:

1. Cash-flow forecasts.
2. Preliminaries schedules.
3. Bills of quantities for uncomplicated commercial and industrial buildings, and plant foundations.
4. Reinforcement schedules.

The examples shown in Figs. 20.6 and 20.7 were produced by a groundworks subcontractor to create quick look-up charts for pricing bills of quantities. The item highlighted in Fig. 20.6 shows the total rate for formwork to beams where the fix and strike time is 1.70 hr and the estimator expects four uses of the shutter. The highlighted cell in Fig. 20.7 is a rate for excavating 1 m of trench for a 150 mm diameter pipe, 1.75 m deep to invert including disposal to a tip 8 km from site.

FORMWORK	**Project:**	LIFEBOAT STATION	**DATE:**	JUN 08

ENTER BASIC DATA									
Carpenter (all-in rate/hr)	£15.50		ASSUMPTIONS						
Labourer (all-in rate/hr)	£12.00								
Plywood cost per sheet	£28.00		Waste %					9.00	..%
Softwood cost per m^3	£250.00		Consumables		making			0.08	..xcraft rate
					fixing			0.07	..xcraft rate
Percentage additions									
LAB for travelling	0.00		Founds	Soffit	Beams	Walls	Columns		
for fluctuations	5.00								
for o/heads and profit	10.00	make	0.80	0.50	1.00	0.95	0.80	..hrs/m^2	
for discount to MC	2.50	timber	0.04	0.04	0.05	0.05	0.04	..m^3/m^2	
Total mark up on labour =	1.18	plant	0.44	4.00	1.10	1.05	0.85	..£/m^2	
MAT/PLT for fluctuations	5.00								
for o/heads and profit	10.00		Lab assist handling			(hrs)		0.15	..hrs/m^2
for discount to MC	2.50								
Total mark up for MAT/PLT =	1.18								

HOURS TO FIX AND STRIKE		1.20	1.30	1.40	1.50	1.60	1.70	2.00
ONE USE	Founds	59.27	62.23	65.18	68.13	71.08	74.41	81.03
	Soffits	58.90	61.84	64.78	67.72	70.67	73.98	80.59
	Beams	65.92	69.07	72.22	75.38	78.53	82.12	88.94
	Walls	65.10	68.22	71.35	74.48	77.61	81.16	87.96
	Columns	61.16	64.17	67.18	70.19	73.20	76.60	83.28
TWO USES	Founds	44.99	47.29	49.58	51.88	54.17	56.47	62.66
	Soffits	46.59	48.91	51.24	53.57	55.89	58.22	64.46
	Beams	49.13	51.50	53.88	56.26	58.64	61.01	67.33
	Walls	48.63	51.00	53.37	55.73	58.10	60.47	66.77
	Columns	46.28	48.60	50.92	53.24	55.56	57.88	64.12
THREE USES	Founds	39.06	41.07	43.08	45.08	47.09	49.09	54.77
	Soffits	41.47	43.50	45.53	47.56	49.59	51.62	57.32
	Beams	42.15	44.19	46.23	48.26	50.30	52.34	58.04
	Walls	41.79	43.83	45.86	47.89	49.92	51.96	57.66
	Columns	40.10	42.12	44.13	46.15	48.17	50.18	55.87
FOUR USES	Founds	35.95	37.79	39.62	41.46	43.29	45.13	50.63
	Soffits	38.79	40.62	42.46	44.29	46.13	47.96	53.47
	Beams	38.49	40.33	42.16	44.00	45.83	47.67	53.17
	Walls	38.20	40.04	41.87	43.71	45.54	47.38	52.88
	Columns	36.86	38.69	40.53	42.36	44.20	46.03	51.54
FIVE USES	Founds	33.95	35.79	37.62	39.46	41.29	43.13	48.63
	Soffits	37.07	38.90	40.74	42.57	44.41	46.24	51.75
	Beams	36.14	37.98	39.81	41.65	43.48	45.32	50.82
	Walls	35.90	37.73	39.57	41.40	43.24	45.07	50.58
	Columns	34.77	36.61	38.44	40.28	42.11	43.95	49.45
SIX USES	Founds	32.62	34.45	36.29	38.12	39.96	41.79	47.30
	Soffits	35.92	37.75	39.59	41.42	43.26	45.09	50.60
	Beams	34.57	36.41	38.24	40.08	41.91	43.75	49.25
	Walls	34.36	36.20	38.03	39.87	41.70	43.54	49.04
	Columns	33.39	35.22	37.06	38.89	40.73	42.56	48.07

Fig. 20.6 *Example of a spreadsheet template for formwork*

DRAINAGE	**Project:**	LIFEBOAT STATION	**Date:**	JUN 08

ENTER BASIC DATA	rates/hr
Margin %	10
Labourer	£12.00
NWRA	£0.20
Excavator	£8.25
Fuel	£4.60
Operator plus rate	£0.40
Pump	£2.00
Compactor	£2.00
16t lorry (all-in)	£26.50

Assumptions (can be changed)

	0-1.5 m	1.75-2.75	3 m+
Rate of dig m^3/hr	12	8	6
Earth support cost	£2.30	£2.65	£3.20
Grade and ram cost	£0.50	£0.55	£0.60
Distance to tip	....	8	km
Tip charges (or work to tip)	....	£13.00	/m^3

DRAIN EXCAVATION			Pipe sizes (mm)				
			100-150	225-300	375-450	525-600	675-750
DEPTH	width	>	0.60	0.70	1.10	1.25	1.45
0.50			6.61	7.29	10.00	11.02	12.38
0.75			9.75	10.74	14.70	16.19	18.18
1.00			12.88	14.19	19.40	21.36	23.97
1.25			16.02	17.64	24.10	26.53	29.76
1.50			19.16	21.09	28.80	31.70	35.55
	width	>	0.70	0.80	1.20	1.40	1.60
1.75			28.13	30.69	40.94	46.06	51.18
2.00			32.09	35.01	46.68	52.52	58.35
2.25			36.05	39.32	52.42	58.98	65.53
2.50			40.00	43.64	58.17	65.43	72.70
2.75			43.96	47.95	63.91	71.89	79.87
	width	>	0.80	0.90	1.30	1.50	1.70
3.00			60.26	65.15	84.72	94.51	104.29
3.25			65.24	70.53	91.71	102.30	112.89
3.50			70.21	75.91	98.70	110.09	121.49
3.75			75.19	81.29	105.69	117.89	130.08
4.00			80.17	86.67	112.68	125.68	138.68

Fig. 20.7 *Example of spreadsheet template for drainage excavation*

The estimator can change any of the data at the top of the page and the total rates change within seconds. These applications show that a spreadsheet can closely mimic the traditional methods used to produce rates, but do so at much greater speed, with clear presentation, but as with all computer methods careful interpretation by an experienced estimator is still required.

Anyone who is used to dealing with figures will soon be charmed with the power of such sophisticated software, and will be able to test various theories to arrive at the best condition or price. There are, however, many dangers awaiting the unwary estimator. The problems arise when:

1. The estimator who builds a spreadsheet model fails to produce a foolproof design, or carries out inadequate checks.
2. Another estimator inadvertently changes data in a model or erases a formula by entering a number in a formula cell.

Often a user is unable to spot mistakes in his own spreadsheet; he is more inclined to believe the results when they are presented on a computer printout. The effect of using inaccurate answers from such calculations could be ruinous in a tender. The following guidelines will help to prevent such errors:

1. Start by planning the general requirements with a sketch showing the labels and layout of the spreadsheet. The optimum size of a spreadsheet will depend on sensible file size for sending to recipients by email, and the data an operator will want to see either on the screen or close to the edges. Information can be broken down into a number of worksheets within a single workbook.
2. Adopt a modular approach whereby the layout will include separate identifiable sections, such as:
   (a) an instructions portion
   (b) an area where the user can change data freely
   (c) the results or summary section.
   These sections could be created in different (but linked) files, on pages of a multiple spreadsheet, or in different parts of the screen display.
3. Protect formulae from accidental erasure or amendment, by putting them in the results area of the spreadsheet. It is also possible to make a formula secure by using the password protection feature found on most versions of the program.
4. Check that the numbers representing money are not only rounded to two decimal places for display purposes but also for subsequent calculations in the model. The reason for this check is that although the spreadsheet has been instructed to show two decimal places it usually keeps a more accurate number in the computer's memory.
5. Carry out simple checks using data from previous manual systems. Other estimators could be asked to test the model to find any bugs or misleading instructions.

Above all, where more than one person is going to use the spreadsheet, keep the design simple.

Spreadsheet programs now offer a safe choice for organizations which recognize the need to introduce computers to their staff cautiously and at low cost. They are powerful in the rapid production of cost information, which is usually in a tabular form. The estimator can create spreadsheets which are an amalgam of his expert knowledge, by holding the production information which he has collected.

Once a format has been created (and saved) for a particular purpose, it is referred to as a template. This is because the layout and formulae will usually be fixed, but the data variables can be changed. Most estimators will be inspired to extend the range of applications and build a valuable selection of templates, to suit their own needs and methods of working. On average an estimator will take about six hours to build a template if he is familiar with the software.

Databases

A database is a computerized filing system, the electronic equivalent of a card index or filing cabinet. A database program allows the user to file information and retrieve it in many ways. Perhaps the most common example is a list of names and addresses. When a list has been built up it is called a *file* of data. A file can be displayed as list on the screen, sorted into order, searched for individual pieces of information and printed. Fig. 20.8 shows a typical *record*, which is the basic building block of a database. Each record contains a number of *fields* of information, such as the name field or postcode field.

A simple database can be set up in few hours. The user firstly designs the layout for the record screen defining the fields which are needed. Secondly, the program can be used to produce printouts in different formats. For example, an address file could be printed with a list of names in the first column and telephone numbers in the next. Before printing lists of data, the program could be asked to sort all the records into alphabetical order; or search for all the suppliers in a particular town or district. The program can also create and store a standard letter so that names and addresses can be merged with text to prepare letters for a selective mailshot, for example. More powerful packages offer much greater scope for manipulating and analysing data and include their own programming language used to develop more sophisticated applications.

The range of applications is large in an estimator's office, from elementary filing tasks to the complex manipulation of data. Examples include:

1. Address lists for suppliers and sub-contractors;
2. Drawing registers;
3. Tender registers;
4. Marketing information;

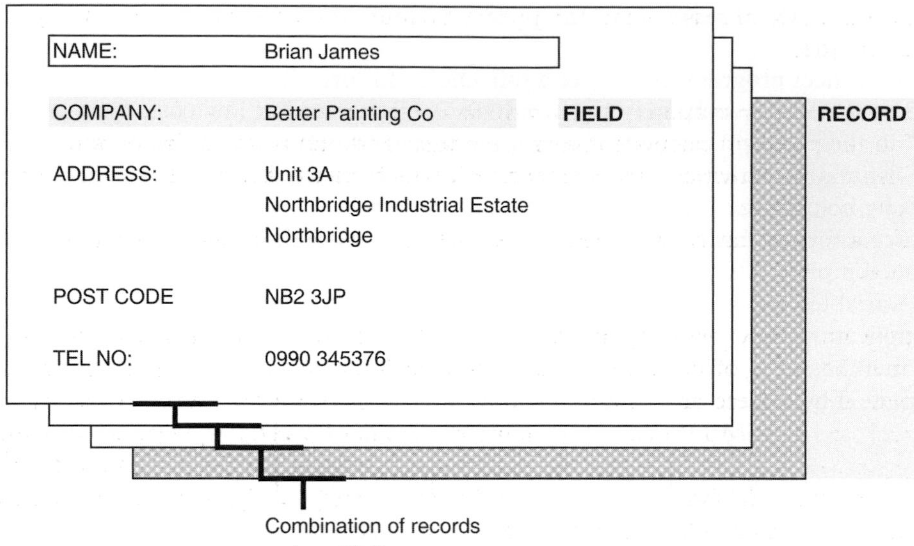

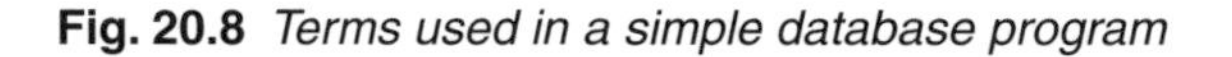

Fig. 20.8 *Terms used in a simple database program*

5. Cost planning data;
6. Bill production and pricing;
7. Personal timesheets.

There are many national databases available to construction organizations using an Internet connection. Usually the user is expected to pay an annual subscription. Estimators seldom develop database applications because the most common database functions (sort and search, for example) are readily provided in spreadsheets. Furthermore, web-based collaborative tools are now the main device for storing data during the tender period.

Planning

It would be quite straightforward to produce tenders using paper, pencil and a calculator. An estimator's program is rarely used during the construction phase but will be an important tool to assess preliminaries and undertake a cash-flow analysis. Where a planning engineer produces the preliminary program, the continuity is likely to be stronger and the program may well form the basis of a control/monitoring document.

There are several reasons for using project planning software to aid manual systems at tender stage:

1. It may be easier to keep complex projects under control.
2. The program will calculate resource loadings and plot resource histograms.
3. The critical activities can be identified, which might affect the workforce levels used for pricing.
4. High-quality charts and clear presentations could impress a client at tender stage.

There are many planning packages available today; some, such as CS Project, give a wide range of powerful features and flexibility to implement the software in different ways. There are others that offer the most common features at lower cost. 'PowerProject' or 'MS Project', for example, both have excellent presentation features and yet are easy to use. The estimator will look for software which will show the overall project duration (starting from submission of tender), highlight critical activities and allow labour and plant resources to be analysed.

Graphics

Graphical software is used for three main purposes:

1. Business graphics – the graphical representation of data drawn from information in a spreadsheet or database.
2. Desktop publishing – the in-house design and publication of leaflets and forms, usually with the aid of a laser printer.
3. Computer-aided design (CAD), usually output to high-quality plotter.

There are many software packages available for desktop computers, to do simple line drawing, rendering and 3D modelling. Industry leaders AutoDesk produce the AutoCAD design package with an architectural add-in, which incorporates the symbol and layering conventions of BS1192. Most packages offer features such as the ability to scale drawings up and down in size, adding text and dimensions, keeping shape libraries, and of interest to an estimator is the ability to attach prices to certain items (attributes) within a drawing.

Google SketchUp is software that can create, modify and share 3D models. It is easier to learn than other 3D modelling programs, which is why it is gaining a large user base. Two versions of the software are available: one which is free for everyone which allows you to build, view and edit 3D models. Google SketchUp Pro adds the ability to share data with other software, get email technical support, and create professional presentations.

Architects and designers use SketchUp for: sketching 3D building models; communicating design ideas with project teams and/or clients; 3D walk-throughs; design presentations; massing model development; and urban planning and design.

Document readers

The most common type of document reader is free software for viewing and printing portable document format (PDF) files. Adobe Reader® version 8.0 for example can also be used for viewing high-fidelity ebooks for Windows, Palm and Pocket PC platforms.

Portable document format is widely used for issuing tender documents on CD-ROM or by email. As a means of communicating formal documents it has the following advantages:

1. Information cannot normally be edited.
2. Pages can be indexed and accessed by hyper-links embedded in text.
3. Members of a team can make comments on a document posted on a web-based system.
4. Password protection prevents unauthorized access.

Evaluation of general purpose and bespoke software for estimating

A swing to Excel is taking place for the following reasons:

- Initial resistance by estimators to use specialist estimating software
- Specific demands of clients for submissions using Excel
- The effect of sector work such as health, defence and education where standard cost plan layouts are used
- The transfer of cost plans to other members of teams
- Flexibility to produce reports for many needs.

Bespoke software is being used because:

- It provides a strong platform for traditional estimates and simple cost plans
- It can be linked to a measurement system
- Useful for 'template' bills of quantities
- Ensures consistent methods are used
- Robust package comparisons
- Effective multi-user system.

Since selecting bespoke software for estimating, contractors are seeing changes:

- Software companies change ownership
- The software is underused when spreadsheets are available in parallel
- Contractors may be paying for more user systems than they need
- Tender submissions generally require Excel format
- For large projects, there is a need to price multiple buildings with high-level cost plans
- Clients are looking for benchmark data in parallel with submissions using spreadsheets.

Some contractors are now:

- Using spreadsheets in preference to bespoke software
- Using bespoke software for their link to measurement systems
- Looking at ways to ensure spreadsheets do not lead to errors
- Investing in a common approach to spreadsheet templates and verification procedures.

What do clients want?

- Many procurement routes impose a cost plan format
- Format depends on which stage of a project is being costed
- Requirement to use Excel spreadsheets
- Clients issue pre-written templates to confirm cost plan build-ups
- Interchange of information is transferable and transparent
- Links are needed to life-cycle costing
- Inputs to internal clients for financial models.

How do contractors maintain estimating consistency?

- Estimating packages provide consistency
- Estimating packages offer a structured system
- Estimating packages provide a central depository for cost plans
- The use of spreadsheets needs to be monitored to identify best practice and possible misuse and errors. More development and discipline is often needed but not put into practice
- Guides to best practice with spreadsheets are needed
- Put in place a regime for data capture.

Computer-aided estimating competencies

- Estimators are not fully trained and many not proficient in the use of estimating packages or spreadsheets.

- Additional training requirements need to be identified and courses provided.
- Expert advice could be employed to improve efficiencies and avoid errors.

Comparison of estimating package and spreadsheet for a cost-planned tender

Stage	Name	Estimating package	Spreadsheet
1	Budget	No	Yes
2	Cost model	No	Yes
3	Cost plan	Yes	Yes
4	Cost plan+	Yes – strong tool for detailed cost plan	Yes
5	BoQ estimate	Yes – strong tool for quote analysis	Yes
6	Handover	Yes – but likely to be converted to a spreadsheet file	Yes
7	Construct	Not suitable for site use (unless contractor installs post-contract software)	Fully compatible with site methods
8	Feedback	Requires re-entry of data	Converts packages to elemental costs
9	Archive	Built-in functionality	Needs agreed format and filing structure/location

Hardware

A basic computer system is made up of the computer itself, a keyboard, monitor and storage devices with access to a printer and a network. Most of these facilities are available in a laptop computer, which may be appropriate for an estimator working at different offices and at home.

The most important part of a computer system is the software, because it is this that dictates what can and cannot be done. On the other hand, now that there is a vast range of software, which can run on a standard microcomputer, people prefer to select the PC first. Of the distinct groups of computer available today, the most popular is the IBM-compatible. In design offices, there is still a loyal band of Apple Mac users who need enhanced graphic production facilities.

There has been much interest in input devices which can save estimators time. The obvious solution would be to send contractors bills of quantities in a standard format, on a CD-ROM, or as file sent electronically with an email. Efforts have been made to use optical character recognition software, but checking and correcting documents can be very time-consuming.

A digitizer is used to take measurements from a drawing. Some of the features of a digitizer are:

1. Automatic scale adjustment – useful if the software with a digitizer can accurately compensate for reduced, photocopied or PDF plans. There needs to be a known dimensions or scale bar on a drawing.
2. Calculation modes to measure lines and areas for irregular shapes and angles.
3. Compatibility with existing equipment and spreadsheets.
4. Automatic transfer of measurements to quantity field in database.
5. Ability to read any scale.

Digitizer software is usually menu-driven using an overlay window over the taking off program. An audit trail can be kept of every measurement so the take-off can be checked manually.

There are three other ways to measure drawings:

1. Using a function available in collaborative web-based files. For example, when a drawing is opened using a web access site, a line can be drawn around a shape to find its area.
2. Drawing software such as AutoCAD has similar functionality and a record can be kept, of polygons for example.
3. A piece of equipment which is still in common use is the electronic planimeter. Its main advantage is portability. It can be taken to the drawings, on site or at home.

Networking

What is the networking revolution and how does it affect estimating today? The aim is to maintain competitive edge by making use of latest information through investing in new technology. A network can be seen as two or more computers linked together, sharing data files, software applications, hardware (including printers and backup devices) and links outside the office.

Every day, there many facets of networking taking place. Electronic mail can be received while the estimator is working on a tender; other estimators can be inputting data for the current estimate and the buyer can access relevant items for enquiries to be sent to suppliers and sub-contractors – electronically. This is all achieved by storing information electronically with a corresponding reduction in storing and handling project documents.

Where networks link computers within a limited area, such as within an estimating department, or head office building, it is called a local area network, or LAN. People share an organization's information in this way using what is more commonly called an 'intranet'. Sometimes the term refers only to the most visible service, the internal *website*. If an organization is linked to outside networks, the transition is seamless, and information can be shared with anyone, regardless of location across the world. By linking the network in this way, a wide area network or WAN is created. The more common term is 'extranet', which can be part of a company's *Intranet* that can be viewed by employees and other users outside the company.

There have been many examples of estimators and buyers creating their own databases which were not available to their colleagues. Clearly access must be for a wider team and networking enables rapid communication throughout the organization. There are some documents, however, which cannot be made freely available. For example, the estimate and subsequent final review decisions are confidential and must not be circulated to a wider audience. There are two reasons: firstly, most tenders are submitted in competition without giving your competitors sight of your pricing levels; and secondly, estimators must avoid any contravention of competition legislation.

Electronic mail is a cheap and quick means of sending messages to colleagues, customers and suppliers. Messages can be left for others to read either because they are not available or the information is urgently required in written form. With a couple of simple menu selections, information can also be copied to a number of people within an organization. Estimators can obtain late specifications or bill pages as an attachment to an email message. This facility is commonly used by estimators to assemble specifications and in-house designs for design-and-build tenders. There may be a project programme to incorporate in a presentation document and CVs can be sent from the personnel department. All this data can be received electronically and assembled on a computer, to produce a presentation document to accompany a tender.

The construction industry has been slow to embrace e-commerce (business carried out by means of networks, mainly the Internet). The UK Government and national trade associations are looking at strategies to educate all members in the supply chain to embrace the benefits of trading electronically. As with most initiatives, the Government has to lead by example, and has started by making the most of its services available on-line. As a major client of the construction industry, the Government has set targets for procuring its goods electronically.

Implementation

Computer-aided estimating is not a single program or technique but the development of opportunities provided by the computer and software providers. This development will take place under the guidance of an information technology strategy produced by the organization's business managers.

A chief estimator needs to look at what the company is doing now and what it hopes to achieve in the future. He might ask himself these questions and put the answers in order of priority:

1. Why use computers?
   - To communicate with clients, consultants and suppliers.
   - To produce post-tender data.
   - To build up an accurate estimate of cost.
   - To allow tender adjustments to be made to an estimate, conveniently and accurately.
   - To reduce manual calculations.
   - To store standard models of bills and cost plans.
2. What are the basic needs?
   - Flexibility with different tender documentation.
   - Flexibility for projects with different timescales.
   - Networking including access to the Internet.
   - Hardware/software compatibility.
   - Uniformity of reports for basic resources, review meetings and post tender allowances.
   - Ability to produce estimates in a standard cost plan format.
3. How do I select a system?
   - Attend a sales exhibition.
   - Ask for technical literature.
   - Find out what the estimators want.
   - Tell a supplier what you want.
   - Consult an independent expert.
   - Speak to other chief estimators.
4. What is *not* wanted?
   - Long, meaningless item code numbers.
   - Too many menus to change resources.
   - Long recalculation times.
5. What *is* wanted?
   - Company-specific reports.
   - Windows.
   - Context-sensitive help windows.
   - Rapid editing and deleting of bill items and prices.
   - Powerful sub-contract comparison system.
   - Checking for unpriced items.
   - Clear reference manual.

One of the most difficult decisions is the selection of software. Should there be a combination of general-purpose and specialist estimating packages? The answer is 'yes' – if a company insists on maintaining a consistent approach using a specialist package there will also be a need to use spreadsheets when exchanging data with other organizations.

The future

Information technology is already an integral part of the estimating process. Tender documents are exchanged electronically and estimates are assembled and submitted using desktop and laptop computers. Collaborative sites reside on the Web which allow users to download information when they need it and deposit bids when the time comes to submit a bid.

Training is as important for the client's advisers as it is for estimators. Now that general-purpose software is available in most offices, the most successful developments are where people's enthusiasm is channelled to speed up the operation of the company's procedures. Since computing skills are better taught through practical examples, training can take place at work. Clearly all estimators must have some computing skills and should feel comfortable using general-purpose software.

Both quantity surveyors and estimators have made a start but have approached the tender process from different standpoints. Quantity surveyors are naturally concerned with describing and quantifying the items of work for a project and the contractor needs to attach resources and prices to it. So quantity surveying software is designed to handle lists of work items in a structured form, whereas the estimator uses lists of resources which he can assign to the work. Coding methods have been tried but no common numbering system has emerged for general use and people resist codes as a way of entering and finding items. The answer might be for consultants to list activities using database software, and the database program would be distributed with the data. The contractor could then select items for sending to sub-contractors and price the rest of the work without the need for his own software. The main benefits would be:

1. The time saved by the contractors entering bills of quantities into their computers.
2. Amendments could be sent on disk or by email.
3. Priced bills could be returned with the tender in the form of a database file which would be transferred to a spreadsheet for checking and analysis of rates, and cash-flow predictions.
4. The database file would be used for valuation purposes during construction.

This idea is not new, and has been pioneered by some consultants, mainly in civil engineering. The main obstacle is the use by contractors of different computer-aided estimating software which is often linked to cost-reporting systems. Contractors would also be reluctant to submit all their calculations used to build up a tender.

Another powerful feature of estimating software, which is seldom used, is the facility to dynamically link money and time. That is, linking the bill of quantities to the programme of work. This link, between these two important factors of a construction project, provides a wealth of information at the fingertips of both management and client.

There are several questions for the future which are constantly raised:

1. *Will computers replace estimators?*
   There have not been any reports of computers successfully replacing estimators but the information produced by estimators can be more comprehensive; and complex changes can be made to the estimate before the tender is submitted. There are many duties of an estimator which computers cannot replace: in particular, the many decisions made at each stage of a tender – site visits, discussions with subcontractors, interpretation of ground conditions, access restrictions, best use of resources, and so on. Clearly the role of the estimator and his assistant is changing to adapt to the use of computers (see Fig. 20.9). The estimator will have time to produce more estimates or look in greater depth at the methods and resources for the contracts he wants to win.
2. *Are computer systems appropriate for operational estimating?*
   At the heart of most computer-aided estimating systems is the facility to price individual items of work, as listed in the bill of quantities. This method is well understood for building work but is difficult to use when a group of items must be priced as a single operation. An alternative approach for some civil engineering projects may be the use of resourced programs generated with planning software. In this way, resources attached to each activity are scheduled and costed without the need to produce unit rates. If a conventional estimating package is used, it will produce resource summaries, which can be compared with a resourced program. For example, the computer might show that an excavator has been allowed for one and a half weeks on a contract that clearly needs an excavator for two weeks. This information was difficult to draw out of the pricing notes using manual methods.

   A leading software package for operational estimating is Candy (Construction Computer Software). This is a single-package, project control system designed for estimating, valuations, planning, cashflow and forecasting components; integrating all aspects of a project's construction process.
3. *How could expert systems aid the estimator?*
   In simple terms, an expert system can be described as having three parts: a user interface, a knowledge base (containing facts, rules and questions) and an inference engine which can draw on the knowledge base to make deductions about a particular problem. There have been few off-the-shelf expert system shells designed for the PC. A shell, comprising the user interface and inference engine, enables the user to input the knowledge part of a system.

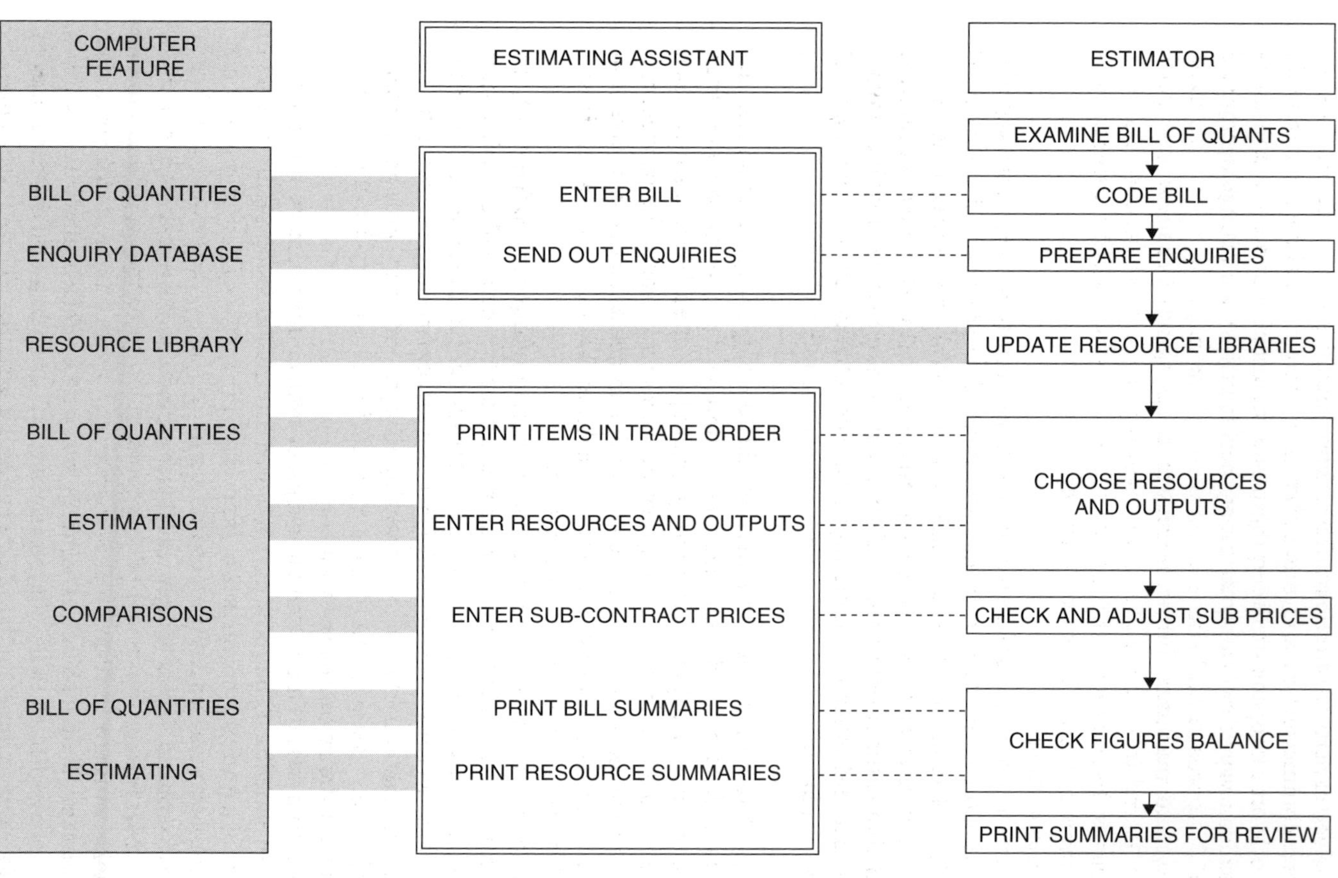

Fig. 20.9 *The roles of the estimator and estimating assistant using computer systems*

In estimating, expert systems have been developed for early cost planning of industrial and commercial buildings. A client or his professional adviser can build up a cost model of a proposed building by answering a series of pre-selected questions, in a similar way to the many fault diagnosis systems used in engineering and medicine. This approach could be used for some of the important decisions made by estimators and managers. The construction industry has been slow to exploit expert systems and it is unlikely that they will replace general-purpose and estimating software. Estimators will continue to use spreadsheets and database systems to schedule and price construction work.

4. *What do estimators want?*
   Large flat screen monitors, unfettered access to the Internet to look up supplier and sub-contractor's details and price lists; bills of quantities will be received electronically and tenders will be submitted in the same way. Journals will print independent test reports of estimating software, and eventually voice recognition systems will allow estimators to talk to their computers, doing away with the typewriter keyboard and other electronic input devices. This could be impractical, of course, in open-plan offices.

Clearly, the implementation of computers to aid the estimating function has not been easy. In fact there is evidence that some contractors have abandoned their attempts at using computer-aided estimating packages. They have, instead, used spreadsheets to produce priced bills of quantities and cost plans. Furthermore, a spreadsheet workbook for a tender commonly contains location factors, inflation indices, benchmark cost plans, sub-contract values, cashflow forecasts as well as the bill of quantities which provides the base information.

For some contractors, estimating software is used to produce site budgets only on those tenders that are successful; that is, after the tender has been submitted. This is not as absurd as it seems. The effort required to get the bill of quantities in the computer is much more worthwhile if the bid is successful and every part of the tender can be allocated to sub-contract or direct work packages. All the adjustments made during the final review meeting can be incorporated in the costs and clear printouts can be produced for the construction team.

It is perhaps ironic that many of the industry's commentators greeted the last two decades of the twentieth century as the time when the construction industry would see an end to the use of bills of quantities and the emergence of computer systems which could replace much of the work of estimators. We now know that both these predictions were wrong. Bills of quantities remain in everyday use and it is still difficult to form a clear computing strategy for estimating.

Estimators will continue to develop their computing skills but they will not have one computer system which will meet all their needs. Experience now tells us that computers will not replace estimators; estimators will always be needed to predict with a reasonable degree of accuracy the costs of construction. What might change is their name and in some instances their status. With the drive towards greater economy, some estimating duties will be carried out by clerks, assistants and specialist buyers. On the other hand, the range of skills required by an estimator has grown. He needs to manage a team which includes quantity surveying, operational and purchasing staff, for projects using a variety of contracts. The aim is to establish the sum of money, time and other conditions required to complete the specified construction work.

The future can be summed up in three particular trends:

1. Flexibility – software must provide for high-level cost plans which are used throughout a project's design stages. Cost plans are now taking the place of bills of quantities and analytical pricing.
2. 'On-demand' – there will be a rapid growth in information services and software accessible on the Web.
3. Location – using collaborative tools and modern communications, staff will not need to be office-based.

Further reading

Aqua Group (1999) *Tenders and Contracts for Building*, 3rd edition, Blackwell

Aqua Group (2002) *Pre-contract Practice and Contract Administration for the Building Team*, Blackwell

Ashworth, A. and Skitmore, R.M. (1982) Accuracy in estimating. Occasional Paper No. 27, the Chartered Institute of Building

Ashworth, A. and Willis, J. (2002) *Pre-contract Studies Development Economics, Tendering and Estimating*, Blackwell Science

Ashworth, A. (2004) *Cost Studies of Buildings*, 4th edition, Prentice Hall

Buchan, R., Grant, F. and Fleming, E. (2003) *Estimating for Builders and Surveyors*, 2nd edition, Elsevier

Chartered Institute of Building (1978) *Information required before estimating: a code of procedure supplementing the code of estimating practice*

Chartered Institute of Building (1981) *The Practice of Estimating* (compiled and edited by P. Harlow)

Chartered Institute of Building (1987) *Code of Estimating Practice*, Supplement No. 1, *Refurbishment and modernization*

Chartered Institute of Building (1988) *Code of Estimating Practice*, Supplement No. 2, *Design and build*

Chartered Institute of Building (1989) *Code of Estimating Practice*, Supplement No. 3, *Management contracting*

Chartered Institute of Building (1993) *Code of Estimating Practice*, Supplement No. 4, Post-tender use of estimating information

Chartered Institute of Building (1997) *Code of Estimating Practice*, 6th edition Co-published with Addison Wesley Longman

Civil Engineering Standard Method of Measurement (1991), 3rd edition, Thomas Telford

Code of Measuring Practice: A Guide for Property Professionals, 6th edition, RICS Property Measurement Group, RICS Books

Co-ordinating Committee for Project Information (1987) *Co-ordinated Project Information for Building Works, a Guide with Examples* (this and other CPI documents may be obtained from BEC Publications, RIBA Publications or Surveyor's Bookshop)

Cook, A.E. (1991) *Construction Tendering: Theory and practice*, Batsford

Cross, D.M.G. (1990) *Builders Estimating Data*, Heinemann-Newnes

Department of the Environment PSA (1987) *Significant Items Estimating* (produced by the Directorate of Building and Quantity Surveying Services), HMSO

Farrow, J.-J. (1984) *Tendering – an Applied Science*, 2nd edition, Occasional Paper, the Chartered Institute of Building

Franks, J. (1998) *Building Procurement Systems: A Client's Guide*, 3rd edition, Chartered Institute of Building, Longman

Griffith, A., Knight, A. and King, A. (2003) *Best Practice Tendering for Design and Build Projects*, Thomas Telford
Kwakye, A.A. (1994) *Understanding Tendering and Estimating*, Gower
Latham, Sir Michael (1994) *Constructing the Team* (Joint review of procurement and contractual arrangements in the United Kingdom construction industry), HMSO
Laxton's Building Price Book Major and Small Works (2008), Elsevier
McCaffer, R. and Baldwin, A. (1991) *Estimating and Tendering for Civil Engineering Works*, 2nd edition, BSP
Ministry of Public Building and Works (1964) *The placing and management of contracts for building and civil engineering work* (Report of the committee. Chairman: Sir Harold Banwell), HMSO
Ministry of Works (1944) *The placing and management of building contracts* (Report of the committee Chairman: Sir Ernest Simon), HMSO
Schedule of dayworks carried out incidental to contract work (2002), Civil Engineering Contractors Association
Sher, W. (1996) *Computer-aided Estimating: A Guide to Good Practice*, CIOB, Longman
Skitmore, R.M. (1989) *Contract Bidding in Construction: Strategic Management and Modelling*, Longman
Smith, A.J. (1995) *Estimating, Tendering and Bidding for Construction*, Macmillan
Smith, R.C. (1986) *Estimating and Tendering for Building Work*, Longman
Standard Method of Measurement of Building Works, 7th edition (1988) BEC and RICS
Whitehead, G. (1990) *Guide to Estimating for Heating and Ventilating Contractors*, HVCA

The Chartered Institute of Buildings Technical Information Papers, 1982–1991:
Paper No. 7 Potter, D. and Scoins, D. (1982) Computer-aided estimating
Paper No. 9 Holes, L.G. and Thomas, R. (1982) A general-purpose data processing system for estimating
Paper No. 11 Harrison, R.S. (1982) Practicalities of computer-aided estimating
Paper No. 15 Skoyles, E.R. (1982) Waste and the estimator
Paper No. 37 Harrison, R.S. (1984) Pricing drainage and external works
Paper No. 39 Braid, S.R. (1984) Importance of estimating feedback
Paper No. 59 Uprichard, D.C. (1986) Computerised standard networks in tender planning
Paper No. 64 Ashworth, A. (1986) Cost models – their history, development and appraisal
Paper No. 65 Ashworth, A. and Elliott, D.A. (1986) Price books and schedules of rates
Paper No. 75 Harrison, R.S. (1987) Managing the estimating function
Paper No. 77 Ashworth, A. (1987) General and other attendance provided for sub-contractors
Paper No. 81 Ashworth, A. (1987) The computer and the estimator
Paper No. 97 Brook, M.D. (1988) The use of spreadsheets in estimating
Paper No. 113 Senior, G. (1990) Risk and uncertainty in lump sum contracts
Paper No. 114 Emsley, M.W. and Harris, F.C. (1990) Methods and rates for structural steel erection
Paper No. 120 Cook, A.E. (1990) The cost of preparing tenders for fixed price contracts
Paper No. 127 Harris, F. and McCaffer, R. (1991) The management of contractor's plant
Paper No. 128 Price, A.D.F. (1991) Measurement of construction productivity: concrete gangs
Paper No. 131 Brook, M.D. (1991) Safety considerations on tendering – management's responsibility

The Chartered Institute of Building, Construction Papers, 1992
Paper No. 2 Massey, W.B. (1992) Sub-contractors during the tender period – an estimator's view

Paper No. 11 Hardy, T.J. (1992) Germany – a challenge for the estimator
Paper No. 16 Milne, M. (1993) Contracts under seal and performance bonds
Paper No. 17 Young, B.A. (1993) A professional approach to tender presentations in the construction industry
Paper No. 19 Harrison, R.S. (1993) The transfer of information between estimating and other functions in a contractor's organization – or the case for going round in circles
Paper No. 23 Emsley, M.W. and Harris, F.C. (1993) Methods and rates for precast concrete erection
Paper No. 32 Morres, R. (1994) Negotiation in construction
Paper No. 33 Harrison, R.S. (1994) Operational estimating
Paper No. 49 Borrie, D. (1995) Procurement in France
Paper No. 54 Pokora, J. and Hastings, C. (1995) Building partnerships
Paper No. 55 Sher, W. (1995) Classification and coding of data for computer-aided estimating systems

The National Joint Consultative Committee for Building (RIBA and BEC)
Standard form of tendering questionnaire select list of contractors (1994)
Code of Procedure for Single Stage Selective Tendering (1989)
Code of Procedure for Two Stage Selective Tendering (1983)
Code of Procedure for Selective Tendering for Design and Build (1985)
Code of Procedure for the Selection of a Management Contractor and Works Contractors (1990)
Code of Procedure for the Letting and Management of Domestic Sub-contract Works (1989)
Guidance Note 1. Joint venture tendering for contracts in the United Kingdom (1985)
Guidance Note 2. Performance bonds (1986)
Guidance Note 3. Fire officers' recommendations (1987)
Guidance Note 4. Pre-tender meetings (1986)
Guidance Note 5. Reproduction of drawings for tender purposes (1990)
Guidance Note 6. Collateral warranties (1990)
Guidance Note 8. Construction Management – Selection and appointment of construction manager and trade contractors (1994)
Guidance Note 9. Charges for admission to approved and select lists and for tender documents (1995)
Procedure note 9 Tendering periods

JCT Practice Note 6 – Series 2 – Main contract tendering (2002) *(Intended to replace the NJCC codes of procedure and the subsequent CIB code of practice listed below).*
Documents produced by working groups of the Construction Industry Board published by Thomas Telford Publishing (1997)

- Briefing the team: a guide to better briefing for clients
- Code of Practice for the selection of main contractors
- Code of Practice for the selection of subcontractors
- Constructing success: code of practice for clients of the construction industry
- Framework for a national register for contractors
- Partnering in the team
- Selecting consultants for the team: balancing quality and price

Index